NIELS BOHR
A Centenary Volume

Edited by

A. P. French and P. J. Kennedy

Harvard University Press
Cambridge, Massachusetts, and London, England

Library of Congress Cataloging-in-Publication Data
Main entry under title:
Niels Bohr: a centenary volume.
  "Works by Niels Bohr": p.
  Includes index.
  1. Bohr, Niels Henrik David, 1885–1962—Addresses,
essays, lectures.   2. Physics—History—Addresses,
essays, lectures.   3. Physicists—Denmark—Biography—
Addresses, essays, lectures.   I. Bohr, Niels Henrik
David, 1885–1962.   II. French, A. P. (Anthony Philip),
1920–   .   III. Kennedy, P. J., 1925–   .
QC16.B63N49   1985   530′.0924   [B]   85-8542
ISBN 0-674-62415-7

Preface

In 1979 the International Commission on Physics Education published a volume to celebrate the centenary of Albert Einstein's birth, a volume that eventually appeared in four languages and reached an extensive international readership. There can be little doubt that Einstein, quite apart from the respect in which he is held by all scientists, also holds a unique place in the public imagination, and the success of the centenary volume owed much to that.

The subject of this companion volume, Niels Bohr, is different in that his name, although known to almost every student of natural science, is not widely recognized by the public. And yet, as we worked on the material for this book, it became overwhelmingly clear to us that here was a man who through his work, his imagination, and his personality had made a unique contribution to science — a contribution that deserved to be far better known. We became convinced, in fact, that Bohr was the one physicist of this century, besides Einstein, whose life and work were so wide-ranging and important that a commemorative book about him might have a broad appeal.

As with the Einstein volume, we have aimed to produce a book that will interest the general public, as well as professional physicists and teachers of science. The articles vary widely in character and in the level of scientific background assumed, and we can hardly expect that all readers will want to go through the book from cover to cover — although perhaps we secretly hope that they will. But most important, we hope to have given in these pages some sense of Bohr as a man, as the acknowledged leader of a group of brilliant intellectual innovators, and as a figure on the world stage. We hope, too, that the historical details, described in many instances by those who participated in them, will give a more general picture of a great era in the development of physics, and a feeling for the diversity of human effort that it involved.

The creation of this book would have been impossible without generous cooperation from many sources. First of all, of course, come those who have given their time and effort to contribute original articles and reminiscences bearing on the various periods and aspects of Bohr's life; we owe them an immense debt of gratitude. Then there are the organizations that provided financial and other support for the preparation of the book. We gratefully acknowledge financial contributions from Unesco, from the

Institute of Physics (U.K.), and from the International Union of Pure and Applied Physics. We also received much help and advice from the Center for History of Physics of the American Institute of Physics (especially Joan Warnow) and from the Niels Bohr Institute (especially Dr. Erik Rüdinger). Special thanks are also due Maurice Ebison, Education Secretary of the Institute of Physics, for his contributions to the stock of quotations from which our selection of marginal quotations was made, and Professor E. J. Burge, Chelsea College, London, who provided (as he did for the Einstein volume) the photographs for a montage of interesting and relevant postage stamps.

For those who are not familiar with the International Commission on Physics Education (ICPE), we should like to mention that this commission, established in 1960, is one of a number of commissions of the International Union of Pure and Applied Physics. The ICPE, with a membership of about twelve persons, each from a different country, has as its main concern the stimulation and promotion of international cooperation in physics education. As with the Einstein centenary volume, royalties from the sale of the present book will go to support the work of the Commission.

We ourselves, during the three years we spent planning and preparing this book, have come to feel an almost personal relationship with Niels Bohr, and an ever-growing admiration for what he achieved. We hope that others, in reading the book, will understand these feelings and perhaps come to share them.

A.P.F.
P.J.K.

Contents

Contents

Contents

Chronology

1885   Niels Bohr is born on October 7, in Copenhagen.

1887   Brother, Harald, is born on April 22.

1903   Bohr leaves Gammelholm School, enters University of Copenhagen.

1905   Albert Einstein publishes special theory of relativity.

1906–1907   Research on surface tension of liquids for Royal Danish Academy competition. Bohr is awarded Gold Medal.

1909   Receives master's degree. Publishes medal-winning work on surface tension.

1911   Receives doctorate. Thesis on electron theory of metals.

1911–1912   Bohr in Cambridge. Meets J. J. Thomson. Ernest Rutherford postulates the nuclear atom.

1912   Bohr joins Rutherford at Manchester; meets George de Hevesy. Appointed assistant at University of Copenhagen. Marries Margrethe Nørlund.

1913   W. H. Bragg, W. L. Bragg, Henry Moseley, and Max von Laue develop x-ray spectroscopy. Bohr's trilogy, "On the Constitution of Atoms and Molecules," is published. Bohr appointed lecturer at University of Copenhagen.

1914–1916   Holds Schuster Readership at Manchester University.

1915   Arnold Sommerfeld generalizes the orbital quantum conditions. Albert Einstein develops general theory of relativity.

1916   Bohr appointed professor of theoretical physics at University of Copenhagen. Hendrik Kramers comes to work with him.

1917   Bohr elected to Royal Danish Academy. Begins planning the Institute for Theoretical Physics.

1918–1922   Publishes three papers under the general title "On the Quantum Theory of Spectra," embodying the development of the correspondence principle. Oskar Klein comes to Copenhagen.

1919   Rutherford discovers nuclear disintegration caused by alpha particles. George de Hevesy comes to Copenhagen.

1920   Bohr visits Berlin. Bonzenfreie Kolloquium. Meets Max Planck, Einstein, and James Franck. Rutherford visits Copenhagen.

1921   Bohr awarded the Hughes Medal of the Royal Society. University Institute for Theoretical Physics ("Bohr Institute") opens.

1922   Bohr visits Cambridge University (Rutherford now Cavendish Professor). Visits University of Göttingen. Bohr Festspiele. Meets Wolfgang Pauli and Werner Heisenberg. Pauli joins institute. Bohr publishes theory of the periodic system. Identification and discovery of hafnium (element 72). Bohr receives Nobel Prize for physics.

1923   First visit to the United States. Meets Albert Michelson and Arthur Compton. Offered Royal Society chair at Cambridge University. John C. Slater arrives in Copenhagen.

1924   Visits to institute by Friedrich Paschen, Max Born, and Heisenberg. Rockefeller International Education Board donates $40,000 for expansion of facilities.

1925   Foundation of matrix mechanics (Heisenberg). Discovery of the exclusion principle (Pauli) and electron spin (Uhlenbeck and Goudsmit).

1926   Heisenberg revisits institute; Kramers and Hevesy leave. Visits by Erwin Schrödinger, P. A. M. Dirac, and Samuel Goudsmit. Bohr elected a Foreign Member of the Royal Society.

1927   Uncertainty principle (Heisenberg) and complementarity argument (Bohr) are formulated. Dirac develops relativistic electron theory. Como conference. Fifth Solvay conference (the Bohr-Einstein debates).

1928   George Gamow arrives at institute; Nevill Mott visits. Birth of liquid-drop model of nucleus (Gamow).

1929   First of annual institute conferences. Hendrik Casimir visits Copenhagen. Ernest Lawrence invents the cyclotron.

1930   Visits to institute by Lev Landau and Rudolf Peierls. Bohr awarded Planck Medal of Deutsche Physikalische Gesellschaft. J. D. Cockcroft and E. T. S. Walton achieve first artificial nuclear transmutations.

1932   Bohr family moves to Carlsberg House of Honor. James Chadwick discovers the neutron.

1933   Bohr works on quantum electrodynamics with Léon Rosenfeld. Develops measurement theory for electromagnetic fields. Begins aid for refugee scientists. Otto Frisch visits institute. Bohr, with his wife, visits the United States, stopping at University of Chicago and California Institute of Technology. Meets Robert Millikan and J. Robert Oppenheimer. Seventh Solvay conference (on nuclear physics).

1934   Frisch joins staff of institute. Next door, the University Institute of Mathematics opens, with Harald Bohr as director. The Bohrs' eldest son, Christian, is drowned in a sailing accident. Irène and Frédéric Joliot-Curie discover artificial radioactivity. Enrico Fermi publishes theory of beta decay.

1936   Bohr develops compound-nucleus theory, with Fritz Kalckar.

1937   Bohr, with wife and son Hans, makes six-month trip around the world. Visits United States, Japan, China, and USSR. Death of Rutherford; Bohr attends funeral at Westminster Abbey.

1938   Installation of cyclotron at Bohr Institute.

1939   Discovery of nuclear fission by Otto Hahn and Fritz Strassmann, and Lise Meitner and Otto Frisch. Reported by Bohr during visit to United States. Bohr and John Wheeler write classic paper on theory of fission. Bohr elected president of Royal Danish Academy of Sciences. World War II begins.

1940   German occupation of Denmark.

1943   Bohr and family escape to Sweden. Bohr, with son Aage, proceeds to Britain and the United States (Washington and Los Alamos).

1944   Meetings with Roosevelt and Churchill.

1945 First atomic bomb exploded. Hiroshima and Nagasaki destroyed. World War II ends. Bohr returns to Denmark in August.

1946 Abraham Pais becomes first postwar foreign member of Bohr Institute.

1949 USSR detonates a fission bomb. Bohr revisits United States.

1950 Open letter to the United Nations.

1952 United States explodes its first hydrogen bomb. Foundation of CERN.

1955 Bohr appointed chairman of Danish Atomic Energy Commission. Death of Einstein. First "Atoms for Peace" conference (Geneva).

1957 Bohr receives the first Atoms for Peace Award. NORDITA established in Copenhagen.

1958 Bohr delivers Rutherford Memorial Lecture.

1961 Bohr works on theory of superconductivity.

1962 Bohr dies on November 18, in Copenhagen.

Bohr's influence on science is only partially expressed in his published work, great as this was. He led science through the most fundamental change of attitude it has made since Galileo and Newton, by the greatness of his intellect and the wisdom of his judgments. But quite apart from their unbounded admiration for his achievements, the scientists of all nations felt for him an affection which has perhaps never been equalled. What he was counted for even more than what he had done.

G. P. Thomson,
Niels Bohr Memorial Lecture, 1964

I. THE MAN AND WHAT HE ACHIEVED

A Short Biography

P. J. Kennedy

Niels Henrik David Bohr was born into a comfortable Danish home on 7 October 1885, the first son of Christian and Ellen Bohr. Eighteen months later his brother, Harald, was born and there began a friendship, filled with affection and respect, that was to last until Harald's death in 1951. This deep friendship was the first of several that Bohr was to inspire and enjoy and that illuminate the record of his eventful life.

Bohr's father was a university professor, a famous physiologist and a lover of science, and his mother (née Ellen Adler) was a generous, intelligent and liberal woman. Niels, Harald, and their older sister, Jenny, grew up in a cultured and stimulating home. From their earliest days they were exposed to a world of ideas and discussion, of conflicting views rationally and good-temperedly examined, and they developed a respect for all who seek deeper knowledge and understanding.

Even in childhood Bohr admired the qualities and abilities of his younger brother, and he saw himself justified when Harald, having obtained his doctorate two years earlier than Niels, went on to become a distinguished mathematician. Nevertheless, both father and brother regarded Niels as "the special one of the family," and they expected some exceptional contribution from him even though his school career was successful without being outstanding. His schooldays did, however, show his increasing interest in physical science and mathematics, as well as other qualities that were to play an important and valuable part in his later scientific career. He formed friendships at school that were to last and grow throughout his life, and he early showed his ability to inspire affection in others and to enable others to express their own best qualities — an ability that was later to contribute in an essential way to the success of the Copenhagen School.

It was at the University of Copenhagen that Niels Bohr's gifts as a

P. J. Kennedy, a coeditor of this book, studied at Oxford University and at Trinity College, Dublin. He is now Senior Lecturer in the Physics Department at the University of Edinburgh. He served for a number of years as Secretary of the International Commission on Physics Education, and has helped organize many of its activities.

3

Niels and Harald Bohr as young students
and in later life. Niels is at right in both
photos.

scientist became recognized. At the age of twenty, he was awarded the Gold
Medal of the Royal Danish Academy of Sciences and Letters for his prize
exercise on the measurement of surface tension by the study of vibrating
fluid jets. This careful and complete piece of research, both experimental
and theoretical, drew upon and extended the work of the renowned Lord
Rayleigh, and served to give Bohr particular insight in his later work on the
liquid-drop model of the nucleus.

Although Bohr worked very hard at his studies, he also played hard, and
no account of his life would be complete without mention of his enthusiasm for sports. In soccer, especially, he was an expert — although not quite
so much as Harald, who won a place on the Danish Olympic team in 1908.
Both brothers became nationally famous for their achievements on the
soccer field.

Both for his master's degree and for his doctorate (1911) Bohr studied the
electron theory of metals. Although these studies were successful, it was in
this work that he began to be aware of the difficulties and limitations of

Niels Bohrs Disputats.

1. Dr. Bohr. 2. Prof. Chrisiansen. 3. Prof. Heegaard.

Det er kun et Aars Tid siden afdøde Professor Bohrs ene Søn, Harald, erhvervede sig den filosofiske Doktorgrad. I Gaar er den lidt yngre Broder, Magister Niels B o h r, fulgt i hans lære Fodspor.

Meningen havde været, at Disputatsen skulde være foregaaet i Annexet; men paa Grund af det amerikanske Professorbesøg var Handlingen i Stedet for kleven henlagt til et bestedent Auditorium paa selve Universitetet. Dette havde til Følge, at Tilhørerne maatte staa langt ud paa Gangen, og det var for saa vidt af det gode, at Opponenterne temmelig hurtig fik sagt, hvad de havde paa Hjrte.

Metallernes Elektrontheori, der er Emnet for Niels Bohrs Afhandling, er jo ogsaa et ret kreditt Emne, og den første af Opponenterne,

Professor Heegaard, bevælede i en Kritik af den sproglige Fremstilling væsentlig ved Udenværkerne. Mere ind paa det reelle kom den anden Opponent, Professor Christiansen, der havde megen Anerkendelse tilovers for Doktorandens Arbejde.

Prof. Christiansen mindede om, at efter H. C Ørsted var Lorentz den Dansker, der havde lært tedst hjemme paa det videnskabelige Omraade. Talen var om, og Professoren havde altid plejet at tage ud til Lorentz, naar han i herhen hørte Spørgsmaal vilde have Besked. Siden Lorentz' Tid havde vi ikke havt nogen Fagmand paa det omhandlede Felt, og Opponenten sluttede derfor med at udtale sin Glæde over, at dette Savn nu med Niels Bohr var afhjulpet.

Report on Bohr's defense of his doctoral thesis, published in the newspaper *Dagbladet*, Copenhagen.

Dear Dad: I'm so glad you liked the notes on Jeans. I'll bring the notebook home and we can go over it. He has been doing the pr[essure] of radiation business lately, and according to him it is the most utter rot. I got an awful lot from a Dane who had seen me asking Jeans questions, and after the lecture came up and talked over the whole thing. He was awfully sound on it, and most interesting. His name was Böhr, or something that sounds like it.

W. L. Bragg (then a student at Cambridge) to W. H. Bragg, 1911

classical physical theory in the description of electron behavior and of the need for some radically different mode of description for atomic processes.

This work foreshadowed his later contributions to quantum theory and, at the same time, stimulated his interest in philosophy and epistemology, an interest he was cultivating through his readings in philosophy and theology, through his attendance at the lectures of the philosopher Harald Høffding and through his membership in Ekliptika, a discussion group made up of philosophy students. He began to recognize the complex role and the limitations of ordinary language in our description of phenomena, and the need to accommodate apparently conflicting aspects in order to form complete descriptions. He was convinced that "there is no point in trying to remove such ambiguities; we must rather recognize their existence and live with them."[1] Such early ideas, which were never lost to Bohr, reappear later in his major contribution to physics and epistemology: the *complementarity* argument.*

Throughout these early years, Niels Bohr maintained with his brother a warm and trusting correspondence in which they exchanged news and discussed their awakening philosophies. Bohr became a prolific and industrious correspondent and left a legacy of interesting and apparently spontaneous letters. In fact, they were far from spontaneous but were, like his scientific papers, written in their final form only after multiple drafts and painstaking revisions. As early as 1911 Bohr began enlisting the aid of an amanuensis — at first his mother and, later on, his wife, his sons, or his colleagues.

In 1910 Bohr met Margrethe Nørlund, and in August 1912 they were married. This began for him yet another durable and fulfilling relationship which was to sustain him until his death. The Bohrs had six sons of whom four survived to adulthood, and Niels Bohr's later years were filled with the pleasure of his many grandchildren. In 1911, in the period between his engagement and his marriage, Bohr made his first visit to Britain, to the Cavendish Laboratory in Cambridge. The head of the Cavendish at the time was J. J. Thomson, the doyen of atomic physics and discoverer of the electron. Bohr had hoped to interest Thomson in his own work on the electron theory of metals but was unable to do so, and he spent his first few months in Britain feeling somewhat frustrated. Then he met and impressed Ernest Rutherford, the professor of physics at the University of Manchester, who had recently devised the new "massive nucleus" model of the atom. Their association, starting as a mutual recognition of worth between the exuberant practical experimenter and the thoughtful young Dane, developed and deepened throughout a quarter-century of friendship and collaboration, gradually approaching the best type of father-and-son relationship.

* At the end of this volume is a glossary containing many scientific terms used in the text that may be unfamiliar to readers. Those having special importance to a particular discussion have been italicized.

Niels Bohr with Margrethe at the time of their engagement and, many years later, in the garden of the Carlsberg House of Honor.

In Manchester, to which he was soon invited, Bohr found a stimulating atmosphere and congenial colleagues. It was there, in the spring and early summer of 1912, during a period of almost continuous work, that he made several important contributions to atomic physics: he helped clarify the nature of radioactive transformation, was probably the first person to recognize the basis of nuclear isotopy, and developed a theory of the energy loss of alpha particles as they passed through matter — a topic that engaged him throughout his life. At the same time, using the Rutherford atomic model, he began to work out the ideas of atomic stability that were to lead to the quantum description of atomic structure. In 1913 Bohr published these last ideas in three articles in the *Philosophical Magazine,* where he set out his startling attempt to combine aspects of classical physics with the concept of Planck's quantum of *action.* This new theory yielded impressive quantitative agreement with measurements in atomic spectra. The three famous papers, the Trilogy,[2] formed the foundation of Bohr's early reputation. His work, although not immediately accepted by everyone, intrigued his contemporaries and made them aware of the need for a new way of describing events at the atomic level. The Bohr atom, although it has been

Niels Bohr with his sons, about 1927.

superseded scientifically, persists even today in the minds of many people as a vivid image of what atoms look like and as a symbol of physics.

Having spent the early years of the First World War in Britain, in 1916 Bohr returned to the University of Copenhagen, where the advantages of teaching theoretical physics as a separate discipline had been recognized by the establishment of a Chair of Theoretical Physics, a chair created especially for him.

Within a few months Bohr made another acquaintance which, like many others, was to develop into a lifelong friendship. Hendrik Kramers, a young Dutchman, eager to find a place removed from the rigors of war, applied to Bohr for a post and received an appointment. In the years that followed, the two men collaborated on scientific papers, planned and ran the new University Institute for Theoretical Physics (later to become the Niels Bohr Institute) and, in the disillusioned years after the Second World War, worked together for the limitation of nuclear weapons and for the extension of international cooperation. In 1918 Bohr published "On the Quantum Theory of Line Spectra." This paper presented a detailed elaboration of the *correspondence principle,* which Bohr had introduced in 1913. In the skilled hands of Bohr and Kramers, this principle proved a powerful tool for elucidating the fine structure of spectra, and for predicting spectral intensities, transition probabilities and selection rules with considerable accuracy. The principle was never so fully understood or exploited by physicists elsewhere.

Soon after the paper on spectra was published, Bohr found himself heavily engaged in the planning — and frequent replanning — of the new University Institute for Theoretical Physics to be built in Copenhagen (and which, despite its name, was to undertake experimental as well as theoretical research). Within a very few years of its opening in 1921, this unpretentious building in the capital city of a small country was to become one of the best-known places in all of physics. The list of visitors over the next two decades — who came for a few weeks, a few months, or perhaps years — reads like a roll-call of the founders of quantum mechanics. There and in their home, Niels and Margrethe Bohr created an atmosphere in which young minds from all over the world wrestled with the deepest problems of their subject. This intensely interactive enterprise, constantly changing in its participants but always led and shaped by the mind and personality of Niels Bohr, came to be known as the Copenhagen School of physics.

In 1922 Bohr made his next major contribution to physics through several papers on the theory of atomic structure and the periodic system of the elements. Later in the same year he was awarded the Nobel Prize in physics "for his services in the investigation of the structure of atoms and of the radiation emanating from them." In his Nobel speech, Bohr surveyed the state of the quantum theory and the progress that had been made in applying it to the problems of atomic structure, but he took pains to point out the limitations and weaknesses of the theory. He was more acutely aware of these and more perturbed by them than were others who had accepted his ideas less critically.

In the years following, Bohr continued to publish work in atomic physics, including a highly controversial paper with Kramers and John Slater. Although Bohr himself had formulated the first quantization rules for the emission process of radiation, this paper revealed his continuing disinclination to accept the concept of the photon. With the arrival of Werner Heisenberg and Wolfgang Pauli in Copenhagen, Bohr's role became less that of an initiator in the progress of quantum physics and more that of a support, a mentor, and a penetrating critic of those who were leading the way. His mind returned to some of the preoccupations of his early days: the physical interpretation of the mathematical formulations of quantum mechanics, the importance of seeking to define the boundary between a measuring apparatus and the measured object, and the place of language in making explicit the outcome of such measurements. These considerations of many years were brought together in his paper "The Quantum Postulate and the Recent Development of Atomic Theory," presented at the Volta

Prof. N. Bohr! To begin, let me introduce myself, by telling that I am a Dutch student in physics and mathematics. I've studied 4 years in Leiden . . . As I didn't like to go to a country that is in war now, I decided to go to Copenhague, and hope to study now mathematical physics . . . Of course I should like very much to come in acquaintance with *you* in the first place, and also with your brother Harald. Therefore I should be very glad, if you would permit me to visit you one of these days.

H. A. Kramers to Niels Bohr, 25 August 1916

I have been very pleased with my collaboration with Dr. Kramers who, I think, is extremely able, and of whom I entertain the greatest expectations. We have worked together on the helium spectrum.

Niels Bohr to C. W. Oseen, 28 February 1917

Postage stamps on the theme of the Bohr atom and atomic energy. The S.G. numbers are the Stanley Gibbons Catalogue numbers for the country in question, and are included for the interest of collectors. This selection was compiled and annotated by E. J. Burge. Stamps are identified from top to bottom, beginning with the far left column.

Belgium	1966	S.G. 1974	European chemical plant, Mol
Greece	1983	S.G. 1632	First International Democritus Congress
Spain	1967	S.G. 1848	Radiology Congress, Barcelona, with portrait of Wilhelm Roentgen
Czechoslovakia	1964	S.G. 1452,3	Czech engineering
Denmark	1963	S.G. 455,6	Fiftieth anniversary of Bohr's atomic theory
Romania	1971	S.G. 3881-4	Centenary of Rutherford's birth
East Germany	1980	S.G. 2214-9	Honoring Frédéric Joliot-Curie, in a series on celebrities' birth anniversaries
Afghanistan	1958	S.G. 437,8	Atoms for Peace
USSR	1967	S.G. 3382-4	World's Fair, Montreal
United Nations	1978	S.G. 298,9	Peaceful uses of atomic energy
Brazil	1963	S.G. 1084	First anniversary of the National Nuclear Energy Commission
Ceylon	1969	S.G. 555-8	Educational centenary
Yugoslavia	1961	S.G. 982	International Nuclear Electronics Conference, Belgrade
United States	1955	S.G. 1072	Atoms for Peace
Turkey	1963	S.G. 2015-7	First anniversary of the opening of the Turkish Nuclear Research Center
Israel	1969	S.G. 431	Twenty-fifth anniversary of the Weizmann Institute of Science
Sweden	1982	S.G. 1134	Honoring Bohr, in a series commemorating Nobel Prize winners in atomic physics
West Berlin	1958	S.G. B175,6	German Catholics' Day

Commemoration Conference held at Como, Italy, in September 1927. In his presentation, Bohr spoke of the epistemological problems of quantum mechanics and set out his complementarity argument — a principle which he was to continue to develop and extend until the end of his life and which he came to believe was his major contribution to our understanding of nature. Bohr believed that this principle had great general usefulness, and he endeavored for many years to apply it to physiology, psychology, and other disciplines.

These new ideas were soon put to stringent test in discussions with Albert Einstein at the Fifth Solvay Conference,[3] which took place a few weeks after the Como gathering. Bohr and Einstein had first met in 1920 and had soon opened a series of discussions which, conducted with mutual pleasure and respect, continued for many years and became part dialogue, part duel, and part crusade. They disagreed on many things, including causality, the meaning of reality, and the incompleteness of quantum-mechanical descriptions, and for nearly twenty-five years each tried, by ingenious argument and subtle logic, to convince the other.

In 1932 the Bohr family moved from their quarters in the institute to a beautiful mansion at Carlsberg. Under the terms of the Carlsberg Foundation, the mansion was intended for the use of an "outstanding citizen of Denmark," and in 1931 the Royal Danish Academy of Science and Letters, of which Bohr was later to serve as president for many years, had offered the mansion to him as his home for life. The house became not only a family home for the Bohrs' children and grandchildren, but also a haven for the many young colleagues and visitors who stayed there, and a convivial meeting place for scientists, artists, and politicians from all over the world.

For the next few years Bohr continued to develop his ideas, both in physics and in epistemology. He published works dealing with the problem of measurement in quantum electrodynamics; an essay entitled "Light and Life," explaining his application of the complementarity argument to biology; and a paper entitled "Can Quantum-Mechanical Description of Physical Reality Be Considered Complete?" — a response to the famous Einstein, Podolsky, and Rosen paper of the same title. During this period Bohr and his brother, who was at that time director of the Mathematics Institute (which had been built adjacent to the Institute for Theoretical Physics) became deeply involved with a Danish group that had been formed to offer support to scientists and other intellectuals who were forced to flee their homeland by the racial policies of the National Socialist government in Germany. Their two institutes acted as a temporary refuge for many of the greatest names in German physics, and the brothers dedicated themselves to finding new posts for those who had suddenly become stateless and unemployed.

Beginning in about 1934, Bohr made his next major contribution to physics through his work in nuclear physics, in particular his theory of the *compound nucleus* and his elaboration of the *liquid-drop model*. This work reached its climax in 1939 with a paper, written jointly with John Wheeler, on the theory of nuclear fission, but Bohr continued to work and publish on these topics well into the 1940s. In the early years of the Second World War he and his institute continued to function, even after the German occupation of Denmark. But Bohr's role changed and, as a public figure and a focus of national admiration and pride, he felt a responsibility to help maintain Danish science and culture under the prevailing conditions. It was in this period that he was asked to write the introduction to the book *Danish Culture in the Year 1940* — a task on which he lavished the same care and attention to detail and language that he gave to his scientific writings.

In 1943 the occupation became more oppressive, and Bohr left Denmark with the encouragement of the British government. Later he traveled by way of Sweden and Britain to the United States, where he joined the Manhattan Project. Although scientifically interested in the progress toward the production of the fission bomb, Bohr turned his attention almost immediately to the political significance of the project and to the need for early and clear recognition of the threat it would pose to postwar stability. On 11 August 1945, a few days after the bombing of Hiroshima, the editorial page of the London *Times* carried an article by Bohr entitled "Energy from the Atom," in which he pleaded for such recognition.[4] In the autumn of 1945 he returned to his family and the institute, and once again took up his role as honored teacher and stimulating friend of a new generation of physicists. He devoted much time and thought to promoting a sane and realistic policy for nuclear arms, and in 1950, in his "Open Letter to the United Nations," he made a heartfelt appeal for world cooperation. He and his longtime friend and collaborator Hendrik Kramers, then chairman of a United Nations committee on nuclear policy, both worked for peace, but their efforts were overtaken by events and Bohr's proposal for openness and free exchange of information was never tried.

In Denmark, Bohr had become a much respected elder statesman and was called upon to guide the government's policy on atomic energy. For the next ten years much of his time was given to the detailed planning and completion of the Danish Atomic Energy Commission's research establishment at Risø. During this period he also completed a treatise on the passage of charged particles though matter and, again with Kramers, lent his name and support to the establishment of a center for European cooperation in science which was to become CERN (the Conseil Européen pour la Recherche Nucléaire). In the early and delicate days of the organization, the division of theoretical physics was housed and grew strong in the Bohr Institute before moving to Geneva.

At the beginning of the 1960s, Bohr and the members of the institute began to plan for a meeting in 1963 to celebrate the fiftieth anniversary of the publication of the original papers on atomic structure. They hoped to renew the intimate atmosphere of the interwar Copenhagen meetings by inviting back members of the institute from those exciting years and giving

Participants in the 1963 golden jubilee conference in Copenhagen (fiftieth anniversary of the Bohr atom).

Front row, from left: Abraham Pais, Felix Bloch, Friedrich Hund, William Houston, Christian Møller, David Dennison, I. I. Rabi, Victor Weisskopf, Aage Bohr, P. A. M. Dirac, Otto Frisch, Oskar Klein, Werner Heisenberg, Patrick Blackett, Richard Courant, Adalbert Rubinowicz.

Second row: Pascual Jordan, J. Holtsmark, Torsten Gustafson, B. Strømgren, E. Werner, G. C. Wick, Ben Mottelsen, L. H. Thomas, Samuel Goudsmit, C. Manneback, Harald Wergeland, André Mercier, Léon Rosenfeld, Guido Beck.

Third row: G. Holm, J. K. Bøggild, M. H. Jørgensen, Jørgen Koch, J. Kistemaker, L. W. Nordheim, P. Kristensen, Ralph Kronig, F. C. Hoyt, H. H. Nielsen, C. F. von Weizsäcker, John Wheeler, N. O. Lassen, N. Svartholm.

Fourth row: S. G. Nilsson, O. B. Nielsen, Thomas Lauritsen, Hilde Levi, Kametaka Ariyama, G. Trumpy, Stefan Rozental, J. C. Jacobsen, C. B. Madsen, M. S. Plesset, K. A. Jessen, Stanley Deser, Torben Huus.

Fifth row: Roald Tangen, Børge Madsen, M. C. Olesen, Høffer Jensen, Aage Petersen, B. Elbek, Max Delbrück, H. B. G. Casimir, J. Rud Nielsen, Lamek Hulthén, Højgaard Jensen, Jens Lindhard, O. Kofoed-Hansen, Mogens Pihl, Roy Poulsen.

Sixth row: Aage Winther, K. G. Brostrøm, Res Jost, Joaquin Luttinger.

them the opportunity to review the past half-century of progress and to guess at the exciting questions for the future.

Before this could happen Niels Bohr died suddenly at his home at Carlsberg, on the afternoon of 18 November 1962. Nevertheless, the reunion he had planned was held, and many of the surviving members of the Copenhagen School returned to exchange their latest opinions and ideas, as he would have wished. Later, at the invitation of Margrethe Bohr and her sons, they gathered once more in the beautiful mansion which they had come to know so well.[5]

A Personal Memoir

James Franck

James Franck, a distinguished atomic physicist, was a friend of Niels Bohr for about forty years. He was born in 1882 and was Director of the Physical Institute at the University of Göttingen from 1920 to 1933. He came to the United States in 1934, and lived there until his death in 1964. In 1925 he shared the Nobel Prize for physics with Gustav Hertz for their famous research on electron excitation of atoms.

Niels Bohr's life exemplifies the fact that truly great scientists are equally great in their subject matter and as human beings.[1] In order to achieve the greatest scientific accomplishments, it is not enough to possess unusual intelligence; one must also have unusual strength of character, diligence, courage, extreme desire for the truth, and the ability to recognize the essential problems and concentrate on them. These are some of the prerequisites for success, and they strongly affect one's whole view of life and how one leads it.

One can use almost any of Bohr's works in order to show how clearly he demonstrated these character traits. I will choose the work of 1913. Half a century later it may seem to many that the concept of discrete quantum states for the atomic electronic system was obvious. It may be thought that if Bohr had not introduced the idea, someone else would have come up with it shortly thereafter. This notion is absolutely wrong. How much courage, independence, and concentration on essentials was necessary is shown by the slowness with which this idea was accepted by the great body of physicists.

In Germany, Bohr's work was little read in the first years after it was published. There was generally a thorough distrust of attempts to construct models of the atom with the limited knowledge at that time, and, in scanning the literature, one sees that only a few people made the effort to read Bohr's work carefully. Both Gustav Hertz and I were initially unable to understand the enormous importance of Bohr's work. We really had every reason to read it carefully, as we had just finished our measurement of the excitation of the 2,537-angstrom line in mercury by electronic collisions.[2] We read Bohr's work before we sent our manuscript to the printer, but decided not to mention it, because we had a problem understanding the

16

Niels Bohr (left) with James Franck, early 1920s.

Another great scientist I must talk about was James Franck, who, together with Gustav Hertz, got a Nobel Prize for first exciting specific states in atoms by bombarding them with electrons of controlled speed. He came to Copenhagen from Göttingen because he was a Jew. He could have stayed under the existing racial laws, which exempted men who had fought in the First World War, but it was unthinkable for him to serve under a regime that persecuted Jews. He had uncommonly fine features, luminous with kindness and obvious interest in your problems; he was the most immediately lovable man I have ever met.

O. R. Frisch
What Little I Remember, 1979

strong ionization in mercury vapor if, as Bohr had concluded, the energy necessary to ionize the atom is much larger than that required merely to excite it. That we judged this difficulty to be big enough for us *not* to accept Bohr's concept shows our misunderstanding of its significance. Many years later Einstein said in a conversation: "I believe that, without Bohr, we would still today know very little about atomic theory." In the beginning even Arnold Sommerfeld, who later spent his whole life in the promotion and development of the Bohr theory and who was immediately ready to accept the Bohr formula for the Balmer series, did not want to extend its application to a model of the atom.

Bohr's work on the correspondence principle and complementarity is another example that reveals especially well the connection between his great success and his character traits. These developments came from Bohr's deep need to make the basis of all the consequences of atomic theory as clear as possible. He could be satisfied with himself only if he arrived at principles that, in a unified and noncontradictory way, clarified the transition from the continuum to the discrete quantized system.

A very large fraction of the best theoretical physicists of our time labeled themselves, proudly and thankfully, as students of Bohr. His pupils were gifted young theorists whose earlier schooling in theoretical physics, and especially in the application of mathematics to the solution of theoretical problems, had been received at other large centers of physics. What Bohr taught them, by example and discussion, was the ability, for which he provided the model, to think through problems to the end, inexorably fighting self-delusion with the courage not to be deterred from tackling seemingly impossible obstacles. The word "teach" can really not be used here, since character traits cannot really be taught; but their importance can be pointed out, thereby awakening them in those people in whom they are, as it were, dormant. Bohr, in a conversation, once made a remark that is instructive about his way of thinking and working: "One must never be satisfied doing only what one can; rather, one must always do what one really cannot."

At the beginning of the Hitler era, Bohr invited to his institute a number of physicists who had to leave their homeland. I was among them, and stayed more than a year as a visiting professor in Copenhagen. In this way I got a far better opportunity than before to observe the influence of Bohr on the group around him. This influence was by no means limited to the colloquia and private discussions at the institute. Bohr's home (where the major part of his own work was done) formed a second center to which came a flood of visitors of the most diverse kinds. In addition to the staff of the institute and short-term guests from abroad, there were colleagues from other parts of the university, important Danish bureaucrats, artists, politicians, and others. Even the Royal Family often came to visit. One could correctly describe Bohr's house as resembling a Greek academy. The conversations and discussions there were by no means limited to physics and natural science; they encompassed philosophy, history, fine arts, religious history, ethical questions, politics, current events, and many other topics. Bohr himself had a large range of interests. He read a great deal, had a good memory, and thought about everything he read and experienced.

Bohr never had the desire to withdraw into the ivory tower of science; rather, he felt it a duty to keep himself informed about the life and activity of human society and, when necessary, to openly express his opinions. Through his example and discussion he helped a large number of people to take such duties seriously and to perform them.

Many people have not understood Bohr's decision to work on the development of the atomic bomb. It was a hard necessity. Physicists throughout the world who read the literature knew that the building of such a bomb was likely to be possible. Should they just fold their hands and wait until Hitler eventually gained possession of one? If the war were lost, what would stop the criminal tyranny that had already enslaved Germany from spreading over the whole world? But as it became clear that Hitler could not develop such a bomb, Bohr raised objections to it. His warnings went unheeded. For years, Bohr seemed to be plagued by the thought that perhaps it was a mistake to seek to influence world politics by logical arguments. Fortunately, his optimistic nature finally gained the upper hand, and he became cheerful and voluble, as we had known him before. He was this way when I saw him for the last time, in June 1962 in Lindau.

Niels Bohr, the Quantum, and the World

Victor F. Weisskopf

Modern scientific progress is often described in terms of revolutions and upheavals, in which one new theory destroys the previous theories. This description overlooks the fact that scientific development is intrinsically evolutionary. Most of the new and so-called revolutionary ideas in modern science were a refinement of the old system of thought, a generalization or an extension. Relativity theory did not do away with Newton's mechanics —which is still used, for example, to calculate the orbits of satellites. Rather, it has extended the applications of Newtonian principles to extreme velocities and has established the general validity of the same concepts in mechanics and electricity.

However, one development in the first quarter of this century did come close to a true revolution in our thinking. This was quantum mechanics. The discoveries of the late nineteenth century had shown that it is impossible to understand the structure of matter, the specific properties of material, using so-called classical physics. A new system of concepts and a new way of dealing with atomic structure had to be introduced, and they revolutionized our ideas of material reality. Even so, these new concepts, such as the *uncertainty principle* and *quantization*, did not vitiate the scientific accomplishments of earlier times. They are a refinement of classical mechanics, necessary to the understanding of very small systems such as atoms or molecules. They do not destroy the validity of classical physics for the description of larger systems.

The quantum revolution caused a major breakthrough in our views about the structure of matter and opened up many new horizons. Within a few years after the formulation of quantum mechanics, problems that had been considered unsolvable for decades—such as the nature of molecular bonds, the structure of metals, and the radiation from atoms—were finally

Victor Weisskopf, a world leader in theoretical nuclear physics, is Institute Professor, Emeritus, at the Massachusetts Institute of Technology. He began his research career at the University of Göttingen, studying under Paul Ehrenfest and Eugene Wigner. In 1937 he emigrated to the United States. During the period 1961–1965 he was Director-General of the European Center for Nuclear Research (CERN) at Geneva.

19

The floodgates were opened by a young Danish physicist, Niels Bohr, through his proposed model of an atom. You must have seen it many times, decorating almost any publication related to atoms: a dot surrounded by several circles, usually foreshortened into intersecting ellipses. That model has now been out of date for half a century. But symbols have long lives: Father Time is still depicted with a sand-glass, not a wristwatch.

O. R. Frisch,
What Little I Remember, 1979

understood. This sudden growth of knowledge and understanding of the structure of matter opened up many new ways of dealing with materials: it led the way to new forms of energy, new kinds of materials, and many new technical possibilities in chemistry, electronics, and nuclear technology. It also began changing our quality of life at an ever-increasing rate, leaving us at odds with our accepted value system when facing the human problems created by all these innovations.

Nothing could be better suited to illustrate this vast development and to illuminate these problems than a study of the life of Niels Bohr. He was a great physicist — one of the greatest; his name ranks with those of Galileo, Newton, Maxwell, and Einstein. His work was at the center of this development, and kept it going for half a century. To a greater degree than any other scientist, he was involved in the human problems of his science, and its impact on society and politics.

Bohr was born in 1885. His life as a scientist began in about 1905 and lasted fifty-seven years. It was in 1905 that Einstein published his papers on special relativity and on the existence of the light quantum; this was only a few years after Planck's discovery of the quantum of action. Bohr had the great luck to be present at the beginning, or perhaps mankind had the great luck of having him at that turning point. What a time to be a physicist! He began when the structure of the atom was still unknown; he ended when atomic physics reached maturity — when the atomic nucleus was put to industrial use for the production of electric power, to medical use in cancer treatment, and also, unfortunately, to military and political use in the most destructive weapon man has ever conceived.

The work of Niels Bohr can be divided into four periods. In each he exerted a tremendous impact on the development of physics. The first period was the decade 1912–1922, from his first meeting with Rutherford until the foundation of his famous Institute of Theoretical Physics in Copenhagen. In this period Bohr introduced the concept of the quantum state and devised an intuitive method of dealing with atomic phenomena. In the second period, from 1922 to 1930, he gathered around him in his new institute a few of the most productive physicists in the world, who, under his leadership, developed the ideas of quantum mechanics. This is the conceptual edifice that replaced his original intuitive method and gives an adequate description of the inner workings of the atom. The third period, 1930–1940, was devoted to the application of the new quantum concepts to electromagnetic phenomena and the exploration of the structure of the atomic nucleus. Then came the Second World War and the last period of his life, in which he acted as the great leader of physics, deeply concerned with the social, political, and human consequences of the new discoveries.

The first period began with the publication, in the year 1913, of Bohr's work on the quantum orbits of the hydrogen atom. This remarkable paper proposed to explain the hitherto mystifying properties of the atom by introducing a completely new concept into physics: that of the *quantum state.* His theory, based upon previous work by Planck and Einstein, applied the idea of the quantum to the structure of the atom. There is hardly any other paper in the literature of physics from which grew so many new theories and discoveries.

This famous paper marked the beginning of a decade of new insights about previously unexplained phenomena: the structure of the spectra of elements, the processes of absorption and emission of light, the reasons for the periodic system of elements, the puzzling sequence of properties of the ninety-two different atomic species. It was the period in which quality, the specificity of chemical substances, was reduced to quantity, to the number of electrons per atom. All this rested on Bohr's assumption of the existence of specific electronic orbits in the atom, the only ones in which electrons were allowed to revolve. It was at that time still a provisional hypothesis. Bohr's contemporaries, however, took it quite literally, although Bohr warned them in his papers and at meetings that this could not be the final

"Physicists' Congress in Copenhagen": report of the first Copenhagen Conference in the newspaper *Politiken,* 9 April 1929. The sketch of Bohr is by the Danish artist Carl Jensen.

explanation, that something fundamental remained to be discovered in order to understand what was going on in the quantization of the atom.

During the second period, the quantum was fully understood. It was a heroic period without any parallel in the history of science, the most fruitful and most interesting one of modern physics. There is no single paper by Bohr himself that characterizes this period, as his 1913 paper characterized the first period. Bohr found a new way of working. He no longer worked alone, but in collaboration with others. It was his great strength to assemble around him the most active, the most gifted, the most perceptive physicists in the world. At the time, Bohr was working at his Institute for Theoretical Physics, in Copenhagen, in the midst of an international group of physicists: Oskar Klein, Hendrik Kramers, Wolfgang Pauli, Werner Heisenberg, Paul Ehrenfest, George Gamow, Felix Bloch, Hendrik Casimir, Lev Landau, John Slater, and many others. It was at that time, and with those people, that the foundations of the quantum concept were established, that the uncertainty relation was first conceived and discussed, that the particle-wave antinomy (*duality*) was for the first time understood. In lively discussions, in groups of two or more, the deepest problems of the structure of matter were explored. One can imagine what atmosphere, what life, what intellectual activity reigned in Copenhagen at that time. Here was Bohr's influence at its best. Here it was that he created his style, the *Kopenhagener Geist,* a style of a very special character that he imposed upon physics. He could be seen, the greatest among his colleagues, acting, talking, living as an equal in a group of young, optimistic, jocular, enthusiastic people, approaching the deepest riddles of nature with a spirit of attack, a spirit of freedom from conventional bonds, and a spirit of joy that can hardly be described. As a very young man, when I had the privilege of working there, I remember that I was taken a little aback by some of the jokes that crept into the discussion — they seemed to me to indicate a lack of respect. I communicated my feelings to Bohr, and he gave me the following answer: "There are things that are so serious that you can only joke about them."

In this great period of physics, Bohr and his associates touched the nerve of the universe. The intellectual eye of man was opened to the inner workings of nature. Once the fundamental tenets of atomic mechanics had been established, it was possible to understand and to calculate almost every phenomenon in the world of atoms — phenomena such as atomic radiation, the chemical bond, the structure of crystals, the metallic state, and many others. Before that time, the world had been perceived as full of different forces: electromagnetic, cohesive, capillary, chemical, and elastic. Then (except for gravity and nuclear forces) all these forces were reduced to one — the electromagnetic force. In the course of only a few years, the basis was laid for a science of atomic phenomena that grew into the vast body of knowledge we have today.

The ideas spurred by Bohr had an enormous impact. Previously, chemis-

Those years, when a unique cooperation of a whole generation of theoretical physicists from many countries created step by step a logically consistent generalization of classical mechanics and electromagnetism, have sometimes been described as the "heroic" era in quantum physics. To everyone following this development, it was an unforgettable experience to witness how, through the combination of different lines of approach and the introduction of appropriate mathematical methods, a new outlook emerged regarding the comprehension of physical experience. Many obstacles had to be overcome before this goal was reached, and time and again decisive progress was achieved by some of the youngest among us.

Niels Bohr,
Rutherford Memorial Lecture, 1958

The 1930 Copenhagen Conference. In the front row are (from left) Oskar Klein, Niels Bohr, Werner Heisenberg, Wolfgang Pauli, George Gamow, Lev Landau and Hendrik Kramers.

try and physics had been far apart. It was chemistry that had been the science of matter and its specific properties. The atom had been a concept of chemistry: the atom of gold, of oxygen, of silver — different specific entities whose existence had been noted but not understood. Physics, on the other hand, had been a science of general properties of motion, of strain and stress, of electric and magnetic fields. Scientists were not yet able to answer the question: "Where do the specific properties of matter come from?"

The specificity of atoms was a great miracle. What prevents nature from producing a gold atom that is slightly different from another? Shouldn't there be intermediate atoms that are not quite gold but halfway to silver? Why can't there be a continuous change from gold to silver? What keeps all atoms of one species so exactly alike? Why are they not altered by the rough treatment they suffer when the material is heated or subjected to other outside influences? These questions became even more acute and disconcerting when Rutherford found that atoms are little solar systems with the atomic nucleus as the sun and with electrons, like planets, circling around

23

it. Such systems should be extremely sensitive to collisions and other perturbing influences.

Bohr saw that there was a connection between these atomic properties and quantum theory. He tried to formulate the situation by postulating the existence of states for atomic systems that are distinct and characteristic for each species. Electrons can assemble around the nucleus in only a few well defined modes — the quantum states — and not in others. The mode that has the lowest energy is the one in which the electrons will invariably assemble under normal conditions. It is a stable configuration because change is possible only if enough energy is supplied to reach the next quantum state, which is a definite step higher in the energy scale. It is this lowest configuration that is responsible for the typical properties of the atoms.

At first this seemed merely a suitable formulation of the strange fact that atoms have specific qualities. It was actually somewhat more than that, because Bohr developed rules for calculating correctly the energy of these quantum states in some simple cases. But the real significance of this new concept emerged when it became clear that it was intimately connected with the dual nature of electrons, whose behavior is sometimes observed as particle motion, sometimes as wave motion. The quantum states of atoms turned out to be nothing other than the specific vibrations of electron waves confined by electric attraction to a space close to the nucleus. This was a most exciting situation — specific atomic states as harmonic vibrations of electron waves under the confining influence of the electric force provided by the nucleus. The characteristic properties of the elements were revealed to be based upon a natural interplay of vibrations. The ancient ideal of "the harmony of the spheres" seemed to be revived.

But this situation was also deeply disturbing. How can it be that electrons exhibit wave and particle properties at the same time? There seems to be an irreconcilable contradiction between electronic particles revolving around the nucleus, and vibrating electronic waves. Obviously, the mere existence of atoms with well defined specific properties is already proof that strange things must go on within the atom in order to make little solar systems exhibit such behavior. The discovery of the wave-particle nature of the electron only reinforces this. The main point in the Bohr approach to this problem lies in the refutation of the following view: Why not solve the whole problem by looking into the structure of the atom? After all, we could make use of the finest means of observation to find out what the detailed structure is like, and this ought to reveal whether the electron is a wave or a particle.

Bohr and Heisenberg recognized that nature is arranged in a way that makes this approach impossible, because no observation of a tiny object can be made without influencing that object. The quantum state has a peculiar way of escaping ordinary observation, because the very act of such observation obliterates the conditions of its existence. The quantum state is a

their real shape only during those attempts at formulation. The interaction between thought and language always fascinated Bohr. He often spoke of the fact that any attempt to express a thought involves some change, some irrevocable interference with the essential idea, and this interference becomes all the stronger as one tries to express oneself more clearly. Here again there is a complementarity, as he frequently pointed out, between clarity and truth — between *Klarheit und Wahrheit,* as he liked to say. This is why Bohr was not a very clear lecturer. He was intensely interested in what he had to say, but he was too much aware of the intricate web of ideas, of all possible cross-connections; this awareness made his talks fascinating but hard to follow.

The work on uranium fission inevitably brought him into a realm where physics and human affairs are hopelessly intertwined. But even before these discoveries, he had been deeply aware of human problems. He was unusually sensitive to the world in which he lived, and realized earlier than many others that atomic physics would play a decisive part in civilization and in the fate of mankind — that science cannot be separated from the rest of the world. World events brought home this point sooner than expected. By the 1930s, the ivory tower of pure science had already been breached. It was the time of the Nazi regime in Germany, and a stream of refugee scientists came to Copenhagen and found help and support from Bohr. He asked some to stay with him at that time; James Franck, George de Hevesy, George Placzek, Otto Frisch, I myself, and many others found a personal and professional haven in Copenhagen. But this was not all: Bohr's institute was the center for everybody in science who needed help, and many a scientist obtained a post somewhere else — in England, in the United States — through Bohr's efforts. Then came the years of war. Denmark was occupied by the Nazis in April 1940, and Bohr developed close connections with the Danish Resistance. He refused to collaborate with the Nazi authorities. Soon he was forced to leave Denmark, escaping to Sweden and then, via Britain, traveling to the United States.

Now the fourth period of his life began. He joined a large group of scientists in Los Alamos who were working on the exploitation of nuclear energy for war purposes. He did not shy away from this most problematic aspect of scientific activity. He faced it squarely as a necessity, but at the same time it was his idealism, his foresight, and his hope for peace that inspired many people at Los Alamos to think about the future and to prepare their minds for the tasks ahead. He believed that, in spite of death and destruction, there is a positive future for this world, transformed by scientific knowledge.

At that time Bohr was actively engaged in a one-man campaign to persuade the leading statesmen of the Western world of the danger and the hope that might come from an atomic bomb. He wanted those in power to make use of this momentous achievement for the creation of a more open world in which scientific development should bring East and West together in a common endeavor. Bohr conferred with Roosevelt, Churchill, and

UNIVERSITETETS INSTITUT
FOR
TEORETISK FYSIK
———

BLEGDAMSVEJ 15-17, KØBENHAVN Ø
TELEFON: TRIA 1616
TELEGRAMADR.: PHYSICUM, KØBENHAVN

DEN October 26th, 1951.

BY AIR-MAIL

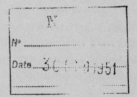

Professor P. Auger,
UNESCO-House,
19, Avenue Kléber,
Paris 16e.

Dear Auger,

From M. Jean Mussard I have received a letter in which, on your instruction, he is sending me some documents relating to the planned European Nuclear Laboratory and asking me for comments on the matter.

I was very happy for our talks at the meeting of the International Union last summer, and I felt that the discussions were most helpful in bringing out different viewpoints as to how one might best proceed with the great European effort with which, in principle, everyone so deeply sympathises. The new idea of Kramers to associate the European Laboratory, in its initial stages, with an existing research institution may, it seems, prove a further step towards relieving the hesitations, as regards commitments to plans and expenditures before everyone has had an opportunity of convincing himself of the soundness of the programme.

In later months we have here thought very much about the matter and of how the project might be organized in its initial stages in case there should be agreement regarding Kramers' suggestion to select the Institute here in Copenhagen as "Pilot institution". On the request of Swedish colleagues we have prepared some tentative notes, of which I enclose a translation, regarding possible procedures in such an eventuality. At a recent meeting of the Swedish Atomic Committee the whole matter was debated and , as you probably know, it was decided to send a letter to the Director General of Unesco expressing the views of Swedish Physicists and authorities.

In case the Unesco consultations should result in a development along such lines the Danish authorities will of course feel it a great responsibility to assist in every way possible. By informal consultations I have learned that the city of Copenhagen is ready to put an appropriate site on the coast line at the disposal for the European Laboratory. The Institute here will also be able to offer hospitality and working conditions for an international expert group as well as facilities for such special scientific and technical research as will be part of the preparations for the major constructions contemplated.

We expect that Kramers' proposal will be closely discussed at the coming Paris meeting and I need hardly add that I should be grateful to hear from you and to learn what you think yourself of the whole matter.

With kindest regards,

Yours sincerely,

Niels Bohr

Facsimile of a letter of 26 October 1951, from Niels Bohr to Pierre Auger, regarding the establishment of a European nuclear laboratory.

other leaders, and he learned quickly the difficulties and pitfalls of diplomatic life (see Part V). His great political concept did not come to fruition; neither did his attempts at raising nuclear technology to an international level in order to avoid a nuclear armament race between powerful nations. Bohr and all others who thought like him—there were a good number all over the world—were deeply disappointed. He ended his efforts for an international understanding on nuclear weapons with his famous letter to the United Nations, written in 1950, in which he made known his thoughts about the need for an open world.

In the last decade of his life, Bohr spent much time organizing international activities in science. He participated in the founding of the Nordic Institute of Atomic Physics (NORDITA) in Copenhagen and he helped create CERN in Geneva—a laboratory in which all European countries could collaborate in the most modern fundamental research in physics. CERN houses one of the world's largest particle accelerators, a symbol of Europe's renaissance in basic science after the United States had taken the lead in this field. This center has become one of the leading laboratories in fundamental physics, where scientists from all over the world collaborate. In many ways, CERN represents Niels Bohr's most significant ideas. It exists today because of Bohr; it was his personality, his stature, and his attention to every detail of its founding and development that made it possible. It is a symbol of the international character of science and is exclusively devoted to basic science. Its results are openly published, and citizens of any country, East or West, are welcome as visitors and collaborators.

In the years following World War II, physics became a major enterprise; large numbers of people and complex machines were needed to carry out physical research. High-energy accelerators made it possible to probe within the nucleus and explore the structure of its constituents, the proton and neutron. Bohr recognized this as a logical continuation of what he and his friends had started. He saw the need for physics on a large scale, on an international scale. In no other human endeavor are the narrow limits of nationality and politics more obsolete and out of place than in the search for more knowledge about the universe.

When Niels Bohr died, an era ended—the era of great scientists who created modern physics. But it was Bohr who helped shape the spirit and the institutions for the continuation of the scientific endeavor into the future.

His death symbolizes his life. Only two weeks before his death he returned from a vacation, fully recovered from a slight stroke he had had a year earlier, and his doctors told him that he could resume work as usual. Two days before his death, he chaired a meeting of the Danish Royal Academy of Science; that weekend he planned to have a party of friends at his house. He was happy and healthy, but when he lay down in the afternoon for a brief rest, he did not wake up. That such a life could be lived in the modern age should be a great encouragement to all of us.[2]

II. THE EARLY YEARS

Bohr's First Theories of the Atom

John L. Heilbron

John L. Heilbron is Professor of History at the University of California, Berkeley, and Director of its Office for History of Science and Technology. He is the author of many works on the history of physics, from the seventeenth to the twentieth century. In this article he shows (contrary to the impression given in elementary textbooks) how difficult and tortuous the path to Bohr's famous model in fact was.

When confronted with an apparent failure in the application of an accepted theory, physicists, like the rest of us, have a choice of strategies. The most obvious, which is also that recommended by armchair methodologists, is to invent an entirely new theory; an example might be Johann Kepler's system of planetary motions. The most likely strategy, because involving the least reconstruction, is to seek the slightest departure from received ideas that will save the phenomena; an example is John Couch Adams's supposition of the existence of a distant planet to account for irregularities in the motions of Uranus. In both cases the strategies worked: Kepler's system, transformed by Newton, became the basis of the world of classical physics; Adams's calculations, and those of his French contemporary Urbain J. J. Leverrier, led to the discovery of Neptune.

They were lucky. Efforts to attribute irregularities in Mercury's motions to an undiscovered planet inside Mercury's orbit failed; no such conservative and minimal adjustment worked, and the little anomaly yielded only to the novel and extravagant theory of general relativity. In contrast, the suggestion by several mathematical physicists of the eighteenth century that a new force, different from Newton's gravity, was necessary to save the motions of the moon proved unnecessary. It is part of the high art of physics to guess correctly whether an apparent failure of theory is an anomaly needing minimal or heroic treatment, or whether it is an anomaly at all.

Perhaps Bohr's greatest strength was his ability to identify, and to exploit, failures in theory. His exercise of this ability amounted to a method. He would collect instances of failure, examine each minutely, and retain those that seemed to him to embody the same flaw. He then invented a hypothesis to correct the flaw, keeping, however, the flawed theory to cover not only parts of experience where it worked, but also parts where neither it nor the new hypothesis, with which it was in contradiction, could account for the phenomena. This juggling made for creative ambiguity as well as for confusion: pushing the contradiction might disclose additional anomalies, and perhaps a better, more inclusive hypothesis. A coherent theory might emerge that would remove the need for cooperation with the flawed theory, and the latter would be restricted to a domain for which it fully sufficed. To work in this way one needs not only creative genius, but also a strong stomach for ambiguity, uncertainty, and contradiction.

Bohr developed this method in steps, beginning with his work on the *electron theory of metals,* on which he wrote his doctoral thesis in 1911, and culminating in the invention and deployment of the correspondence principle in about 1918. It appears nascent in his atomic theories of 1912–13, and triumphant in the struggle that led to the invention of *matrix mechanics.* Like most things, however, it had its time and place. During the 1930s it prompted him to spy revolution in novelties that yielded to slight and natural alterations of the prevailing quantum theory. The main subject of this essay is the nascent method as it appears in Bohr's atomic theories of 1912–13.[1] I begin, however, with an account of his doctoral thesis and a survey of the state of atomic theory when he switched his attention to it in 1912.[2]

THE ELECTRON THEORY OF EVERYTHING

At the turn of the century several important physicists, building on J. J. Thomson's evidence of the ubiquity of the electron and H. A. Lorentz's theory of electron behavior, tried to explain all physical phenomena as consequences of the interactions of electrons among themselves and with "molecules," or collections of electrons. The first outstanding success of the program came in the theory of metals. Thomson, Lorentz, Paul Drude, and others obtained promising agreement with experiment on the assumption that electrons move through metals as do ions through a dilute solution or molecules through a perfect gas. For example, in 1900 Drude deduced that the ratio of thermal (κ) to electrical (σ) conductivity should be the same for all metals, and directly proportional to the absolute temperature, T. His expression agreed with previous empirical generalizations, and with measurement to within a factor of 1.5. Drude then tried his hand at thermoelectric effects; he obtained encouraging indications that a more refined electron theory might account for these also. Lorentz provided such a theory in 1905, treating the free electrons in a metal by the statistical methods worked out for gases. He obtained for κ/σ a value two-thirds of Drude's, which worsened the empirical fit. He advised anyone worried about agreement between theory and experiment to recalculate κ/σ using some special hypothesis about the dependence of electron mean free path on velocity.

One whom it worried was Bohr. In his master's thesis of 1909 he showed that the theoretical value of κ/σ could indeed be raised to the experimental by supposing the mean free path to change with velocity, an assumption he thought equivalent to introducing a force between electrons and metal molecules. The introduction of a central attraction diminishing as the pth power of the distance, the chief innovation of Bohr's doctoral thesis, freed Lorentz's theory from the assumption that electrons interact with metal molecules only when striking them. Bohr brought κ/σ into agreement with experiment with $p = 3$. A few other standard problems in the theory of metals also yielded to him. But several did not, notably two not treated by Drude or by Lorentz in his theory of 1905: heat radiation and magnetism.

In 1903 Lorentz had formulated a theory applicable to radiation whose vibration period is much longer than the average time interval between successive collisions of an electron with metal molecules. His result under this restriction coincided with the long-wave limit of *Planck's radiation formula*. But Planck's formula rested on assumptions very different from those of Drude and Lorentz. Could the agreement be pushed further? In 1907 Thomson thought he had done so. Bohr disagreed. His lengthy calculations showed that Lorentz's derivation held only at the limit considered. Bohr concluded that Thomson's program was hopeless. "The cause of failure is very likely this: that the electromagnetic theory does not agree with the real conditions in matter."[3]

In considering *magnetism,* Bohr followed the theory that Paul Langevin had presented in 1905. Langevin had observed that the electric field set up while an external magnetic field rises or decays, or during reorientation of electron orbits, will cause a change in orbital frequency. The change amounts to a current whose magnetic moment, μ_{dia}, opposes the external field. Langevin concluded that all bodies are diamagnetic. Paramagnetism could arise, therefore, only in substances whose atoms contain moments μ_{para} that can align with the magnetic field and swamp the universal diamagnetism. Langevin noticed that perfect alignment is prevented by thermal agitation, and that in equilibrium the angles at which the moments stand to the field should distribute according to the *Maxwell-Boltzmann law.* Calculation of paramagnetic susceptibility on this hypothesis gave a theoretical expression inversely proportional to absolute temperature, in agreement with *Curie's law.*

Although Langevin's theory had considerable plausibility, Bohr condemned it as untenable, at least with respect to diamagnetism. If mechanical thermal equilibrium is to prevail, he said, the change of velocity induced during the buildup of the magnetic field must quickly equalize among the electrons in a molecule after the field is established. If so, diamagnetism cannot arise from undamped motions of bound electrons. Since Bohr also argued that free electrons cannot cause diamagnetism, because the field cannot alter their distribution in space or velocity, it appeared that the electron theory of metals could not give an account of a fundamental property of matter in its domain.

To harvest the fruit of Langevin's theory without the worm, one might declare the sizes and moments of atoms immune from the averaging demanded by statistical mechanics. Such a declaration, which Bohr deemed necessary, implies the existence of what he called "forces in nature of a kind completely different from the usual mechanical sort," forces that might freeze the structure of atoms and molecules so as to legitimize Langevin's approach and fix the sizes of atoms.

Bohr took from his doctoral thesis the conclusion that the electron theory of metals could easily be brought into agreement with experiment in the cases of κ/σ and of related problems that did not involve the internal structure of atoms, but that it could not give a satisfactory account of heat radiation and magnetism without an additional, radical, extramechanical hypothesis. He probably already associated this hypothesis with the non-classical quantum postulate that Einstein and others had detected as the foundation of Planck's radiation theory. By the spring of 1911 Bohr had pinpointed serious flaws in the prevailing electron theory, and had an inkling of the hypothesis needed to repair it. But he had no precise idea how or where to introduce the hypothesis into the theory.

Bohr spent the academic year 1911–12 in England, engaged in postdoctoral study. He began at Cambridge, where he hoped to discuss problems of the theory with Thomson, to publish his thesis in English, and to find a way of attacking magnetism. But Thomson had ceased to work on the theory of metals, and he was temperamentally averse to the close collaboration and constant conversation that Bohr needed to develop ideas. In the spring of 1912 Bohr moved to Ernest Rutherford's laboratory at the Univer-

Niels Bohr (left) with Max Planck.

sity of Manchester, in order to learn about the experimental side of radioactivity. He soon took an interest in the theory of the nuclear model that Rutherford had then just proposed. Its interest for Bohr lay as much in its imperfections as in its successes. It precisely expressed, and perhaps therefore could model, the results of Bohr's dissertation: in special cases, such as the scattering of alpha particles, the nuclear atom accounted for the phenomena with specific assumptions about the number of charges on the nucleus; but it suffered from radical mechanical instability, which could be overcome only by the sort of rigidification by quantum mechanical fiat that Bohr had come to recognize as necessary from his study of the electron theory of metals.

THEORIES OF ATOMIC STRUCTURE AROUND 1910

The prevailing approach to atomic structure when Bohr was finishing his dissertation derived from Thomson's model and the problems that Thomson had studied with its aid. Thomson had dealt with the greatest uncertainty in atomic theory — ignorance about the positive constituent of the atom — by supposing that bound electrons circulate in coplanar rings within a sphere that acts as if it were filled uniformly with a resistanceless positive charge. This arrangement has the great advantage over the Saturnian atom — in which the electron rings surround a central positive nucleus — of mechanical stability. The Saturnian model had suggested itself to the first physicists who attempted to picture an atom containing electrons; but it was dropped after the discovery that it is not stable against small displacements of the electrons in the plane of their orbits. Thomson's atom, like the Saturnian, eventually collapses from loss of energy by radiation. But, as Thomson showed, the loss in both cases can be made negligible: the larger the number of electrons equally spaced around a ring, the smaller their total radiation. As for the ultimate collapse, it offered an easy explanation of radioactive disintegration.

At first, Thomson supposed that the electrons provided all or most of the mass of the atom. Hence, their number n in an atom of atomic weight A would be about $1,000A$. To check this hypothesis, he devised theories of the scattering of X rays and beta rays by the electrons in his model atoms. Experiments done at the Cavendish Laboratory showed that he had vastly overestimated the electron populations of atoms: n was revealed to be about $2A$, not $1,000A$. Obtaining a more accurate relation between n and A became a major goal of Thomson's research program. In 1910 the Cavendish's best result was $n = 3A$. A little later, Rutherford found that $n = A/2$ by analyzing alpha-particle scattering.

A second line in Thomson's program for atomic theory was to explain the periodic properties of the elements. He made it plausible that periodicity could be a consequence of electromagnetic forces alone by examining the mechanical stability of a ring of electrons against small displacements

You can imagine it is fine to be here, where there are so many people to talk with . . . , and this with those who know most about these things; and Professor Rutherford takes such a lively and effective interest in all that he believes there is something in. In late years he has worked out a theory of the structure of atoms, which seems to be quite a bit more firmly founded than anything one has had hitherto.

Niels Bohr (in Manchester) to his
brother Harald,
12 June 1912

37

around their equilibrium orbit. A single ring of two to seven electrons within a neutralizing positive sphere is stable; the eighth electron must go to the center to achieve stability. The ninth also goes inside, where it and the eighth form a ring of two. Thomson showed that in general the requirement of mechanical stability implies a unique distribution of electrons into rings for each total n of atomic electrons. And he pointed out strong analogies between the properties of certain model atoms and the chemical behavior of elements in the second and third periods of Mendeleev's table.

The third line concerned the building of molecules, the binding together of model atoms. In the vexed case of a diatomic molecule of an elementary gas like hydrogen or oxygen, Thomson argued that a transfer of charge between the initially identical constituents takes place. His illustration of the process is characteristic of his method, which differed fundamentally from Bohr's. Imagine that each atom may be likened to a sealed flask partially filled with water and suspended by a spring. The weak electrical interaction, when the atoms are close together, may be represented by a siphon connecting the flasks. The slightest displacement of one flask relative to the other will cause water to flow through the siphon, increasing the displacement; and the disparity will increase until the air pressure above the water in the lower flask equals the liquid pressure driving the siphon. The flow of water may be taken as transfer of charge between identical model atoms, and the transfer as chemical binding.

In June 1912, when Bohr went to work full time on the theory of the nuclear atom, his first objective was to obtain solutions, on the basis of this new model, to the problems on which Thomson had made progress: the nature of radioactivity and chemical periodicity, and the binding of molecules. It is noteworthy that in Bohr's agenda of 1912, as in Thomson's a decade earlier, explanation of series spectra does not appear. This neglect contrasts sharply with the concern of the few physicists who around 1910 tried to introduce the quantum into the atom. Two examples deserve attention.

The first, the handiwork of Arthur Eric Haas, a doctoral student at the University of Vienna, concerned a single electron oscillating in a neutralizing Thomson sphere of radius a. A simple calculation using Newtonian mechanics shows that the motion is harmonic and that the frequency of oscillation f is independent of the amplitude: $f^2 = e^2/4\pi^2 ma^3$. In what he took to be the spirit of quantum theory, Haas set $f = W/h$, W being an amount of energy and h Planck's constant; since the only unique finite energy in the system occurs when the amplitude equals the radius, he proposed $hf = e^2/a$. Eliminating f from the preceding equations, Haas had what he called "an electrodynamical interpretation of Planck's quantum of action"—namely, $h = 2\pi e(ma)^{1/2}$.[4]

Lorentz took an immediate interest in what he called Haas's "risky hypothesis." In lectures at Göttingen in 1910, he recommended it for having "connected the riddles of the energy elements with the question of the nature and action of positive electricity." In discussions at the Solvay Conference of 1911, called to consider the pressing problems of radiation and the quantum, he insisted on a link between "the size of the constant h and the dimensions of atoms (positive Thomson spheres)." The Solvay participants discussed several ways other than Haas's for introducing the quantum into the atom, and they considered, without reaching agreement, whether the quantum to be employed was Planck's h or $h/2\pi$ or $h/4$. Their papers and proceedings appeared as a book in 1912.[5] It gave Bohr, who read it, something to think about. It also inspired J. W. Nicholson, author of the other early quantized atom we are to consider, to hit on a method of quantization that Bohr came later to adapt.

Nicholson had studied single-ring models containing a few electrons; the most interesting of these theoretical atoms, "nebulium," had four electrons. He offered striking evidence in favor of the existence of this "protoelement" in the stars. Small vibrations of electrons perpendicular to the plane of their ring can be stable, unlike vibrations in the plane. Nicholson computed the frequencies $f_\perp$ of the perpendicular vibrations, and compared them with the frequencies $v_?$ of unassigned lines in nebular spectra. Now, $f_\perp$ depends upon the charge on the nucleus, n, the number of electrons in the ring, p, and the ring radius, a. With a free choice of a, Nicholson made nine of eleven nebular $v_?$'s agree with as many $f_\perp$'s for normal and ionized "nebulium." A few months later, astrophysicists found a new $v_?$ that agreed perfectly with an unassigned $f_\perp$, and in the meantime recognized that the two "nebular" lines that Nicholson could not accommodate belonged to a terrestrial element.

The counsel from Solvay inspired Nicholson to try to fix the frequency f of the unperturbed rotation of the ring by a quantum rule. He computed the total energy E of each of his model atoms, using the parameters that experience had forced upon him. He then formed E/f, and learned that in all cases it equaled a whole number of quanta. After studying the Solvay proceedings, Nicholson required that the *angular momentum* of his models be an integral multiple of $h/2\pi$. That worked, too. He ended with the following picture of a radiating atom: "We are led to suppose [by the quantum theory] that lines of a series may not emanate from the same atom, but from atoms whose internal angular momenta have, by radiation or otherwise, run down by various discrete amounts from some standard value."

BOHR'S ATOMIC THEORIES, 1912–1914

Bohr was drawn to the problem of atomic structure by his critical reaction to a paper by Rutherford's mathematical physicist, C. G. Darwin (whom his good friend Bohr called "the grandson of the real Darwin"), on the slowing of alpha particles in their passage through matter. Bohr objected that Darwin had neglected resonance effects that enhance energy transfer when the time of flight of the particle past an atom roughly equals the natural period of oscillation of some of its electrons. In trying to calculate the effects, Bohr rediscovered the mechanical instability of the Saturnian model; to proceed further, he needed to introduce a hypothesis that would allow him to calculate mechanical quantities from mechanically unstable models. The hypothesis, or rather fiat: any circular orbit satisfying the condition

$$T/f = K \tag{1}$$

is stable against mechanical perturbations and radiation loss. Here T and f represent the kinetic energy and frequency of the orbiting electron, and K is analogous to Planck's h. The kinetic energy T replaces the total energy W so as to take account of two major differences between Rutherford's model and a Planck oscillator: W is negative, and f depends on T. Several considerations, other than a loose analogy to Planck's procedure, prompted or justified writing a prescription of stability in the form (1). The formula provided, as Bohr observed, a resolution of the problem of atomic size. And it allowed him at last to express mathematically the conviction to which he had been led by his attempts to save the electron theory of metals.

Bohr's chief line of research from June 1912 to February 1913 was the exploitation of condition (1) in the service of the nuclear model and the standard problems addressed in Thomson's theory of atomic structure. In June or July 1912, in a memorandum drawn up for discussion with Rutherford, Bohr began by reinterpreting Thomson's main accomplishment: the reduction of chemical periodicity to atomic structure. He calculated the total energy W_p of an electron in a single ring of p electrons stabilized by his quasi-quantum fiat, and discovered that W_p changes from negative at $p = 7$ to positive at $p = 8$. On this reckoning an electron can leave a single ring containing more than seven others. The eighth electron goes into orbit outside the ring of seven; placing it inside does not make the ring capable of holding more, in contrast to Thomson's model.

Whereas Thomson embroiled all atomic electrons in chemical and optical behavior, arrived at chemical periods of steadily increasing length, and offered no neat distinction between the mechanics of ionization and radioactivity, Rutherford's model afforded a firm base for understanding the populations of at least the first two periods, and for distinguishing the regions implicated in chemistry and spectroscopy (the electronic structure) from those responsible for radioactivity (the nuclear black box). Bohr thus arrived at the concept of isotopy on his own, and could recognize immediately as "in complete accord with my ideas" the electrochemistry of the radioelements as summarized by his friend George de Hevesy in 1912. Much therefore depended on Bohr's demonstration that W_p changes sign between $p = 7$ and $p = 8$. The proposition is, however, altogether wrong; it rests on a numerical error—a doubling of the potential energy—that Bohr soon discovered. No doubt the need to find an alternative to Thomson's theory of periodicity encouraged his productive mistake.

Things are going rather well, for I believe I have found out a few things; but, to be sure, I have not been so quick to work them out as I was so stupid to think. I hope to have a little paper ready and to show it to Rutherford before I leave, and I am therefore so busy, so busy.

Niels Bohr to his brother Harald,
17 July 1912

Although Bohr failed to explain chemical periodicity in his first tilt with atomic structure, he retained the result to which his error led him: that an electron added to an atom with a saturated ring or rings goes outside, not inside, the existing structure. With this proposition and Rutherford's approximation $n \approx A/2$, Bohr could go far beyond Thomson and state precisely the number of electrons in any normal atom. Since the nuclear model requires $n_{He} = 2$, it follows that $n_H = 1$, $n_{Li} = 3$, and so on, each neutral atom containing a number of orbiting electrons equal to the number Z of its element in the periodic table, counting from hydrogen as one. Bohr spent much time in 1912 and 1913 assigning ring structures to light atoms and searching for a principle that would account for the periodicity in the table of elements.

In the problem of diatomic molecules, Bohr worked from calculation and a model (a ring of binding electrons coaxial with the two nuclei), whereas Thomson had made do with analogy. The simplest case is the neutral hydrogen molecule, H_2. The electrons, always diametrically opposite each other, circulate under a net electrical attraction that provides the

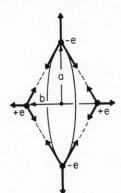

The H_2 molecule according to Bohr's 1913 theory.

needed centripetal force. For the nuclei (two protons) to stand in equilibrium under the electric forces alone, we must have $b = a/\sqrt{3}$. For each electron, ordinary mechanics requires that

$$ma(2\pi f)^2 = \frac{e^2}{a^2} X. \qquad (2)$$

For the hydrogen molecule, this model gives $X = 1.049$. This may be compared with the result for a single electron orbiting a single central charge: for the hydrogen atom $X = 1$, and for singly ionized helium $X = 4$. Combining equations (1) and (2) gives the following results:

$$-W = \frac{Xe^2}{2a} = \text{kinetic energy of orbiting electron} = \frac{\pi^2 m e^4 X^2}{2K^2} \qquad (3)$$

$$a = \text{orbital radius} = \frac{K^2}{\pi^2 m e^2 X} \qquad (4)$$

41

$$f = \text{orbital frequency} = \frac{\pi^2 m e^4 X^2}{2K^3} \qquad (5)$$

Even without knowing K, Bohr could deduce from their relative energies that H_2^+ should dissolve spontaneously into a hydrogen atom and a hydrogen nucleus, and that miscegenated molecules such as $H \cdot He$ do not form. These results were encouraging, and scarcely precedented.

Bohr returned to Copenhagen in the autumn of 1912 to teach, and to wrestle with dispersion and magnetism. He began to worry that he might be anticipated. Others had found pieces to his puzzle: the Solvay symposiasts, the several guessers at isotopy, and, in January 1913, in the unlikely person of a Dutch lawyer, Antonie van den Broek, an independent discoverer of the doctrine of atomic number. Van den Broek reasoned from two approximations: Rutherford's result, $n \approx A/2$, and the average change in atomic weight, $\Delta A \approx 2$, from one element to the next in the periodic table.

Sketches by Bohr of his models for hydrogen and helium.

From the two approximations he inferred an exact law, $\Delta n = 1$; from Thomson's experiments on positive rays, he took $n_H = 1$, and from Rutherford's on alpha particles, $n_{He} = 2$.[6] He declared that "each element must [therefore] have an inner charge equal to $n[e]$ — i.e., the nth element must have n intra-atomic charges of the same sign" that determine all its characteristics. "I am afraid I must hurry if [my work] is to be new when it appears," Bohr wrote in February 1913, after reading van den Broek. "The question is indeed such a burning one."

RADIATIVE TRANSITION

As the old questions were burning, a new one ignited. By the end of 1912 Bohr had met Nicholson's model of the atom, which, like his, was quantized and nuclear. But Nicholson's atoms spoke the language of spectra, whereas Bohr's were mute; and nebulium had the same electronic population as the doctrine of atomic number demanded for beryllium. "I therefore thought at first that the one [model] or the other was altogether wrong." The conflict did not last long for Bohr. On New Year's Day 1913, he wrote to his brother: "[My] calculations would be valid for the final, chemical state of the atoms, whereas Nicholson's would deal with the atoms' sending out radiation, when the electrons are in the process of losing energy before they have occupied their final positions." Having settled with Nicholson, Bohr returned to his own problems. "I do not at all deal with the question of calculation of the frequencies corresponding to the lines in the visible spectrum," he wrote to Rutherford. "I have only tried, on the basis of the simple hypothesis, which I used from the beginning, to discuss the constitution of the atoms and molecules in their 'permanent' state."

Radiation appeared on Bohr's agenda in February 1913, when H. M. Hansen, Copenhagen's expert on spectroscopy, asked or advised him to explain the Balmer formula for the frequencies v_n of the lines of the visible spectrum of atomic hydrogen:

$$v_n = R\left(\frac{1}{4} - \frac{1}{n^2}\right), \tag{6}$$

where R is the *Rydberg constant*. "As soon as I saw Balmer's formula," Bohr recalled, "the whole thing was immediately clear to me." Clarification may have dawned along the following lines. In Balmer's formula, frequency appears as a difference, which, in accordance with the rough contemporary practice, could be connected with an energy difference by multiplying by h. Thus, Rh/n^2 would represent an energy, possibly the energy of one of Nicholson's radiating or radiative states. But Bohr's equation (3) for energy $-W_n$ contains in its denominator K^2, where K is the same kind of quantity as h; compare this with equation (1). By putting $K = \alpha h n$, where α is a pure numerical factor, equation (3) as applied to the hydrogen atom ($X = 1$) takes on the form Rh/n^2 with $Rh = \pi^2 m e^4 / 2\alpha^2 h^2$. Using the then most recent values of the constants, Bohr would have found $\alpha = \frac{1}{2}$, or $K_n = nh/2$. He

43

From top: J. J. Balmer (1825–1898); the longer-wavelength lines in the Balmer spectrum; the Balmer lines in the ultraviolet, up to the series limit.

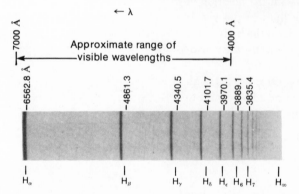

Line	Color	λ (Å)
H_α	Red	6563
H_β	Turquoise	4861
H_γ	Blue	4341
H_δ	Violet	4102
H_ϵ	Extreme violet	3970

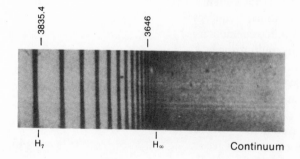

then had as the condition defining the nonpermanent or Nicholson states

$$(T/f)_n = nh/2, \tag{7}$$

from which he could compute $-W_n$, reverse his argument, and obtain the Rydberg constant R of equation (6) as a product of the atomic constants:

$$R = \frac{2\pi^2 me^4}{h^3}.$$

To make a theory of this game, he needed a justification of equation (7). He supplied three:

a. The first part of Bohr's three-part paper on atomic structure published in 1913 opens with an appeal to Planck's radiation theory. In falling from infinity into an excited or Nicholson state — say, the nth — under the attraction of a bare nucleus of charge Ze, an electron will radiate away energy E_n at frequency ν_n, where, "from Planck's theory," $E_n = nh\nu_n$. Bohr offered the following connection between this arbitrary adaptation of Planck and the desired expression $T_n = nhf_n/2$. A planetary electron bound by an inverse-square force has as total energy W_n the negative of its kinetic energy, $-W_n = T_n$; the energy required to remove it again to infinity, $-W_n$, equals the amount it lost by radiation during the binding, wherefore $E_n = -W_n = T_n = nh\nu_n$. It remains to show that $\nu_n = f_n/2$. Bohr put the average of the initial mechanical frequency, $f = 0$, and the final f_n, equal to the radiated frequency. Here he reached an abyss. On both ordinary and Planck radiation theory, the frequencies in the light from an atomic radiator are just those of the electrons producing it; on Bohr's argument, invented to introduce a factor of two, mechanical and radiated frequencies no longer coincide. Pushing his hybrid theory had revealed an anomaly more serious than any with which he had started.

Having achieved $T_n = nhf_n/2$, Bohr presented the equivalents of equations (3) – (5) with $K = nh/2$ and $X = Z$:

$$\left. \begin{aligned} -W_n &= \frac{2\pi^2 me^4 Z^2}{n^2 h^2} \\[1em] f_n &= \frac{4\pi^2 me^4 Z^2}{n^2 h^2} \\[1em] a_n &= \frac{e^2}{(-2W_n)} = \frac{n^2 h^2}{4\pi^2 me^2 Z^2} \end{aligned} \right\} \tag{8}$$

The formulas for the several spectral series of neutral atomic hydrogen ($Z = 1$) follow from energy balance:

$$\nu_{n,m} = \frac{W_n - W_m}{h} = \frac{2\pi^2 me^4}{h^3} \left(\frac{1}{m^2} - \frac{1}{n^2} \right) \tag{9}$$

The Balmer formula results from $m = 2$, $n = 3, 4, \ldots$

The daring (not to say scandalous) character of Bohr's quantum postulate cannot be stressed too strongly: that the frequency of a radiation emitted or absorbed by an atom did not coincide with any frequency of its internal motion must have appeared to most contemporary physicists well-nigh unthinkable.

Léon Rosenfeld, introduction to a reprint of the Trilogy, 1963

Although evidently ad hoc, Bohr's "derivation" of his fundamental condition $T_n = nhf_n/2$ had some precedents in the widespread view that series spectra arise during electron capture, and in then recent reformulations of the quantum postulate by Planck and others. But the unintelligible outcome of Bohr's calculation, which, in the case of the Balmer formula, takes the form $v_n = f_2 - nf_n/2$ ("the frequency of the nth Balmer line equals the mechanical frequency of the second orbit less $n/2$ times that of the nth") had no precedent.

b. After submitting his ad hoc condition on the radiation, $E_n = nhv_n$, and its justification, intended to connect his work closely with Planck's, Bohr discarded both in favor of a condition on the orbits and one-quantum emission, $hv = \Delta W$. The restriction is just $(T/f)_n = nh/2$, now stated without argument. As Bohr observed, it is equivalent to Nicholson's condition that the orbital angular momentum be an integral multiple of $h/2\pi$.

c. The third derivation, the deepest of all, begins by stipulating asymptotic agreement at large values of n between radiated frequencies as calculated by ordinary mechanics and by Bohr's theory. The rationale for the stipulation was that with large n and big radii, the consequences of the restriction arising from atomic binding should disappear: the agreement concerns the numerical values of spectral frequencies, not their methods of production. Stipulating therefore that $v_{n,n-1} \approx f_n \approx f_{n-1}$ for $n \gg 1$, and taking as condition on the states $T_n = \beta(n)hf_n$, $\beta(n)$ an unknown function, Bohr observed that the Balmer formula demands $\beta(n) = \alpha n$, with α a constant. That $\alpha = \frac{1}{2}$ emerges from the asymptotic condition

$$v_{n,n-1} = \frac{\pi^2 m e^4}{2h^3}\left[\frac{1}{\alpha^2(n-1)^2} - \frac{1}{\alpha^2 n^2}\right] \approx \frac{\pi^2 m e^4}{\alpha^2 h^3 n^3} \approx f_n \approx \frac{\pi^2 m e^4}{2\alpha^3 h^3 n^3}.$$

(The equations for v and f follow from Equations (3)–(5), with $K = \alpha nh$.) At the end of 1913, in a lecture to the Danish Physical Society, Bohr showed how to derive R without imposing any condition on the orbit. From the Balmer formula, read as an energy equation, he took $-W_n = -Rh/n^2$; from the usual relations among mechanical quantities of the orbit, $f_n = (-2W_n^3/m)^{1/2}/\pi e^2$, whence $f_n^2 = 2R^3 h^3/\pi^2 m e^4 n^6$. Asymptotically,

$$(v_{n,n-1})^2 = R^2\left[\frac{1}{(n-1)^2} - \frac{1}{n^2}\right]^2 \approx \frac{4R^2}{n^6},$$

which, when equated to the above value of f_n^2, gives $R = 2\pi^2 m e^4/h^3$ once again. By this time Bohr had decided that it was "misleading" to use his original analogy to Planck's oscillator. He had greatly refined the hypothesis with which he repaired the mechanical instability of the nuclear atom.

In one particular, Bohr's mode of spectral analysis enabled him to make a striking prediction. Spectroscopists had detected lines whose frequencies approximately fitted the series formula

$$v_n = R\left[\frac{1}{4} - \frac{1}{(n/2)^2}\right],$$

which is known as the *Pickering series,* and had attributed them to hydrogen on analogy to the Balmer formula. Bohr had no place for half integers, since in his view the running term of the spectral formulas for hydrogen numbered the possible orbits of the single electron; a half-integral orbit would correspond to an inadmissible half-quantum. He accordingly rewrote Pickering's series as $v = 4R(1/4^2 - 1/n^2)$, and attributed it not to hydrogen but to ionized helium, which, with a nuclear charge of two, should have an effective Rydberg constant four times that of hydrogen. English spectroscopists confirmed Bohr's conjecture by finding Pickering lines in helium carefully cleansed of hydrogen. Given the great precision of wavelength measurements, however, the agreement with experiment was not very good; the Pickering formula had never agreed with experiment nearly as well as the Balmer formula did. Bohr cleared up this business in October 1913 in a master stroke that destroyed the Rydberg's universality and made theorists take him seriously. He had neglected, he said, the small motion of the heavy nucleus in estimating the electron's energy in the *stationary states.* Repairing this omission in the way taught in elementary mechanics, which amounts to replacing m by $m' = m/(1 + m/m_Z)$, m_Z being the mass of the nucleus, Bohr came up with a formula for the true hydrogen Rydberg: $R_H = (m'_H/m)R$. Theory required $R_{He}/R_H = 4.00163$. Spectroscopists looked, and reported $R_{He}/R_H = 4.0016$. The impression made by this extraordinary confirmation — in which refinements required by the flawed mechanical theory were invoked to confirm the nonmechanical quantum postulate — may be gauged from Hevesy's description (in a waywardly spelled letter to Bohr) of Einstein's reaction to the news that the Pickering series belongs to helium. "When he heard this, he was extremely astonished and told me: 'Than the frequency of the light does not depand at all on the frequency of the electron . . . And this is an *enormous achiewement.* The theory of Bohr must be then wright.'"

CONFIRMATIONS AND EXTENSIONS

The success with ionized helium capped Bohr's own initiatives. Further support for his views came in 1914, from three lines of inquiry taken up without reference to his theories or even to the general problem of atomic structure. One, begun as a study of the nature of X rays, continued as an exploration of characteristic X-ray spectra, and ended in a search for new elements. The explorer, H. G. J. Moseley, whom Bohr had met at Manchester, discovered that the X-ray lines of highest frequency emitted by the metals he examined satisfied the equation $v = (3R/4)(Z - 1)^2$, Z being the atomic number of the metal. The equation resembled Balmer's, and the unsuccessful theory that Moseley invented to derive it resembled Bohr's. The relations among them were made clear in 1915 by Walther Kossel, who interpreted X-ray emission as the transition of a vacancy in the electronic structure from the inner to the outer reaches of the atom.

A second corroboration emerged from the experiments of James Franck

Niels Bohr (right) with Arnold Sommerfeld in Lund, Sweden, in 1919.

Dear Colleague: I thank you very much for sending me your extremely interesting work, which I had already studied in the Philosophical Magazine. The problem of expressing the Rydberg-Ritz constant by Planck's h has been for some time in my thoughts. A few years ago I talked about it to Debye. Although I am for the present still rather sceptical about atom models in general, nevertheless the calculation of this constant is indisputably a great achievement . . . Are you also going to use your atom model for the Zeeman effect? I wanted to work on it. Perhaps I can hear of your plans in greater detail through Mr. Rutherford, whom I hope to see in October.

Arnold Sommerfeld to Niels Bohr,
4 September 1913

and Gustav Hertz, who had started measuring what they thought were ionization potentials in gases before Bohr took up the study of atomic structure. Bohr was able to reinterpret the values they gave for "ionization potentials" as energies of the lowest *excited states* of the gas atoms; and he was able to extract, from their finding that an atom does not exchange energy with an electron whose energy is insufficient to "ionize" it, a demonstration of the existence of discrete electronic orbits or atomic stationary states. The third corroboration was squeezed from the discovery in 1913 by Johannes Stark that a spectral line can be split into several by a strong electric field. Since physicists at that time argued that on classical theory the effect ought not to have been detectable, Stark's results immediately challenged quantum theory, and several physicists showed that, with one or another adjustment of Bohr's quantum hypothesis, the observed splitting could be calculated.

In 1915 Arnold Sommerfeld began to rework Bohr's hodgepodge theory into a formal structure by imposing conditions on the electron orbits and by making the deduction of the energy levels of the atoms a straightforward algorithm. Bohr improved Sommerfeld's approach by bringing out connections between quantum formalism and the *Hamilton-Jacobi theory* of classical mechanics. This effort, which occupied him from 1916 to 1919, sharpened the contradiction between the main elements in his theory, and, in a few cases, showed how to fashion or calculate quantum-theoretical quantities from a knowledge of "corresponding" classical concepts. Characteristically, as he took the demands of classical mechanics more and more seriously in order to fix the fulcrum on which his correspondence principle would turn, he also restricted still further the reach of "ordinary concepts" (as he called the ideas of classical science) in atomic physics. Advance was possible there, Bohr wrote in the autumn of 1914, only by "departing from the usual considerations to an even greater extent than has [yet] been necessary." His chief guide and compass for this departure — for tracking down the true quantum theory that would replace his contradictory and makeshift one — were to be the usual considerations and ordinary concepts that no longer sufficed.

The Theory of the Periodic System

Helge Kragh

Helge Kragh studied physics and chemistry at the University of Copenhagen and continued his studies at the Niels Bohr Institute. He teaches in the Institute of Mathematics and Physics at Roskilde University, Roskilde, Denmark.

Although Niels Bohr's atomic theory of 1913 was essentially a theory of the hydrogen atom, it also promised a deeper understanding of the structure of more complicated atoms. On the basis of earlier atomic models, such as J. J. Thomson's, it was not possible to construct realistic models of atoms. But with Bohr's theory the dream of many chemists and physicists — to explain the nature and interrelationship of the chemical elements by means of atomic theory — seemed within the reach of realization. The idea that the periodic system of elements can, in principle, be deduced from the mechanics of atoms was old. It went back to Lothar Meyer and Dmitri Mendeleev, who first established the periodic system in 1869.

Bohr's theory of 1913 resulted in a flow of more or less speculative proposals concerning the arrangement of electrons in the various elements. Bohr himself had attacked the problem in the second part of his Trilogy, in which he presented tentative structures for part of the periodic system. "It will be attempted," he wrote, "to obtain indications of what configurations of the electrons may be expected to occur in the atoms." Bohr pictured atoms as systems of concentric electron rings and proposed the configurations shown in the table. Although most of these structures were wrong, he clearly indicated the fundamental explanation of the periodic system: the similarity between elements of the same chemical group is caused by their having the same number of electrons in the outer shells. In Thomson's earlier explanation, which was not based on the nuclear atom, the similarity was assumed to depend on the inner electron structures.

After his *tour de force* of 1913 Bohr turned his interest to the general structure of quantum theory and, in particular, the development of what came to be known as the correspondence principle. "For the present I have stopped speculating on atoms," he wrote to Henry Moseley in 1913. Other

Bohr's arrangement of the electrons in light atoms, as proposed in 1913.

H (1)	F (4,4,1)	Cl (8,4,4,1)
He (2)	Ne (8,2)	Ar (8,8,2)
Li (2,1)	Na (8,2,1)	K (8,8,2,1)
Be (2,2)	Mg (8,2,2)	Ca (8,8,2,2)
B (2,3)	Al (8,2,3)	Sc (8,8,2,3)
C (2,4)	Si (8,2,4)	Ti (8,8,2,4)
N (4,3)	P (8,4,3)	V (8,8,4,3)
O (4,2,2)	S (8,4,2,2)	Cr (8,8,4,2,2)

scientists attempted to construct models of the higher atoms, guided by Bohr's theory and the vast amount of chemical and spectroscopic data. By 1920 there were several proposals as to the arrangement of electrons throughout the entire periodic system — proposals, for example, by Rudolf Ladenburg in Germany, L. Vegard in Norway, Irving Langmuir in the United States, and C. R. Bury in England. However, none of these proposals was satisfactory or generally accepted. Whether based on chemical or physical methods, all involved a great deal of speculation and very little theoretical justification. Bohr wanted to understand the reasons for the periodicity of the elements in terms of the general principles of quantum theory. In his new theory of atomic structure, developed in 1921 – 1923, he attacked this difficult problem in a highly original and largely successful way. Although Bohr's theory was not the first attempt to account for the periodic system in terms of atomic structure, it was the first theory of the complete periodic system which built consistently on the new quantum theory.

A NEW ATOMIC THEORY

From 1914 to 1916 Bohr held the Schuster Readership in physics at the University of Manchester, while waiting to be appointed a professor at Copenhagen. He obtained the professorship in 1916 and started to work, under unsatisfactory conditions, in the buildings of the Institute of Polytechnics. Bohr realized that the inadequate facilities at the institute were preventing theoretical physics from developing in Denmark and proposed in 1917 that the university should establish its own institute for fundamental physics. Planning the new institute turned out to be slow and troublesome, and Bohr expended much time and energy to bring into existence what eventually became Universitetets Institut for Teoretisk Fysik, better known as the "Bohr Institute." It was inaugurated on 3 March 1921, with Bohr, of course, as its director. He engaged himself whole-heartedly in the establishment and running of the institute. The administrative burdens, together with his scientific endeavors, led him to a state of collapse in the

51

POLITIK[

Kjøbenhavn. Torsdag den 3. Marts 1921.

Universitetets Atom-Institut, der indvies i Dag

Institutets store Laboratorium og Medlemmer af Staben.

Ved en Højtidelighed indvies i Eftermiddag Universitetets Institut for teoretisk Fysik, populært kaldet Atom-Institutet, hvis Grundlægger og Leder er den kun 35 Aar gamle, allerede verdensberømte Videnskabsmand, Professor Niels Bohr.

Paa det store Billede ses det Rum, hvor de videnskabelige Forsøg med Atomerne foretages efter den af Professor Franck opfundne Metode.

I Hjørnet ser vi enkelte af Institutets Medarbejdere. I øverste Række fra venstre til højre: Ungareren, Professor Hevesy, Docent H. M. Hansen og Professor Niels Bohr. Nederst Tyskeren, Professor Franck og Hollænderen, Dr. Kramers.

Vi henviser i øvrigt til omstaaende Artikel.

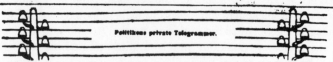

Politikens private Telegrammer.

Lloyd George afgiver i Dag en Erklæring, som vil afgøre Evropas foreløbige Skæbne.

The opening of the Niels Bohr Institute, as reported in the newspaper *Politiken,* Copenhagen.

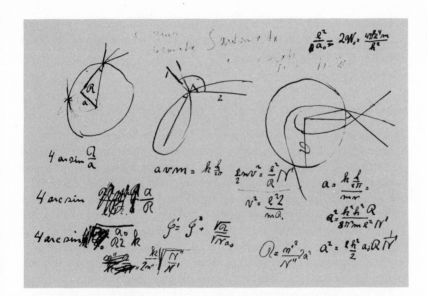

Rough notes concerning the structure of higher atoms, made by Bohr in 1920.

spring of 1921. Nevertheless, it was at that time that he developed his new theory of the structure of atoms.

Bohr first presented his new ideas in a lecture before the Physical Society in Copenhagen in December 1920. In March and September 1921, he published the essential features of the theory in two short papers in *Nature*. These papers aroused great interest among physicists in England and Germany, who urged Bohr to publish a more detailed and complete version of his theory. Bohr's use of the correspondence principle in particular puzzled his colleagues, as shown by the following excerpts from letters: "I have read your *Nature* letter with eager interest . . . Of course I am now even more interested to know how you saw it all in terms of correspondence" (Ehrenfest to Bohr, 27 September 1921). "Everybody is eager to know whether you can fix the 'rings of electrons' by the correspondence principle or whether you have recourse to the chemical facts to do so" (Rutherford to Bohr, 26 September 1921).

A more complete version of Bohr's theory was published in an extraordinary sixty-four-page article entitled "The Structure of the Atoms and the Physical and Chemical Properties of the Elements," which appeared in *Zeitschrift für Physik* in March 1922. The article gained a wide readership and was soon translated into English and French. German physicists obtained a firsthand impression of Bohr's theory in June 1922, when he delivered a series of lectures at the University of Göttingen. These lectures, later known as the "Bohr festival," were an important event in the history of atomic physics (see the reminiscence by Friedrich Hund in this volume). The young German physicists, such as Pauli, Hund, Alfred Landé, Walther Gerlach, Pascual Jordan, and Heisenberg, were fascinated by Bohr's lectures and personality. They became convinced that the quantum theory of atoms was the hottest area of current physics and that Bohr was its prophet. Pauli later said that when he met Bohr in Göttingen, "a new phase of my life began."

FROM HYDROGEN TO URANIUM, AND BEYOND

At the time that Bohr's theory emerged there were, basically, two different conceptions as to the arrangement of electrons in the atom. According to the static model, proposed by Gilbert Lewis and Irving Langmuir and favored by most chemists, the electrons were arranged in shells, each electron or pair of electrons occupying a stationary position. However, most physicists preferred a dynamic model based on Bohr's model of 1913. According to this view, the electrons moved rapidly in circular (or elliptic) orbits in a ringlike structure. Bohr thought that both models were inadequate. He dissociated himself in particular from the chemists' static atom, which, he pointed out, could not be justified by quantum theory.

The basic features of Bohr's new picture of the atom can be summarized as follows:

a. The electrons move around the nucleus in elliptic orbits, grouped according to their *quantum numbers*. In classifying the orbits, Bohr used the principal quantum number n ($n = 1, 2, 3, \ldots$) and the azimuthal quantum number k ($k = 1, 2, \ldots, n$). Electrons with $n = 1$ were called K electrons, those with $n = 2$ were called L electrons, and so on. A group of z electrons with specific n and k quantum numbers was written by means of the notation $(n_k)^z$. The two quantum numbers specify the geometric orbit. If $k = n$, the orbit is circular, whereas k smaller than n indicates a Kepler ellipse whose eccentricity increases with $n - k$. The so-called inner quantum number (j), used in the classification of spectra, played an insignificant role in Bohr's scheme.

b. The structure of an atom of atomic number Z can be considered as the result of the successive addition of Z electrons to a bare nucleus. According to Bohr's so-called *Aufbauprinzip* ("construction principle"), the addition of electron number p to a partially completed atom with $p - 1$ bound electrons will leave the quantum numbers of the $p - 1$ electrons unchanged. When, in this building-up process, a new atom is formed, the principal quantum number of the last captured electron will differ from that of the already bound electrons in the outer shell only if the atom being formed belongs to a new period of the periodic system. Thus, in each new period n increases by one unit. For example, although helium has the structure $(1_1)^2$ the new electron in lithium will be in the quantum state 2_1 — that is, it moves elliptically around the two crossed 1_1 circular orbits of helium.

c. The valence electrons of most metals move in elliptic n_1 orbits which, because of their eccentricity, penetrate the stable shell of eight electrons that

My early research work, in the early 1920s, was based on Bohr orbits, and was completely unsuccessful. I was taking the Bohr orbits as physically real and trying to build up a mathematics for them. The Bohr orbits apply to individual electrons, and for an atom containing several electrons one required a theory of Bohr orbits in interaction. I worked hard on this problem, along with other people. One sees how futile such work was. Heisenberg showed that one needed a completely new mathematics, involving noncommutative algebra. The Bohr orbits were an unsound physical concept and should not be used as the basis for such a theory. I learnt my lesson then. I learnt to distrust all physical concepts as the basis for a theory. Instead one should put one's trust in a mathematical scheme, even if the scheme does not appear at first sight to be connected with physics. One should concentrate on getting an interesting mathematics.

P. A. M. Dirac, "The Mathematical Foundations of Quantum Theory," 1978

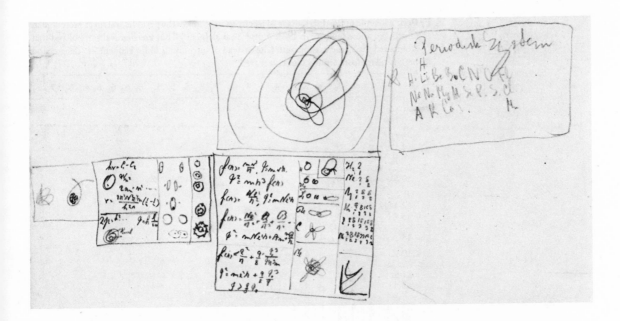

Bohr's notes of 1921 for his new atomic theory, embodying penetrating electron orbits and a new scheme for the distribution of electrons.

characterizes the noble gases. The idea of penetrating orbits was independently published by Erwin Schrödinger in 1921, but only Bohr applied it systematically to the periodic system. Bohr showed that in order to account for the optical spectrum of sodium, its valence electron must be assumed to penetrate the neon core and thus be classified as a 3_1 electron. Most earlier theories had assumed the electron to be in a 1_1 state, and Schrödinger had proposed the state 2_1.

d. The structure of the so-called *transition group* elements was explained as a result of the penetration effect. Whereas the two valence electrons in calcium are both in a state 4_1, corresponding to the two 3_1 electrons in magnesium, the next electron will not be in the N group but in the state 3_3 — that is, belonging to the M group. From scandium to copper, the M group is completed from $(3_1)^4(3_2)^4$ in calcium to $(3_1)^6(3_2)^6(3_3)^6$ in copper. Bohr's explanation of the transition groups, including the rare earths, thus resembled the older ideas of intermediate shells proposed by Rudolf Ladenburg and others. However, Bohr's explanation was based on a model in which the concept of electron shells had no proper meaning.

Elaborating on these and other concepts and making use of the empirical knowledge of the elements, Bohr was able to assign definite structures to the atoms of the entire periodic system (see table on next page). Bohr had considerable confidence in his theory. He believed that he had established the true principles for describing the structure of atoms and that a physical explanation of the periodic system, as dreamt of by Mendeleev, was within his reach. He did not hesitate to construct hypothetical transuranium elements, announcing to his audience in Göttingen that "one could go

Bohr's scheme of the distribution of electrons in atoms of the periodic system as presented in Göttingen and elsewhere in 1922. Notice that element 72 has no chemical symbol and that Bohr used the name cassiopeium (Cp) instead of lutecium (Lu) for element 71. The version here is taken from Bohr's Nobel lecture.

	1_1	2_1	2_2	3_1	3_2	3_3	4_1	4_2	4_3	4_4	5_1	5_2	5_3	5_4	5_5	6_1	6_2	6_3	6_4	6_5	6_6	7_1	7_2
1 H	1																						
2 He	2																						
3 Li	2	1																					
4 Be	2	2																					
5 B	2	2	(1)																				
10 Ne	2	4	4																				
11 Na	2	4	4	1																			
12 Mg	2	4	4	2																			
13 Al	2	4	4	2	1																		
18 A	2	4	4	4	4																		
19 K	2	4	4	4	4		1																
20 Ca	2	4	4	4	4		2																
21 Sc	2	4	4	4	4	1	(2)																
22 Ti	2	4	4	4	4	2	(2)																
29 Cu	2	4	4	6	6	6	1																
30 Zn	2	4	4	6	6	6	2																
31 Ga	2	4	4	6	6	6	2	1															
36 Kr	2	4	4	6	6	6	4	4															
37 Rb	2	4	4	6	6	6	4	4			1												
38 Sr	2	4	4	6	6	6	4	4			2												
39 Y	2	4	4	6	6	6	4	4	1		(2)												
40 Zr	2	4	4	6	6	6	4	4	2		(2)												
47 Ag	2	4	4	6	6	6	6	6	6		1												
48 Cd	2	4	4	6	6	6	6	6	6		2												
49 In	2	4	4	6	6	6	6	6	6		2	1											
54 X	2	4	4	6	6	6	6	6	6		4	4											
55 Cs	2	4	4	6	6	6	6	6	6		4	4				1							
56 Ba	2	4	4	6	6	6	6	6	6		4	4				2							
57 La	2	4	4	6	6	6	6	6	6		4	4	1			(2)							
58 Ce	2	4	4	6	6	6	6	6	6	1	4	4	1			(2)							
59 Pr	2	4	4	6	6	6	6	6	6	2	4	4	1			(2)							
71 Cp	2	4	4	6	6	6	8	8	8	8	4	4	1			(2)							
72 —	2	4	4	6	6	6	8	8	8	8	4	4	2			(2)							
79 Au	2	4	4	6	6	6	8	8	8	8	6	6	6			1							
80 Hg	2	4	4	6	6	6	8	8	8	8	6	6	6			2							
81 Tl	2	4	4	6	6	6	8	8	8	8	6	6	6			2	1						
86 Em	2	4	4	6	6	6	8	8	8	8	6	6	6			4	4						
87 —	2	4	4	6	6	6	8	8	8	8	6	6	6			4	4					1	
88 Ra	2	4	4	6	6	6	8	8	8	8	6	6	6			4	4					2	
89 Ac	2	4	4	6	6	6	8	8	8	8	6	6	6			4	4	1				(2)	
90 Th	2	4	4	6	6	6	8	8	8	8	6	6	6			4	4	2				(2)	
118?	2	4	4	6	6	6	8	8	8	8	8	8	8	8		6	6	6				4	4

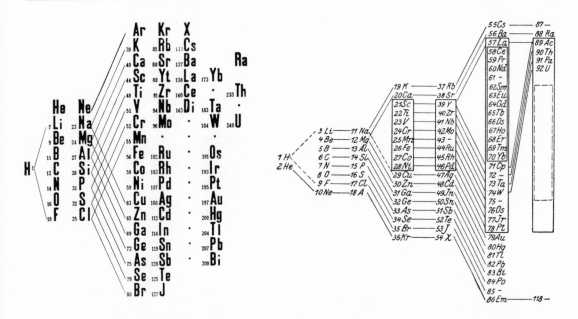

on . . . and build some one hundred or one thousand elements"! The element with atomic number 118, still unknown today, was predicted to be a noble gas with eight Q electrons in its outer shell (see table).

In lectures in Göttingen, Cambridge, and Copenhagen, Bohr illustrated his theory with a new version of the periodic system, sometimes known as the Bohr-Thomsen table. Bohr's version was based on a much earlier table proposed by Danish chemist Julius Thomsen in 1895: Bohr simply modernized Thomsen's system, with which he had been acquainted since his student days. In Thomsen's system the chemical groups are vertical, contrary to the standard version, and the transition groups fit into the system in a natural way. In his lectures on atomic theory during the period in question, Bohr also made use of large plates that pictured the electron orbits in the atoms of selected elements. The orbits were drawn to scale and colored in red and black. The plates, with their visual qualities and beautiful geometric forms, impressed the audience—in particular the picture of radium (see next page), in which all eighty-eight orbits were meticulously drawn. The original plates have unfortunately disappeared.

The periodic systems of Julius Thomsen (left) and Bohr. Thomsen's diagram is an unpublished version from 1898, made for his lecture room. It hung for many years in front of successive audiences of Danish chemistry students—among them Niels Bohr, who as a student attended classes in physics, mathematics, astronomy and chemistry.

METHOD AND SCIENTIFIC STYLE

Bohr's method for building up atoms cannot really be characterized by means of standard labels such as "empiricism" and "deductivism." Perhaps the most suitable label is "eclecticism," since Bohr applied a peculiar mixture of empirical evidence and theoretical reasoning, tied together with a more intuitive understanding. The most important components of Bohr's method are reconstructed in the figure on page 59.

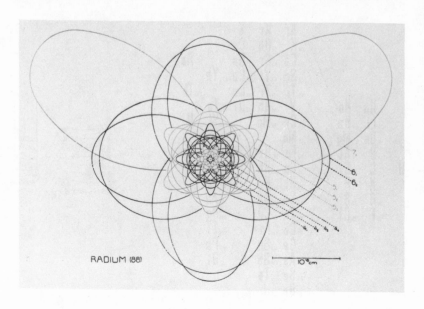

RADIUM (88)

10^{-8} cm

The eighty-eight electron orbits of radium in Bohr's model.

Bohr's theory relied heavily on chemical and spectroscopic evidence, as does atomic theory in general. However, it was the use of general concepts that distinguished his theory from all other atomic theories and supplied it with Bohr's personal imprint. Most research papers in physics are objective and rather bloodless accounts which reveal nothing about the personality of their authors; in a sense they are, and are intended to be, anonymous. Bohr's works on the periodic system are very different. They are permeated by the spirit of Bohr's thinking. They could not possibly have been written by any other physicist.

The general principles, as applied by Bohr, were used in a philosophical rather than a mathematical or physical way. That is, they were not stated quantitatively but were qualitative considerations of an intuitive kind. Bohr's use of the correspondence principle, perhaps the most characteristic component of the theory, illustrates this point.

Bohr worked out the correspondence principle as a powerful tool of quantum theory in about 1918. In the early twenties he used it as a means of choosing the correct quantum state of an electron which is added to an atom not yet complete. According to Bohr, the correspondence principle offered "for the first time a rational theoretical basis . . . for the discussion of the arrangement of the orbits of the electrons." What Bohr had in mind was that the building up of atoms was ruled by the correspondence principle. If the capturing process of a free electron is assumed to end in a state n_k, this can take place only if the process is consistent with the correspondence principle. But when is a hypothetical capturing process consistent with the correspondence principle and, in this sense, rational?

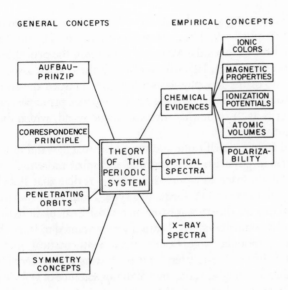

GENERAL CONCEPTS   EMPIRICAL CONCEPTS

Bohr's own explanation was markedly vague. "Among the processes that are conceivable and that according to the quantum theory might occur in the atom, we shall reject those whose occurrence cannot be regarded as consistent with a correspondence of the required nature." Throughout his construction of the periodic system, Bohr applied correspondence arguments in a way which creates the impression that he had reached his results by means of elaborate calculations. For example, he stated that if lithium is formed by adding a third electron to the helium electron structure, the 1_1 orbit cannot be reached through a continuous alteration of the 2_1 orbit. "A closer investigation," he wrote, shows that a transition from L to K will "not exhibit a correspondence with a harmonic component in the motion of the atom." However, Bohr's "closer investigation"—a favorite phrase in his idiosyncratic terminology—did not imply a mathematical deduction from quantum theory or, for that matter, calculations at all. Such calculations—for example, of transition probabilities based on Fourier coefficients—were later performed by Frank Hoyt, an American physicist working at Bohr's institute, but his laborious calculations turned out to be unfruitful. Bohr realized instinctively that the atom was too complicated a system to be treated by means of the mathematical machinery of current quantum theory. He had little confidence in the "mathematical chemistry" that Sommerfeld, Max Born, and other German theorists regarded as an ideal. Instead, Bohr relied on a sort of intuitive understanding of what went on during the formation of atoms. Because of the intuitive nature of Bohr's explanations, they could not help appearing obscure or even incomprehensible. Substantial parts of Bohr's theory were in a sense nonobjective,

The conceptual structure behind Bohr's theory of the periodic system.

59

Pauli, the great critic and puritan, did not like this attempt at an explanation. He sensed already in 1921 that a major principle was hidden in these regularities. As an interesting testimony to Pauli's attitude, I quote some remarks that I found scribbled on the margins of a book in the Pauli library at CERN. The book contains Bohr's famous paper on the *Aufbauprinzip*. During his discussion of the adding of the eleventh electron to the closed shell of ten electrons, Bohr remarks: "We must expect that the eleventh electron goes into the third orbit." *("Wir müssen erwarten dass das 11.Electron (Na) in die 3.Bahn geht.")* Pauli, obviously annoyed by this statement, writes hastily in the margin with two exclamation marks, "No reason to *expect* anything; you concluded it from the spectra!!" *("Wir müssen es nicht erwarten aber wir wissen es aus den Spectren!!")*

V. F. Weisskopf, "Three Steps in the Structure of Matter," 1970

because the reasoning applied by Bohr could not be reproduced or understood by other physicists. Bohr's idiosyncratic use of the correspondence principle is difficult to justify. More often than not, the principle seems to act as a deus ex machina. Hendrik Kramers, Bohr's closest associate at the time, was one of the few physicists who really grasped the spirit of the theory. He once recollected how the correspondence principle appeared to most physicists as "a somewhat mystical magic wand, which did not act outside Copenhagen."

Not until Bohr himself came and waved the magic wand in Göttingen did the German physicists — by training and spirit more inclined to the mathematical approach to atomic structure — realize that Bohr's theory was not of the mathematical-deductive type. From Bohr's writings they had believed that the theory was based on extensive calculations of penetration and correspondence effects. Naturally they wanted to learn the details of what they thought were Bohr's secret mathematical methods. As Kramers said thirteen years later: "It is interesting to recollect how many physicists abroad thought, at the time of the appearance of Bohr's theory of the periodic system, that it was extensively supported by unpublished calculations which dealt in detail with the structure of the individual atoms, whereas the truth was, in fact, that Bohr had created and elaborated with a divine glance a synthesis between results of a spectroscopical nature and of a chemical nature."

During the Göttingen lecture series in 1922, German physicists realized that methodologically Bohr's theory was of an entirely different kind from

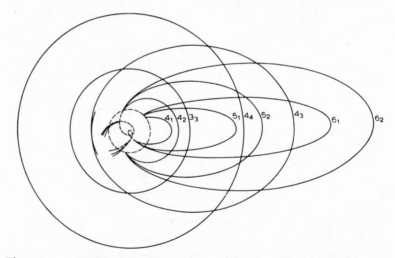

The various stages of the capture of the nineteenth electron in the potassium atom, as shown by Bohr in his Nobel lecture and on other occasions in 1922. The electron ends up in a 4_1 orbit that penetrates into the closed shell with eighteen electrons, symbolized by a dotted ring. The elliptic 4_1 orbit precesses about the nucleus.

what they had expected and were used to. Many years later, Heisenberg recalled his impressions as follows: "It could very distinctly be felt that Bohr had not reached his results through calculations and proofs but through empathy and inspiration and that it was now difficult for him to defend them in front of the advanced school of mathematics in Göttingen."

The German physicists' immediate response, characterized by enthusiasm as well as a certain lack of critical attitude, was no doubt due to the enormous authority that Bohr had in the community of German atomic physicists. They endowed Bohr with an almost godlike status and were ready to believe, without reservation or closer examination, that he could do miracles in the area of atomic structure. Sommerfeld's response to Bohr's first announcement of his theory — that "it evidently represents the greatest advance in atomic structure since 1913"—was echoed by most of his colleagues.

ELEMENT NUMBER 72

In 1921 the exact place of the *rare-earth* elements in the periodic system was still a matter of discussion among chemists and physicists. Most versions of the periodic system implied that the rare-earth group included the unknown element 72. After the development of X-ray spectroscopy, pioneered by Moseley, it was known that only a few elements remained to be discovered, in particular those of atomic numbers 43, 61, and 72. The result of Bohr's theory with respect to the rare earths was therefore a topic of considerable interest.

According to Bohr, the rare earths were characterized by a building up of the N group, from $(4_1)^6(4_2)^6(4_3)^2$ at cerium to $(4_1)^8(4_2)^8(4_3)^8(4_4)^8$ at element 71, lutecium. The unknown element 72 therefore had to be a homologue of zirconium and not a rare earth. Using rather obscure arguments involving correspondence and symmetry, Bohr wrote in Göttingen the electron structure for element 72 as shown in the table on page 56.

Bohr's confidence in his theory, including its prediction of element 72 as a member of the fourth subgroup, increased during the summer and fall of 1922, when the technique of X-ray spectroscopy was successfully applied to confirm the theory. The confirmation was primarily due to Dirk Coster, a Dutch physicist who worked with Manne Siegbahn at Siegbahn's laboratory in Lund, Sweden. Bohr invited Coster to the new institute in Copenhagen where they jointly investigated the relationship between Bohr's theory and the X-ray data from all known elements. Coster concluded that the data were in "beautiful agreement" with the theory. Bohr and Coster were also able to interpret the data for the rare earths in terms of the theory. They concluded that the measured X-ray absorption edges of the elements were well accounted for if the building up of the 4_4 quantum state started at element 58 and was completed at element 71.

During 1922 the question of element 72 took a new and dramatic turn.

In May, Alexandre Dauvillier, who worked at Maurice de Broglie's private laboratory in Paris, investigated a sample that the French chemist Georges Urbain had isolated in 1911 from the earth ytterbium. On the basis of chemical analysis Urbain claimed to have discovered a new rare-earth element, which he placed as number 72 in the periodic system. Urbain named the element celtium, after France, but the discovery was not generally accepted. Dauvillier was able to detect in Urbain's sample two weak X-ray lines, which he interpreted as proof that element 72 was a constituent of the sample. "It is now unquestionable that the element of atomic number 72 is actually celtium, [which has] conclusively won its place among the chemical elements," the two French scientists wrote.

The announcement of the discovery of celtium was naturally considered most inconvenient in Copenhagen. If Dauvillier and Urbain were right, element 72 belonged to the rare earths, which contradicted Bohr's theory. Bohr was not willing to accept the French claim at its face value and asked Coster about his opinion. "The question [of celtium] is of paramount interest, since, as you know, the ideas of atomic structure seem to require

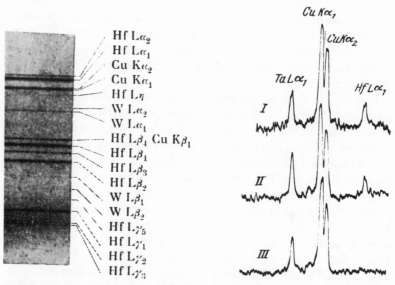

An early X-ray emission spectrum of hafnium. On the left is an almost complete L spectrum, taken in 1923 from a sample containing 95 percent hafnium. The L lines originate from electrons jumping from a higher energy state to the state characterized by $n = 2$. The strong copper lines are X rays emitted from the anticathode of the X-ray tube. On the right is a photometric record of an early hafnium sample, probably from January 1923. The intensity of the X rays is plotted against the wavelength. The three curves are from three different stages of chemical precipitation with sodium phosphate. In curve III the hafnium content has vanished. The tantalum line of constant intensity is a calibration standard from which the amount of hafnium can be estimated.

Plaque commemorating the discovery of hafnium: "Element no. 72, discovered 1922 in Copenhagen." This plaque was presented to the City of Copenhagen on the occasion of its eight-hundredth anniversary (1967) and is now in the town museum.

that a substance with atomic number 72 must have essentially different properties from the ones shown by the rare earths." Coster rejected celtium and assured Bohr that Dauvillier's measurements were generally inaccurate. Consequently, Bohr chose to ignore celtium and to stick to his theory.

Bohr himself had not intended to search for an alternative to celtium. It was Coster and George de Hevesy, the Hungarian-born chemist who worked at Bohr's Institute, who took the initiative. Guided by Bohr's theory, they prepared samples of zirconium minerals and investigated their X-ray spectra. After a few weeks' work they were able to identify two fairly strong X-ray lines, which they ascribed with certainty to atomic number 72. Coster and Hevesy's discovery was made just in time for Bohr to announce it in his Nobel lecture in Stockholm on 11 December 1922.

Continued investigations at Bohr's institute soon confirmed the existence of considerable amounts of element 72 in zirconium minerals. Within a year the new element had been isolated, and Bohr and his associates discussed a name for it. Hevesy and Coster proposed to call it hafnium, referring to the Latin name for Copenhagen. Although Bohr preferred to keep a low profile in the discovery of hafnium, there is no doubt that he played an important role, both as director of the institute and as originator of the theory which predicted the main properties of the element. Congratulating Hevesy on the discovery, Rutherford wrote: "I am sure that it is highly gratifying to Bohr, even if he tries to suppress it."

The announcement of the discovery of hafnium led immediately to a priority dispute with Urbain and Dauvillier in Paris. To complicate the matter, yet another candidate for element 72 was advanced by British chemist Alexander Scott, who claimed to have discovered it in 1918. Scott's name for the element was oceanium. Although oceanium turned out not to be a serious competitor for hafnium, the rivalry with Paris gave rise to a long and bitter priority conflict. Even though Bohr was not directly involved in the conflict, it disturbed him very much. At the height of the priority debate he wrote to Rutherford: "You can hardly imagine how great a comfort your kind letter was to us all in this terrible muddle about the new element, in which we quite innocently have dropped. We had never dreamt of any competition with the chemists in the hunt for new elements, but wished only to prove the correctness of the theory. In the letter from Urbain which the editor of *Nature* kindly sent us for possible comments, he tries however to shift the whole matter, paying no regard to the important scientific discussion of the properties of the element 72, but tries only to claim a priority for announcing a detection of such an element."

THE SIGNIFICANCE OF BOHR'S THEORY

Bohr's theory of the periodic system quickly became generally accepted among physicists. The success was, however, short-lived, as evidence contradicting the details of the theory soon accumulated. The evidence came in particular from X-ray spectroscopy and chemistry. For example, Dauvillier and Louis de Broglie[1] showed that the X-ray data were difficult to reconcile with Bohr's proposal of four 2_1 electrons. They argued that the particles should be classified as two 2_1 electrons and two 2_2 electrons. Partly on the basis of Dauvillier and de Broglie's proposal, Edmund Stoner developed in 1924 a complete set of electron distributions which differed from Bohr's. In particular, Stoner incorporated the inner quantum number j, which he interpreted as the angular momentum of the individual electrons. Each n_k sublevel was further subdivided as n_{kj}, where $j = k$ or $k - 1$; each of the n_{kj} sublevels was assumed to be populated with a maximum of $2j$ electrons. In this way the complete L level contains two electrons with $k = 1$ and six electrons with $k = 2$ — that is, $(2_1)^2(2_2)^6$ as opposed to Bohr's $(2_1)^4(2_2)^4$.

Stoner's system was soon recognized as superior to Bohr's. It agreed better with the details of X-ray spectroscopy and also with the chemical evidence. The system received strong support from J. D. Main-Smith, who independently worked out a complete system of electron distributions, mainly guided by chemical considerations. Although Stoner and Main-Smith's system was empirically superior to Bohr's, it lacked the theoretical justification which characterized Bohr's system. This foundation was supplied in 1925, when Pauli introduced his famous *exclusion principle*. Pauli's work made possible a satisfactory understanding of the electron distributions throughout the periodic system. The exclusion principle, or *Pauli Verbot,* has proved to be the magic key which can unlock the secrets of the periodic system. In its standard version it states that two or more electrons in the same atom cannot have the same set of quantum numbers. The contributions of Stoner, Main-Smith, and Pauli took place before quantum mechanics was introduced, but the new atomic physics did not change their results materially. Even today the results of Stoner, Main-Smith, and Pauli are taught in textbooks, although, of course, differently phrased and interpreted.

When Main-Smith and Stoner proposed their scheme in 1924, Bohr was involved in other areas such as radiation theory and the application of the correspondence principle to dispersion. His attitude to the new concept of electron configurations was rather conservative. Although he did not stick to the details of his own theory, his first impression of Stoner's work was negative. In a letter to Coster of December 1924 he admitted the "formal beauty and simplicity of his [Stoner's] classification"; but he also stressed that "from quantum-theoretical points of view, it cannot mean a final solution to the problem, since we do not yet possess any possibility of connecting the classification of levels in a rational manner with a quantum-theoretical analysis of electron orbits." Bohr was uncompromising in his demand that electron orbits must be specified in the sense of the correspondence principle, and he was unable to see how this could be done with Stoner and Main-Smith's scheme.

Only with the advent of Pauli's work did Bohr admit that the new scheme was also superior to his own when considered from the point of view of quantum theory. Still, Bohr was not completely happy with Pauli's theory, because it did not bring the completion of the electron shells into relation with his beloved correspondence principle. Pauli's work rested on the *Aufbauprinzip* but not on the other concepts on which Bohr had founded his theory. On the contrary, Pauli rejected Bohr's extensive use of correspondence and symmetry. In a letter of December 1924 he wrote to Bohr: "I have often told you that I think that the correspondence principle has really nothing to do with the problem of the completion of the electron groups in the atom . . . I believe that everything really is much simpler; one does not at all have to talk about harmonic interplay."

If Bohr's theory of the periodic system was only a parenthesis in the development of physics, why does it still deserve attention? For one thing, despite its short lifetime it was an important step toward the presently accepted theory. Bohr's theory provided the first theoretically founded scheme of the constitution of all the elements. When Stoner, Main-Smith, and Pauli developed their ideas, they did so against the background of Bohr's theory, which served as a source of inspiration and criticism. Furthermore, the main results of Bohr's theory are still valid. His version of the periodic system and the distribution of electrons according to their principal quantum numbers has not been changed by the later development.

More important, perhaps, the very fact that the theory is the intellectual child of Niels Bohr makes it historically significant. If one wants to understand the physical thinking which was peculiar to Bohr, the theory of the periodic system is a profitable work to study. Indeed, his writings concerning the periodic system are throughout impregnated with his personal style. For example, it is remarkable that in his voluminous articles from 1921 to 1923 there are virtually no mathematical formulas. Lengthy arguments in involved sentences, filled with general and often vague considerations, became the hallmark of Bohr's philosophical approach to physics. This style, the first full example of which was given with the theory of the periodic system, can also be found in many of Bohr's later works. His work on radiation theory of 1924 (the Bohr-Kramers-Slater theory), his contributions to the interpretation of quantum mechanics, and his theory of nuclear reactions (developed from 1936 to 1939) have much of the same general nature. In his *Autobiographical Notes* (1946), Einstein praised Bohr's "unique instinct and tact" which expressed "the highest form of musicality in the sphere of thought." Indeed, to label Bohr's theory of the periodic system as "musical" is a good characterization.

Obviously, Bohr's personal style, his idiosyncratic vocabulary, and his belief in a complementarity between clarity *(Klarheit)* and truth *(Wahrheit)* made some of his works difficult to appreciate for physicists outside the Copenhagen circle. Paul Dirac once expressed this difference in mental attitude as follows: "While I am very much impressed by what Bohr said, his arguments were mainly of a qualitative nature . . . what I wanted was statements which could be expressed in terms of equations, and Bohr's work very seldom provided such statements." Dirac was not referring to Bohr's theory of the periodic system, which was already a past chapter when he first met Bohr, but his statement is valid for that theory as well.

Bohr's theory was not, as was his atomic theory of 1913, destined to revolutionize physics. The approach to atomic physics which in a few years created the new quantum mechanics was based on ideas which in many ways contradicted essential parts of Bohr's theory. The atomic theory of 1921–1923 can justly be seen as conservative, the culmination of the semimechanical atomic models which during the first two decades of the century were developed with increasing confidence. In Bohr's mind there apparently was little doubt about the reality of the electron orbits or the fertility of the semimechanical model concept upon which his theory relied. Probably he believed that his models represented fairly realistically the actual constitution of the atoms. A literal or at least semiliteral interpretation was certainly the message which his lectures and articles signaled to his colleagues in physics. Bohr's use of large plates, with the atomic models carefully drawn, strongly suggests that he did not seriously doubt the reality of well-defined electron orbits.

It has been said that Bohr's atomic theory of 1913 was a late product of Victorian physics. This is an unorthodox view of a truly revolutionary theory but there is some truth in it, and it applies to the theory of the periodic system also. Before the advent of quantum mechanics, Bohr was not ready to follow the new radical approach to atomic structure which had already been hinted at by Heisenberg and Pauli. According to Pauli, one should be able to deduce the periodic system from a theory which was based solely on observable physical quantities and which precluded electron orbits and visualizable atomic models. He considered the exclusion principle as a step toward this goal and presented it as a purely formal principle. In December 1924 he wrote to Bohr: "For weak men, who need the crutch of the idea of unambiguously defined electron orbits and mechanical models, the rule can be grounded as follows: 'If more than one electron have the same quantum numbers in strong fields, they would have the same orbits and therefore collide.'" Pauli certainly did not regard Bohr as a "weak man," but he may well have been alluding sarcastically to Bohr's theory of the periodic system, which did not live up to the antimodel standards heralded by Pauli and Heisenberg. It was left to younger physicists to recognize that atomic models, as developed and used by Bohr and others, had to be transcended in order to build a fruitful new atomic theory. When the first steps were taken along this path — by Heisenberg, Pauli, Dirac, Jordan, and Born — Bohr quickly abandoned his old approach and devoted himself to supplying quantum mechanics with a rational foundation.[2]

Bohr and Rutherford

Mark Oliphant

Sir Mark Oliphant was born in Australia and, like Rutherford before him, came to study at Cambridge University on an 1851 Exhibition scholarship. In the 1930s he worked closely with Rutherford, who (in Bohr's words) "found in Marcus Oliphant a collaborator and friend whose general attitude and working power reminds us so much of his own." He was Director of the Research School of Physical Sciences at Canberra, 1950–1963, and Governor of South Australia, 1971–1976. He is now retired.

During the short period he spent in Manchester, Bohr formed a close relationship with Rutherford, a very personal relationship which grew with time, even beyond Rutherford's death. After World War I, Bohr made frequent visits to Cambridge, where he and his wife stayed with the Rutherfords in a very informal manner. Bohr spoke to colloquia and to the Kapitza Club.[1]

Rutherford, though always inspiring, was not a great lecturer — "To 'Er' was Rutherford!" Bohr was much worse. His failing was that he used too many words to express any idea, wandering about as he spoke, often inaudibly. I remember him speaking for over two hours to the Kapitza Club. But Rutherford, when present, would call a halt to Bohr's discourse, teasing him by remarking that his sense of time was as uncertain as the principle he was discussing.

When word reached us at the Galvani celebrations in Italy, in 1937, that Rutherford had died, Bohr addressed a silent gathering on the passing of his great master and friend. His emotion moved us all, especially those of us who shared his deep personal grief.

I had long talks with Bohr in Washington, when success in the nuclear weapon project was imminent. Perhaps I should say that he talked to me! He brought draft after draft of a letter he proposed sending to the president, concerning the possible first use of the weapon. He appealed, in his complex jargon, for it to be used in the interests of peace, not war. He referred continually to Rutherford, as the father of nuclear physics, and to what Rutherford's views would have been had he been still alive.

Rutherford formed much closer relationships with some of his colleagues and research students than with others, and this appeared to have nothing

The Bohrs (right) with the Rutherfords (Lady Rutherford on the extreme left) and Mrs. Oliphant, around 1930. The photograph was taken by Marcus Oliphant.

to do with similarity of outlook or personality. James Chadwick, a somewhat dour man of few words, loved Rutherford as did Bohr. Rutherford played golf with friends, and talked enthusiastically with colleagues about whatever subject was under discussion, but with none was he as close as with Bohr, Chadwick, and a few others. Despite his flippant remarks about the uncertainty principle, Rutherford's sympathies were with Bohr rather than with Einstein, over causality in particle physics. But the old physics was generally to the fore when his beloved alpha particles were involved. "Bohr, my boy, you are too complacent about ignorance."

Bohr shared with Rutherford the belief that instinct played as great a part in the development of physics as did laborious experimenting or mathematical calculation, but both demanded testing of speculation by observation. Each was able to make vivid visual models of his thoughts, models which helped the experimenter greatly.

The Bohrs spent happy holidays with the Rutherfords at their holiday cottages in northern Wales and Wiltshire. Such holidays were periods of complete relaxation — they walked, met scattered neighbors, sawed wood for the fire, or just did nothing.

It is strange that two people so close to each other should have been so different in all but their love of physics as the basic science of nature, and their deep concern for all humanity. Rutherford, who called himself a liberal, was one of the most politically conservative people I have ever known. He admired Stanley Baldwin greatly, upholding Baldwin's actions on the abdication of Edward VIII. Bohr made no political judgments, other than those involving cruelty or human misery. In the words of the Russell-Einstein manifesto, he remembered his humanity and forgot the rest.[2]

Bohr, Göttingen, and Quantum Mechanics

Friedrich Hund

As a result of the fortunate appointments of Max Born and James Franck to the faculty in 1921, the University of Göttingen was well prepared to seize and extend the new ideas in physics. Franck, while in Berlin, had heard Bohr lecture and had already gained a deep impression of Bohr's way of thinking about the atom. And Born, in his researches on the structure of crystal lattices, had realized the importance of quantum theory; now his field of interest shifted from the dynamics of crystal lattices toward fundamental questions of quantum theory. The scientific surroundings were favorable: David Hilbert was vividly interested in physics; Richard Courant's research in mathematics met the requirements of the physicists; Carl Runge had been a spectroscopist; Robert Pohl, Ludwig Prandtl, and Emil Wiechert, renowned worldwide in their specialties, participated in fundamental physics. The discussions in the weekly physics colloquia and in the seminars were at a correspondingly high level.

All this was nourished by the seven lectures Bohr gave during two weeks in 1922.[1] The weather was fine and the intellectual life in the city was fascinating, but the consequences of the past war were still troublesome. The term *Bohr-Festspiele* ("Bohr festival"), probably coined by Franck, reflects to some extent the enchantment radiated by the lectures. Each lecture, filling a long evening (at that hour one was hungry in 1922!) and delivered in a low voice and in very cautious language, showed Bohr's fundamental approach and gave an excellent overview of the connection between the spectra radiated by the atoms and the physical and chemical properties of the elements expressed in the periodic table. Important people joined in the event, among them Sommerfeld, Ehrenfest, Landé, and Pauli. Sommerfeld brought along a young blond student, looking almost like a

Friedrich Hund is Professor Emeritus at the University of Göttingen, where he first became Assistant Professor in 1925. He is noted for his fundamental work in theoretical spectroscopy, and has also written an authoritative history of quantum theory.

Participants in the "Bonzenfreie Kolloquium," held in Berlin in 1920. Bohr had come to lecture in Berlin, but some of the younger physicists felt that the senior people (the "Bonzen") had monopolized the discussion. ("Bonze," literally a Buddhist priest, was a term used for a VIP in a pejorative sense.) Bohr was persuaded to conduct a colloquium for the younger people, with "bigwigs" strictly excluded. Front row, from left: Otto Stern, James Franck, Bohr, Lise Meitner, Hans Geiger, and Peter Pringsheim. Behind Meitner is Otto Hahn. Behind Hahn and to the right is George de Hevesy.

schoolboy: Werner Heisenberg. Heisenberg joined boldly in the discussions.

The deep impression the lectures made on the Göttingen physicists, and the enormous respect Bohr gained among us younger people, came less from a real acceptance of the line of approach to atomic physics given in the correspondence principle than from optimism that the time was now ripe for really understanding the properties of matter. Explaining the numbers 2, 8, 8, 18, 18, 32 of the periodic table looked like a promising beginning, but we also noted a deeper significance. We knew Bohr's theory from

Participants in the "Bohrfestspiele" of 1922. Seated in front is Max Born. Standing are (from left) Carl Oseen, Bohr, James Franck, and Oskar Klein.

Sommerfeld's book,[2] but now we experienced directly a somehow deeper, freer, and apparently more powerful comprehension, both of the fundamental difference between atomic physics and ordinary physics, and of their connection. A seminar led by Born and Hilbert concerned itself with these matters. When, a few months after this brilliant performance, Bohr received the Nobel Prize, he became a kind of idol.

During Bohr's lectures, Born had the opportunity to notice the talents of Heisenberg, and he brought him to Göttingen. From autumn 1922 until Easter 1926 (except for a stay at Munich and a stay at Copenhagen), Heisenberg was a member of the Göttingen circle.

Born noticed that Bohr's ideas did not yet form a consistent theory, and

In speaking of the beginnings of
quantum mechanics in Göttingen,
we must undoubtedly start with the
Bohr Festival in 1922. At the
instigation of Hilbert and the physicists Franck, Born, and Pohl, the
university had asked the Dane Niels
Bohr to give a comprehensive series
of lectures on his theory. Guests
were invited from outside, including
Sommerfeld from Munich, and the
whole gathering, as one of the first
after the great economic stringency
of the postwar period, bore the
marks of a joyous new beginning; a
new beginning in the international
relations of science, but also in the
tasks of the newborn atomic physics.
Not only this, but Göttingen, in
gorgeous summer weather, was re-
splendent with gardens and flowers,
and so, despite the difficulty of the
subject, the atmosphere and the
excitement of the students who filled
most of the auditorium gave the
lectures such a festive air that the
name Bohr Festival soon made the
rounds—in allusion to the Handel
Festivals in the Göttingen civic
theater, which had just begun.

Werner Heisenberg,
Tradition in Science, 1983

he began to direct his own efforts toward such a goal. With Heisenberg's help, he demonstrated that the special assumptions hitherto used led to incorrect results when applied to the helium atom. In a lecture course and in the book, *Atommechanik I,* resulting from it, Born explored the limits of currently used principles. Then he tried to adapt the mathematical methods more closely to the peculiarities of the atom. The decisive step to a quantum mechanics was then taken by Heisenberg in the summer of 1925. This step consistently exploited Bohr's correspondence principle; Heisenberg himself perceived it this way.

Looking back, we can easily reconstruct the path from Bohr's initial attack on the physics of the atom to the Göttingen quantum mechanics of 1925. The theory to be created had to account for the stability of atoms and the combination principle of spectra. In the simplest case — the frequencies of a classical system — the generally energy-dependent fundamental frequency $v(E)$ and the overtones $\tau v(E)$ "correspond" to the quantum theoretical frequencies $[E(n) - E(n - \tau)]/h$. The quantum-theoretical energies $E(n)$ must thus be chosen in such a manner that the quantum-theoretical frequencies go over to the classical as the system becomes larger and larger and approximates a classical and macroscopic system. Since the quantum frequency takes the form $(1/h)(dE/dn)$ for large n, one could equate this expression to the classical dependence $\tau v(E)$. Bohr successfully did this in 1913 with the hydrogen atom. The three simple examples of this procedure are the harmonic oscillator, the rigid rotator, and the hydrogen atom:

Oscillator: $\qquad E(n) = hvn; \qquad \dfrac{dE(n)}{h\, dn} = v$

Rotator: $\qquad E(n) \sim n^2; \qquad \dfrac{dE(n)}{dn} \sim n \sim \sqrt{E}$

H atom: $\qquad E(n) \sim -\dfrac{1}{n^2}; \qquad \dfrac{dE(n)}{dn} \sim \dfrac{1}{n^3} \sim |E|^{3/2}$

On the right side, one sees the classical forms of dependence: v independent of E, $v \sim \sqrt{E}$, $v \sim |E|^{3/2}$. The multiplicative factors omitted here can be determined.

Since it was evident that this procedure could not be generally correct, one had to be content with the convergence of the quantum frequency to the classical in the limit of large n. This procedure did not, however, lead to a unique result. In 1925 Heisenberg was able to extract still more from the correspondence principle. The oscillation contained in the classical motion

$$\chi_\tau(E) e^{2\pi i \tau v(E)t}$$

had to correspond to the quantum "transition amplitudes"

$$\chi(n, n - \tau) e^{2\pi i v(n, n - \tau)t}$$

Calculations with the classical amplitudes use the trivial "combination principle"

$$\rho v + (\tau - \rho)v = \tau v.$$

For calculations with the quantum amplitudes, Heisenberg found rules adapted to the quantum combination principle

$$v(n, n - \rho) + v(n - \rho, n - \tau) = v(n, n - \tau),$$

the combination principle for spectra. This quantum mechanics, conceived by Heisenberg and mathematically developed by Born, Heisenberg, and Jordan, was a sharpened form of Bohr's correspondence principle. It was a modification (no longer pictorial) of classical mechanics, taken far enough to include the combination principle of spectra.

Heisenberg identified three influences from which his quantum mechanics resulted—namely, Sommerfeld's keen attack on facts previously not understood, Born's demand for a mathematically consistent theory, and Bohr's deep insight into the essentials of the physics of atoms.

Bohr's importance for the contribution that Göttingen was able to make to quantum theory cannot be overestimated. In Göttingen this was attested by the first volumes of the series of monographs *Struktur der Materie,* which were edited by Born and Franck after 1925. The introduction at the beginning of the series promised "a close connection between experience and theory in the spirit of Bohr." "Bohr's correspondence principle" was the guide for the first volume; the foreword to the second volume saw "no conflict with the concepts of Bohr"; the content of the third volume was regarded as "clear evidence for the fundamental concepts of Bohr's theory"; the fourth volume was patterned directly on Bohr's considerations.

The completion of quantum mechanics still awaited Schrödinger's development of de Broglie's ideas. And of course a crucial contribution to the physical understanding of Schrödinger's wave function, again developed in Göttingen, was the probability interpretation by Max Born.[3]

The Trilogy

Dear Harald,

Perhaps I have found out a little about the structure of atoms. Don't talk about it to anybody, for otherwise I couldn't write to you about it so soon. If I should be right it wouldn't be a suggestion of the nature of a possibility (i.e., an impossibility, as J. J. Thomson's theory) but perhaps a little bit of reality. It has grown out of a little information I got from the absorption of α-rays (the little theory I wrote about last time). You understand that I may yet be wrong; for it hasn't been worked out fully yet (but I don't think so); also, I do not believe that Rutherford thinks that it is completely wild; he is a man of the right sort, and he would never say that he was convinced of something that was not fully worked out. Believe me, I am eager to finish it in a hurry, and to do that I have taken off a couple of days from the laboratory (this is also a secret).

This was intended only as a little greeting from

your Niels

who is longing very much to talk with you.

—Letter from Niels Bohr to his brother,
19 June 1912[1]

In the spring of 1912 Bohr began studies that culminated in a set of three epoch-making papers, commonly called the Trilogy, in which he developed his new theory of atomic structure and its application to atoms and molecules.

In his Rutherford Memorial Lecture (1958) Bohr recalled the beginnings of that work, which grew directly out of Rutherford's discovery of the nucleus:

Early in my stay in Manchester in the spring of 1912 I became convinced that the electronic constitution of the Rutherford atom was governed throughout by the quantum of action . . . I rapidly became absorbed in the general theoretical implications of the new atomic model and especially in the possibility it offered of a sharp distinction as regards the physical and chemical properties of matter, between those directly originating in the atomic nucleus itself and those primarily depending on the distribution of the electrons bound to it at distances very large compared with nuclear dimensions.

While the explanation of the radioactive disintegrations had to be sought in the intrinsic constitution of the nucleus, it was evident that the ordinary physical and chemical characteristics of the elements manifested properties of the surrounding electron system. It was even clear that, owing to the large mass of the nucleus and its small extension compared with that of the whole atom, the constitution of the electron system would depend almost exclusively on the total electric charge of the nucleus. Such considerations at once suggested the prospect of basing the account of the physical and chemical properties of every element on a single integer, now generally known as the atomic number, expressing the nuclear charge as a multiple of the elementary unit of electricity.[2]

On 6 March 1913 Bohr wrote to Rutherford, enclosing a draft of the first paper in the series. In his reply, dated March 20, Rutherford reacted positively but raised a prickly question: "There appears to me one grave difficulty in your hypothesis, which I have no doubt you fully realise, namely, how does an electron decide what frequency it is going to vibrate at when it passes from one stationary state to the other? It seems to me that you would have to assume that the electron knows beforehand where it is going to stop." Rutherford continued: "There is one criticism of a minor character which I would make in the arrangement of the paper. I think in your endeavour to be clear you have a tendency to make your paper much too long . . . I do not know if you appreciate the fact that long papers have a way of frightening readers." Bohr had in the meantime compounded the problem (although all ended well), as he recalled in his Rutherford Memorial Lecture:

The second point raised with such emphasis in Rutherford's letter brought me into a quite embarrassing situation. In fact, a few days before receiving his answer, I had sent Rutherford a considerably extended version of the earlier manuscript, the additions especially concerning the relation between emission and absorption spectra and the asymptotic correspondence with the classical physical theories. I therefore felt the only way to straighten matters was to go at once to Manchester and talk it all over with Rutherford himself. Although Rutherford was as busy as ever, he showed an almost angelic patience with me, and after discussions through several long evenings during which he declared he had never

Some algebraic formulae caught my eye . . . It was part of a paper by a Mr. N. Bohr of whom I had never heard . . . I sat down and began to read. In half an hour I was in a state of excitement and ecstasy, such as I have never experienced before or since in my scientific career. I had just finished a year's work revising a book on *Modern Electrical Theory*. These few pages made everything I had written entirely obsolete. That was a little annoying, no doubt; but the annoyance was nothing to the thrill of a new revelation, such as must have inspired Keats's most famous sonnet. And I had so nearly missed the joy of discovering this work for myself and rushing up to the laboratory to be the first to tell everyone else about it! Twenty years have not damped my enthusiasm.

Norman Campbell, quoted in obituary by L. Hartshorn, 1949

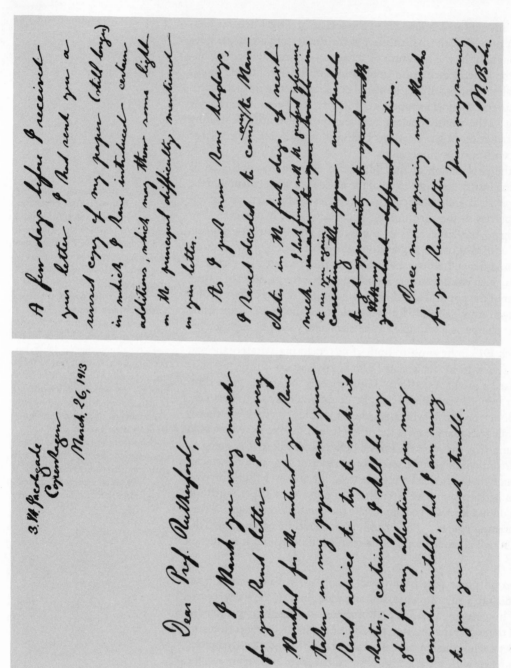

Letter from Bohr to Rutherford, 26 March 1913 (draft). After the words "I have decided to come over to Manchester in the first days of next week," Bohr had added: "in order to spare you trouble in correcting the paper and perhaps to get opportunity to speak with you about different questions." But he struck out this phrase and it did not appear in the final copy.

thought I should prove so obstinate, he consented to leave all the old and new points in the final paper.[3]

After this tussle (typical of Bohr's doggedness and determination), the second and third papers of the Trilogy evoked relatively little objection from Rutherford, and he duly submitted them on Bohr's behalf to the *Philosophical Magazine*. The original articles comprise seventy-one printed pages. Here we reproduce some of the most important parts — although the whole set of papers is well worth reading, even today, for the picture they give of the comprehensiveness of Bohr's endeavors, and of his mastery of the subject matter.

A.P.F.

THE

LONDON, EDINBURGH, AND DUBLIN

PHILOSOPHICAL MAGAZINE

AND

JOURNAL OF SCIENCE.

[SIXTH SERIES.]

JULY 1913.

I. *On the Constitution of Atoms and Molecules.*
By N. Bohr, Dr. phil. Copenhagen*.

Introduction.

IN order to explain the results of experiments on scattering of α rays by matter Prof. Rutherford† has given a theory of the structure of atoms. According to this theory, the atoms consist of a positively charged nucleus surrounded by a system of electrons kept together by attractive forces from the nucleus; the total negative charge of the electrons is equal to the positive charge of the nucleus. Further, the nucleus is assumed to be the seat of the essential part of the mass of the atom, and to have linear dimensions exceedingly small compared with the linear dimensions of the whole atom. The number of electrons in an atom is deduced to be approximately equal to half the atomic weight. Great interest is to be attributed to this atom-model; for, as Rutherford has shown, the assumption of the existence of nuclei, as those in question, seems to be necessary in order to account for the results of the experiments on large angle scattering of the α rays‡.

In an attempt to explain some of the properties of matter on the basis of this atom-model we meet, however, with difficulties of a serious nature arising from the apparent

* Communicated by Prof. E. Rutherford, F.R.S.
† E. Rutherford, Phil. Mag. xxi. p. 669 (1911).
‡ See also Geiger and Marsden, Phil. Mag. April 1913.

instability of the system of electrons: difficulties purposely avoided in atom-models previously considered, for instance, in the one proposed by Sir J. J. Thomson*. According to the theory of the latter the atom consists of a sphere of uniform positive electrification, inside which the electrons move in circular orbits.

The principal difference between the atom-models proposed by Thomson and Rutherford consists in the circumstance that the forces acting on the electrons in the atom-model of Thomson allow of certain configurations and motions of the electrons for which the system is in a stable equilibrium; such configurations, however, apparently do not exist for the second atom-model. The nature of the difference in question will perhaps be most clearly seen by noticing that among the quantities characterizing the first atom a quantity appears—the radius of the positive sphere—of dimensions of a length and of the same order of magnitude as the linear extension of the atom, while such a length does not appear among the quantities characterizing the second atom, viz. the charges and masses of the electrons and the positive nucleus; nor can it be determined solely by help of the latter quantities.

The way of considering a problem of this kind has, however, undergone essential alterations in recent years owing to the development of the theory of the energy radiation, and the direct affirmation of the new assumptions introduced in this theory, found by experiments on very different phenomena such as specific heats, photoelectric effect, Röntgen-rays, &c. The result of the discussion of these questions seems to be a general acknowledgment of the inadequacy of the classical electrodynamics in describing the behaviour of systems of atomic size†. Whatever the alteration in the laws of motion of the electrons may be, it seems necessary to introduce in the laws in question a quantity foreign to the classical electrodynamics, *i. e.* Planck's constant, or as it often is called the elementary quantum of action. By the introduction of this quantity the question of the stable configuration of the electrons in the atoms is essentially changed, as this constant is of such dimensions and magnitude that it, together with the mass and charge of the particles, can determine a length of the order of magnitude required.

This paper is an attempt to show that the application of the above ideas to Rutherford's atom-model affords a basis

* J. J. Thomson, Phil. Mag. vii. p. 237 (1904).
† See f. inst., 'Théorie du rayonnement et les quanta.' Rapports de la réunion à Bruxelles, Nov. 1911. Paris, 1912.

for a theory of the constitution of atoms. It will further be shown that from this theory we are led to a theory of the constitution of molecules.

In the present first part of the paper the mechanism of the binding of electrons by a positive nucleus is discussed in relation to Planck's theory. It will be shown that it is possible from the point of view taken to account in a simple way for the law of the line spectrum of hydrogen. Further, reasons are given for a principal hypothesis on which the considerations contained in the following parts are based.

I wish here to express my thanks to Prof. Rutherford for his kind and encouraging interest in this work.

PART I.—BINDING OF ELECTRONS BY POSITIVE NUCLEI.

§ 1. *General Considerations.*

The inadequacy of the classical electrodynamics in accounting for the properties of atoms from an atom-model as Rutherford's, will appear very clearly if we consider a simple system consisting of a positively charged nucleus of very small dimensions and an electron describing closed orbits around it. For simplicity, let us assume that the mass of the electron is negligibly small in comparison with that of the nucleus, and further, that the velocity of the electron is small compared with that of light.

Let us at first assume that there is no energy radiation. In this case the electron will describe stationary elliptical orbits. The frequency of revolution ω and the major-axis of the orbit $2a$ will depend on the amount of energy W which must be transferred to the system in order to remove the electron to an infinitely great distance apart from the nucleus. Denoting the charge of the electron and of the nucleus by $-e$ and E respectively and the mass of the electron by m, we thus get

$$\omega = \frac{\sqrt{2}}{\pi} \frac{W^{\frac{3}{2}}}{eE\sqrt{m}}, \quad 2a = \frac{eE}{W} . \quad . \quad . \quad . \quad (1)$$

Further, it can easily be shown that the mean value of the kinetic energy of the electron taken for a whole revolution is equal to W. We see that if the value of W is not given, there will be no values of ω and a characteristic for the system in question.

Let us now, however, take the effect of the energy radiation into account, calculated in the ordinary way from the acceleration of the electron. In this case the electron will

no longer describe stationary orbits. W will continuously increase, and the electron will approach the nucleus describing orbits of smaller and smaller dimensions, and with greater and greater frequency; the electron on the average gaining in kinetic energy at the same time as the whole system loses energy. This process will go on until the dimensions of the orbit are of the same order of magnitude as the dimensions of the electron or those of the nucleus. A simple calculation shows that the energy radiated out during the process considered will be enormously great compared with that radiated out by ordinary molecular processes.

It is obvious that the behaviour of such a system will be very different from that of an atomic system occurring in nature. In the first place, the actual atoms in their permanent state seem to have absolutely fixed dimensions and frequencies. Further, if we consider any molecular process, the result seems always to be that after a certain amount of energy characteristic for the systems in question is radiated out, the systems will again settle down in a stable state of equilibrium, in which the distances apart of the particles are of the same order of magnitude as before the process.

Now the essential point in Planck's theory of radiation is that the energy radiation from an atomic system does not take place in the continuous way assumed in the ordinary electrodynamics, but that it, on the contrary, takes place in distinctly separated emissions, the amount of energy radiated out from an atomic vibrator of frequency ν in a single emission being equal to $\tau h\nu$, where τ is an entire number, and h is a universal constant*.

Returning to the simple case of an electron and a positive nucleus considered above, let us assume that the electron at the beginning of the interaction with the nucleus was at a great distance apart from the nucleus, and had no sensible velocity relative to the latter. Let us further assume that the electron after the interaction has taken place has settled down in a stationary orbit around the nucleus. We shall, for reasons referred to later, assume that the orbit in question is circular; this assumption will, however, make no alteration in the calculations for systems containing only a single electron.

Let us now assume that, during the binding of the electron, a homogeneous radiation is emitted of a frequency ν, equal to half the frequency of revolution of the electron in its final

* See f. inst, M. Planck, *Ann. d. Phys.* xxxi. p.758 (1910); xxxvii. p. 642 (1912); *Verh. deutsch. Phys. Ges.* 1911, p. 138.

orbit; then, from Planck's theory, we might expect that the amount of energy emitted by the process considered is equal to $\tau h\nu$, where h is Planck's constant and τ an entire number. If we assume that the radiation emitted is homogeneous, the second assumption concerning the frequency of the radiation suggests itself, since the frequency of revolution of the electron at the beginning of the emission is 0. The question, however, of the rigorous validity of both assumptions, and also of the application made of Planck's theory, will be more closely discussed in § 3.

Putting

$$W = \tau h \frac{\omega}{2}, \qquad\qquad\qquad (2)$$

we get by help of the formula (1)

$$W = \frac{2\pi^2 m e^2 E^2}{\tau^2 h^2}, \quad \omega = \frac{4\pi^2 m e^2 E^2}{\tau^3 h^3}, \quad 2a = \frac{\tau^2 h^2}{2\pi^2 m e E} \cdot \quad (3)$$

If in these expressions we give τ different values, we get a series of values for W, ω, and a corresponding to a series of configurations of the system. According to the above considerations, we are led to assume that these configurations will correspond to states of the system in which there is no radiation of energy; states which consequently will be stationary as long as the system is not disturbed from outside. We see that the value of W is greatest if τ has its smallest value 1. This case will therefore correspond to the most stable state of the system, i. e. will correspond to the binding of the electron for the breaking up of which the greatest amount of energy is required.

Putting in the above expressions $\tau = 1$ and $E = e$, and introducing the experimental values

$$e = 4\cdot7 \,.\, 10^{-10}, \quad \frac{e}{m} = 5\cdot31\,.\,10^{17}, \quad h = 6\cdot5\,.\,10^{-27},$$

we get

$$2a = 1\cdot1\,.\,10^{-8}\ \text{cm.}, \quad \omega = 6\cdot2\,.\,10^{15}\ \frac{1}{\text{sec.}}, \quad \frac{W}{e} = 13\ \text{volt.}$$

We see that these values are of the same order of magnitude as the linear dimensions of the atoms, the optical frequencies, and the ionization-potentials.

The general importance of Planck's theory for the discussion of the behaviour of atomic systems was originally pointed out by Einstein*. The considerations of Einstein

* A. Einstein, *Ann. d. Phys.* xvii. p. 132 (1905); xx. p. 199 (1906); xxii. p. 180 (1907).

have been developed and applied on a number of different phenomena, especially by Stark, Nernst, and Sommerfield. The agreement as to the order of magnitude between values observed for the frequencies and dimensions of the atoms, and values for these quantities calculated by considerations similar to those given above, has been the subject of much discussion. It was first pointed out by Haas*, in an attempt to explain the meaning and the value of Planck's constant on the basis of J. J. Thomson's atom-model, by help of the linear dimensions and frequency of an hydrogen atom.

Systems of the kind considered in this paper, in which the forces between the particles vary inversely as the square of the distance, are discussed in relation to Planck's theory by J. W. Nicholson†. In a series of papers this author has shown that it seems to be possible to account for lines of hitherto unknown origin in the spectra of the stellar nebulæ and that of the solar corona, by assuming the presence in these bodies of certain hypothetical elements of exactly indicated constitution. The atoms of these elements are supposed to consist simply of a ring of a few electrons surrounding a positive nucleus of negligibly small dimensions. The ratios between the frequencies corresponding to the lines in question are compared with the ratios between the frequencies corresponding to different modes of vibration of the ring of electrons. Nicholson has obtained a relation to Planck's theory showing that the ratios between the wave-length of different sets of lines of the coronal spectrum can be accounted for with great accuracy by assuming that the ratio between the energy of the system and the frequency of rotation of the ring is equal to an entire multiple of Planck's constant. The quantity which Nicholson refers to as the energy is equal to twice the quantity which we have denoted above by W. In the latest paper cited Nicholson has found it necessary to give the theory a more complicated form, still, however, representing the ratio of energy to frequency by a simple function of whole numbers.

The excellent agreement between the calculated and observed values of the ratios between the wave-lengths in question seems a strong argument in favour of the validity of the foundation of Nicholson's calculations. Serious

* A. E. Haas, *Jahrb. d. Rad. u. El.* vii. p. 261 (1910). See further, A. Schidlof, *Ann. d. Phys.* xxxv. p. 90 (1911); E. Wertheimer, *Phys. Zeitschr.* xii. p. 409 (1911), *Verh. deutsch. Phys. Ges.* 1912, p. 431; F. A. Lindemann, *Verh. deutsch. Phys. Ges.* 1911, pp. 482, 1107; F. Haber, *Verh. deutsch. Phys. Ges.* 1911, p. 1117.

† J. W. Nicholson, *Month. Not. Roy. Astr. Soc.* lxxii. pp. 49, 139, 677, 693, 729 (1912).

objections, however, may be raised against the theory. These objections are intimately connected with the problem of the homogeneity of the radiation emitted. In Nicholson's calculations the frequency of lines in a line-spectrum is identified with the frequency of vibration of a mechanical system in a distinctly indicated state of equilibrium. As a relation from Planck's theory is used, we might expect that the radiation is sent out in quanta; but systems like those considered, in which the frequency is a function of the energy, cannot emit a finite amount of a homogeneous radiation; for, as soon as the emission of radiation is started, the energy and also the frequency of the system are altered. Further, according to the calculation of Nicholson, the systems are unstable for some modes of vibration. Apart from such objections—which may be only formal (see p. 23)—it must be remarked, that the theory in the form given does not seem to be able to account for the well-known laws of Balmer and Rydberg connecting the frequencies of the lines in the line-spectra of the ordinary elements.

It will now be attempted to show that the difficulties in question disappear if we consider the problems from the point of view taken in this paper. Before proceeding it may be useful to restate briefly the ideas characterizing the calculations on p. 5. The principal assumptions used are :

(1) That the dynamical equilibrium of the systems in the stationary states can be discussed by help of the ordinary mechanics, while the passing of the systems between different stationary states cannot be treated on that basis.

(2) That the latter process is followed by the emission of a *homogeneous* radiation, for which the relation between the frequency and the amount of energy emitted is the one given by Planck's theory.

The first assumption seems to present itself; for it is known that the ordinary mechanics cannot have an absolute validity, but will only hold in calculations of certain mean values of the motion of the electrons. On the other hand, in the calculations of the dynamical equilibrium in a stationary state in which there is no relative displacement of the particles, we need not distinguish between the actual motions and their mean values. The second assumption is in obvious contrast to the ordinary ideas of electrodynamics, but appears to be necessary in order to account for experimental facts.

In the calculations on page 5 we have further made use

of the more special assumptions, viz. that the different stationary states correspond to the emission of a different number of Planck's energy-quanta, and that the frequency of the radiation emitted during the passing of the system from a state in which no energy is yet radiated out to one of the stationary states, is equal to half the frequency of revolution of the electron in the latter state. We can, however (see § 3), also arrive at the expressions (3) for the stationary states by using assumptions of somewhat different form. We shall, therefore, postpone the discussion of the special assumptions, and first show how by the help of the above principal assumptions, and of the expressions (3) for the stationary states, we can account for the line-spectrum of hydrogen.

§ 2. Emission of Line-spectra.

Spectrum of Hydrogen.—General evidence indicates that an atom of hydrogen consists simply of a single electron rotating round a positive nucleus of charge e^*. The reformation of a hydrogen atom, when the electron has been removed to great distances away from the nucleus—*e.g.* by the effect of electrical discharge in a vacuum tube—will accordingly correspond to the binding of an electron by a positive nucleus considered on p. 5. If in (3) we put $E = e$, we get for the total amount of energy radiated out by the formation of one of the stationary states,

$$W_\tau = \frac{2\pi^2 m e^4}{h^2 \tau^2}.$$

The amount of energy emitted by the passing of the system from a state corresponding to $\tau = \tau_2$ to one corresponding to $\tau = \tau_1$, is consequently

$$W_{\tau_2} - W_{\tau_1} = \frac{2\pi^2 m e^4}{h^2}\left(\frac{1}{\tau_2^2} - \frac{1}{\tau_1^2}\right).$$

If now we suppose that the radiation in question is homogeneous, and that the amount of energy emitted is equal to $h\nu$, where ν is the frequency of the radiation, we get

$$W_{\tau_2} - W_{\tau_1} = h\nu,$$

* See f. inst. N. Bohr, Phil. Mag. xxv. p. 24 (1913). The conclusion drawn in the paper cited is strongly supported by the fact that hydrogen, in the experiments on positive rays of Sir J. J. Thomson, is the only element which never occurs with a positive charge corresponding to the loss of more than one electron (comp. Phil. Mag. xxiv. p. 672 (1912)).

and from this

$$\nu = \frac{2\pi^2 m e^4}{h^3}\left(\frac{1}{\tau_2^2} - \frac{1}{\tau_1^2}\right) \quad . \quad . \quad . \quad . \quad . \quad (4)$$

We see that this expression accounts for the law connecting the lines in the spectrum of hydrogen. If we put $\tau_2=2$ and let τ_1 vary, we get the ordinary Balmer series. If we put $\tau_2=3$, we get the series in the ultra-red observed by Paschen* and previously suspected by Ritz. If we put $\tau_2=1$ and $\tau_2=4, 5,..$, we get series respectively in the extreme ultra-violet and the extreme ultra-red, which are not observed, but the existence of which may be expected.

The agreement in question is quantitative as well as qualitative. Putting

$$e = 4\cdot7 \cdot 10^{-10}, \quad \frac{e}{m} = 5\cdot31 \cdot 10^{17}, \quad \text{and} \quad h = 6\cdot5 \cdot 10^{-27},$$

we get

$$\frac{2\pi^2 m e^4}{h^3} = 3\cdot1 \cdot 10^{15}.$$

The observed value for the factor outside the bracket in the formula (4) is

$$3\cdot290 \cdot 10^{15}.$$

The agreement between the theoretical and observed values is inside the uncertainty due to experimental errors in the constants entering in the expression for the theoretical value. We shall in § 3 return to consider the possible importance of the agreement in question.

It may be remarked that the fact, that it has not been possible to observe more than 12 lines of the Balmer series in experiments with vacuum tubes, while 33 lines are observed in the spectra of some celestial bodies, is just what we should expect from the above theory. According to the equation (3) the diameter of the orbit of the electron in the different stationary states is proportional to τ^2. For $\tau=12$ the diameter is equal to $1\cdot6 \cdot 10^{-6}$ cm., or equal to the mean distance between the molecules in a gas at a pressure of about 7 mm. mercury; for $\tau=33$ the diameter is equal to $1\cdot2 \cdot 10^{-5}$ cm., corresponding to the mean distance of the molecules at a pressure of about $0\cdot02$ mm. mercury. According to the theory the necessary condition for the appearance of a great number of lines is therefore a very small density of the gas; for simultaneously to obtain an

* F. Paschen, *Ann. d. Phys.* xxvii. p. 565 (1908).

intensity sufficient for observation the space filled with the gas must be very great. If the theory is right, we may therefore never expect to be able in experiments with vacuum tubes to observe the lines corresponding to high numbers of the Balmer series of the emission spectrum of hydrogen; it might, however, be possible to observe the lines by investigation of the absorption spectrum of this gas (see § 4).

It will be observed that we in the above way do not obtain other series of lines, generally ascribed to hydrogen; for instance, the series first observed by Pickering* in the spectrum of the star ζ Puppis, and the set of series recently found by Fowler† by experiments with vacuum tubes containing a mixture of hydrogen and helium. We shall, however, see that, by help of the above theory, we can account naturally for these series of lines if we ascribe them to helium.

A neutral atom of the latter element consists, according to Rutherford's theory, of a positive nucleus of charge $2e$ and two electrons. Now considering the binding of a single electron by a helium nucleus, we get, putting $E=2e$ in the expressions (3) on page 5, and proceeding in exactly the same way as above,

$$\nu = \frac{8\pi^2 m e^4}{h^3}\left(\frac{1}{\tau_2^2} - \frac{1}{\tau_1^2}\right) = \frac{2\pi^2 m e^4}{h^3}\left(\frac{1}{\left(\frac{\tau_2}{2}\right)^2} - \frac{1}{\left(\frac{\tau_1}{2}\right)^2}\right).$$

If we in this formula put $\tau_2=1$ or $\tau_2=2$, we get series of lines in the extreme ultra-violet. If we put $\tau_2=3$, and let τ_1 vary, we get a series which includes 2 of the series observed by Fowler, and denoted by him as the first and second principal series of the hydrogen spectrum. If we put $\tau_2=4$, we get the series observed by Pickering in the spectrum of ζ Puppis. Every second of the lines in this series is identical with a line in the Balmer series of the hydrogen spectrum; the presence of hydrogen in the star in question may therefore account for the fact that these lines are of a greater intensity than the rest of the lines in the series. The series is also observed in the experiments of Fowler, and denoted in his paper as the Sharp series of the hydrogen spectrum. If we finally in the above formula put $\tau_2=5, 6,...$, we get series, the strong lines of which are to be expected in the ultra-red.

* E. C. Pickering, *Astrophys. J.* iv. p.369 (1896); v. p. 92 (1897).
† A. Fowler, *Month. Not. Roy. Astr. Soc.* lxxiii. Dec. 1912.

While there obviously can be no question of a mechanical foundation of the calculations given in this paper, it is, however, possible to give a very simple interpretation of the result of the calculation on p. 5 by help of symbols taken from the ordinary mechanics. Denoting the angular momentum of the electron round the nucleus by M, we have immediately for a circular orbit $\pi M = \dfrac{T}{\omega}$, where ω is the frequency of revolution and T the kinetic energy of the electron; for a circular orbit we further have $T = W$ (see p. 5) and from (2), p. 5, we consequently get

$$M = \tau M_0,$$

where

$$M_0 = \frac{h}{2\pi} = 1 \cdot 04 \times 10^{-27}.$$

If we therefore assume that the orbit of the electron in the stationary states is circular, the result of the calculation on p. 5 can be expressed by the simple condition: that the angular momentum of the electron round the nucleus in a stationary state of the system is equal to an entire multiple of a universal value, independent of the charge on the nucleus. The possible importance of the angular momentum in the discussion of atomic systems in relation to Planck's theory is emphasized by Nicholson*.

The great number of different stationary states we do not observe except by investigation of the emission and absorption of radiation. In most of the other physical phenomena, however, we only observe the atoms of the matter in a single distinct state, i. e. the state of the atoms at low temperature. From the preceding considerations we are immediately led to the assumption that the "permanent" state is the one among the stationary states during the formation of which the greatest amount of energy is emitted. According to the equation (3) on p. 5, this state is the one which corresponds to $\tau = 1$.

.
.

Proceeding to consider systems of a more complicated constitution, we shall make use of the following theorem, which can be very simply proved:—

"In every system consisting of electrons and positive nuclei, in which the nuclei are at rest and the electrons move in circular orbits with a velocity small compared with the

---

* J. W. Nicholson, *loc. cit.* p. 679.

velocity of light, the kinetic energy will be numerically equal to half the potential energy."

By help of this theorem we get—as in the previous cases of a single electron or of a ring rotating round a nucleus—that the total amount of energy emitted, by the formation of the systems from a configuration in which the distances apart of the particles are infinitely great and in which the particles have no velocities relative to each other, is equal to the kinetic energy of the electrons in the final configuration.

In analogy with the case of a single ring we are here led to assume that corresponding to any configuration of equilibrium a series of geometrically similar, stationary configurations of the system will exist in which the kinetic energy of every electron is equal to the frequency of revolution multiplied by $\dfrac{\tau}{2}h$ where τ is an entire number and h Planck's constant. In any such series of stationary configurations the one corresponding to the greatest amount of energy emitted will be the one in which τ for every electron is equal to 1. Considering that the ratio of kinetic energy to frequency for a particle rotating in a circular orbit is equal to π times the angular momentum round the centre of the orbit, we are therefore led to the following simple generalization of the hypotheses mentioned on pp. 15 and 22.

"*In any molecular system consisting of positive nuclei and electrons in which the nuclei are at rest relative to each other and the electrons move in circular orbits, the angular momentum of every electron round the centre of its orbit will in the permanent state of the system be equal to $\dfrac{h}{2\pi}$, where h is Planck's constant*"*.

In analogy with the considerations on p. 23, we shall assume that a configuration satisfying this condition is stable if the total energy of the system is less than in any neighbouring configuration satisfying the same condition of the angular momentum of the electrons.

As mentioned in the introduction, the above hypothesis will be used in a following communication as a basis for a theory of the constitution of atoms and molecules. It will be shown that it leads to results which seem to be in conformity with experiments on a number of different phenomena.

---

* In the considerations leading to this hypothesis we have assumed that the velocity of the electrons is small compared with the velocity of light. The limits of the validity of this assumption will be discussed in Part II.

The foundation of the hypothesis has been sought entirely in its relation with Planck's theory of radiation; by help of considerations given later it will be attempted to throw some further light on the foundation of it from another point of view.

April 5, 1913.

•   •   •

On the Constitution of Atoms and Molecules.
*By N. Bohr, Dr. phil. Copenhagen.**

PART II.—SYSTEMS CONTAINING ONLY A SINGLE NUCLEUS †.

§ 1. *General Assumptions.*

FOLLOWING the theory of Rutherford, we shall assume that the atoms of the elements consist of a positively charged nucleus surrounded by a cluster of electrons. The nucleus is the seat of the essential part of the mass of the atom, and has linear dimensions exceedingly small compared with the distances apart of the electrons in the surrounding cluster.

As in the previous paper, we shall assume that the cluster of electrons is formed by the successive binding by the nucleus of electrons initially nearly at rest, energy at the same time being radiated away. This will go on until, when the total negative charge on the bound electrons is numerically equal to the positive charge on the nucleus, the system will be neutral and no longer able to exert sensible forces on electrons at distances from the nucleus great in comparison with the dimensions of the orbits of the bound electrons. We may regard the formation of helium from α rays as an observed example of a process of this kind, an α particle on this view being identical with the nucleus of a helium atom.

On account of the small dimensions of the nucleus, its internal structure will not be of sensible influence on the constitution of the cluster of electrons, and consequently will have no effect on the ordinary physical and chemical properties of the atom. The latter properties on this theory

* Communicated by Prof. E. Rutherford, F.R.S.
† Part I. was published in Phil. Mag. xxvi. p. 1 (1913).

will depend entirely on the total charge and mass of the nucleus; the internal structure of the nucleus will be of influence only on the phenomena of radioactivity.

From the result of experiments on large-angle scattering of α-rays, Rutherford* found an electric charge on the nucleus corresponding per atom to a number of electrons approximately equal to half the atomic weight. This result seems to be in agreement with the number of electrons per atom calculated from experiments on scattering of Röntgen radiation†. The total experimental evidence supports the hypothesis ‡ that the actual number of electrons in a neutral atom with a few exceptions is equal to the number which indicates the position of the corresponding element in the series of elements arranged in order of increasing atomic weight. For example on this view, the atom of oxygen which is the eighth element of the series has eight electrons and a nucleus carrying eight unit charges.

We shall assume that the electrons are arranged at equal angular intervals in coaxial rings rotating round the nucleus. In order to determine the frequency and dimensions of the rings we shall use the main hypothesis of the first paper, viz.: that in the permanent state of an atom the angular momentum of every electron round the centre of its orbit is equal to the universal value $\frac{h}{2\pi}$, where h is Planck's constant.

We shall take as a condition of stability, that the total energy of the system in the configuration in question is less than in any neighbouring configuration satisfying the same condition of the angular momentum of the electrons.

If the charge on the nucleus and the number of electrons in the different rings is known, the condition in regard to the angular momentum of the electrons will, as shown in § 2, completely determine the configuration of the system, *i.e.*, the frequency of revolution and the linear dimensions of the rings. Corresponding to different distributions of the electrons in the rings, however, there will, in general, be more than one configuration which will satisfy the condition of the angular momentum together with the condition of stability.

In § 3 and § 4 it will be shown that, on the general view of the formation of the atoms, we are led to indications of

* Comp. also Geiger and Marsden, Phil. Mag. xxv. p. 604 (1913).
† Comp. C. G. Barkla, Phil. Mag. xxi. p. 648 (1911).
‡ Comp. A. v. d. Broek, *Phys. Zeitschr.* xiv. p. 32 (1913).

the arrangement of the electrons in the rings which are consistent with those suggested by the chemical properties of the corresponding element.

In § 5 it will be shown that it is possible from the theory to calculate the minimum velocity of cathode rays necessary to produce the characteristic Röntgen radiation from the element, and that this is in approximate agreement with the experimental values.

In § 6 the phenomena of radioactivity will be briefly considered in relation to the theory.

· · ·

§ 3. Constitution of Atoms containing very few Electrons.

As stated in § 1, the condition of the universal constancy of the angular momentum of the electrons, together with the condition of stability, is in most cases not sufficient to determine completely the constitution of the system. On the general view of formation of atoms, however, and by making use of the knowledge of the properties of the corresponding elements, it will be attempted, in this section and the next, to obtain indications of what configurations of the electrons may be expected to occur in the atoms. In these considerations we shall assume that the number of electrons in the atom is equal to the number which indicates the position of the corresponding element in the series of elements arranged in order of increasing atomic weight. Exceptions to this rule will be supposed to occur only at such places in the series where deviation from the periodic law of the chemical properties of the elements are observed. In order to show clearly the principles used we shall first consider with some detail those atoms containing very few electrons.

For sake of brevity we shall, by the symbol $N(n_1, n_2, \cdots)$, refer to a plane system of rings of electrons rotating round a nucleus of charge Ne, satisfying the condition of the angular momentum of the electrons with the approximation used in § 2. $n_1, n_2, \cdots$ are the numbers of electrons in the rings, starting from inside. By $a_1, a_2, \cdots$ and $\omega_1, \omega_2, \cdots$ we shall denote the radii and frequency of the rings taken in the same order. The total amount of energy W emitted by the formation of the system shall simply be denoted by $W[N(n_1, n_2, \cdots)]$.

N = 1.   Hydrogen.

In Part I. we have considered the binding of an electron by a positive nucleus of charge e, and have shown that it is possible to account for the Balmer spectrum of hydrogen on the assumption of the existence of a series of stationary states in which the angular momentum of the electron round the nucleus is equal to entire multiples of the value $\frac{h}{2\pi}$, where h is Planck's constant. The formula found for the frequencies of the spectrum was

$$\nu = \frac{2\pi^2 e^4 m}{h^3}\left(\frac{1}{\tau_2^2} - \frac{1}{\tau_1^2}\right),$$

where τ_1 and τ_2 are entire numbers. Introducing the values for e, m, and h used on p. 479, we get for the factor before the bracket $3\cdot1.10^{15}$*; the value observed for the constant in the Balmer spectrum is $3\cdot290.10^{15}$.

For the permanent state of a neutral hydrogen atom we get from the formula (1) and (2) in § 2, putting F=1,

1(1). $a = \dfrac{h^2}{4\pi^2 e^2 m} = 0\cdot55.10^{-8}$, $\omega = \dfrac{4\pi^2 e^2 m}{h^3} = 6\cdot2.10^{15}$,

$$W = \frac{2\pi^2 e^4 m}{h^2} = 2\cdot0.10^{-11}.$$

These values are of the order of magnitude to be expected.

For $\dfrac{W}{e}$ we get $0\cdot043$, which corresponds to 13 volts; the value for the ionizing potential of a hydrogen atom, calculated by Sir J. J. Thomson from experiments on positive rays, is 11 volts*. No other definite data, however, are available for hydrogen atoms. For sake of brevity, we shall in the following denote the values for a, ω, and W corresponding to the configuration 1(1) by a_0, ω_0 and W_0.

* This value is that calculated in the first part of the paper. Using the values $e=4\cdot78.10^{-10}$ (see R. A. Millikan, Brit. Assoc. Rep. 1912, p. 410), $\frac{e}{m}=5\cdot31.10^7$ (see P. Gmelin, Ann. d. Phys. xxviii. p. 1086 (1909)) and A. H. Bucherer, Ann. d. Phys. xxxvii. p. 597 (1912)), and $\frac{e}{h}=7\cdot27.10^{16}$ (calculated by Planck's theory from the experiments of E. Warburg, G. Leithäuser, E. Hupka, and C. Müller, Ann. d. Phys. xl. p. 611 (1913)) we get $\frac{2\pi^2 e^4 m}{h^3}=3\cdot26.10^{15}$ in very close agreement with observations.

* J. J. Thomson, Phil. Mag. xxiv. p. 218 (1912).

At distances from the nucleus, great in comparison with a_0, the system 1(1) will not exert sensible forces on free electrons. Since, however, the configuration :

1(2) $a = 1.33 a_0$, $\omega = 0.563 \omega_0$, $W = 1.13 W_0$,

corresponds to a greater value for W than the configuration 1(1), we may expect that a hydrogen atom under certain conditions can acquire a negative charge. This is in agreement with experiments on positive rays. Since $W[1(3)]$ is only 0.54, a hydrogen atom cannot be expected to be able to acquire a double negative charge.

N = 2.   *Helium.*

As shown in Part I, using the same assumptions as for hydrogen, we must expect that during the binding of an electron by a nucleus of charge 2e, a spectrum is emitted, expressed by

$$\nu = \frac{2\pi^2 m e^4}{h^3} \left(\frac{1}{\left(\frac{\tau_2}{2}\right)^2} - \frac{1}{\left(\frac{\tau_1}{2}\right)^2} \right),$$

This spectrum includes the spectrum observed by Pickering in the star ζ Puppis and the spectra recently observed by Fowler in experiments with vacuum tubes filled with a mixture of hydrogen and helium. These spectra are generally ascribed to hydrogen.

For the permanent state of a positively charged helium atom, we get

2(1) $a = \frac{1}{2} a_0$, $\omega = 4 \omega_0$, $W = 4 W_0$.

At distances from the nucleus great compared with the radius of the bound electron, the system 2(1) will, to a close approximation, act on an electron as a simple nucleus of charge e. For a system consisting of two electrons and a nucleus of charge 2e, we may therefore assume the existence of a series of stationary states in which the electron most lightly bound moves approximately in the same way as the electron in the stationary states of a hydrogen atom. Such an assumption has already been used in Part I. in an attempt to explain the appearance of Rydberg's constant in the formula for the line-spectrum of any element. We can, however, hardly assume the existence of a stable configuration in which the two electrons have the same angular momentum round the nucleus and move in different orbits. In such a configuration the electrons would be so near to each other that the deviations

from circular orbits would be very great. For the permanent state of a neutral helium atom, we shall therefore adopt the configuration :

2(2) $a = 0.571 a_0$, $\omega = 3.06 \omega_0$, $W = 6.13 W_0$.

Since $W[2(2)] - W[2(1)] = 2.13 W_0$,

we see that both electrons in a neutral helium atom are more firmly bound than the electron in a hydrogen atom. Using the values on p. 488, we get

$2.13 \cdot \frac{W_0}{e} = 27$ volts   and $2.13 \frac{W_0}{h} = 6.6 \cdot 10^{15} \frac{1}{sec.}$;

these values are of the same order of magnitude as the value observed for the ionization potential in helium, 20.5 volt*, and the value for the frequency of the ultra-violet absorption in helium determined by experiments on dispersion $5.9 \cdot 10^{15} \frac{1}{sec.}$ †.

* J. Franck u. G. Hertz, *Verh. d. Deutsch. Phys. Ges.* xv. p. 34 (1913).
† C. and M. Cuthbertson, Proc. Roy. Soc. A. lxxxiv. p. 13 (1910). (In a previous paper (Phil. Mag. Jan. 1913) the author took the values for the refractive index in helium, given by M. and C. Cuthbertson, as corresponding to atmospheric pressure; these values, however, refer to double atmospheric pressure. Consequently the value there given for the number of electrons in a helium atom calculated from Drude's theory has to be divided by 2.)

. . . . .

. . . . .

§ 5. *Characteristic Röntgen Radiation.*

According to the theory of emission of radiation given in Part I, the ordinary line-spectrum of an element is emitted during the reformation of an atom when one or more of the electrons in the outer rings are removed. In analogy it may be supposed that the characteristic Röntgen radiation is sent out during the settling down of the system if electrons in inner rings are removed by some agency, e.g. by impact of cathode particles. This view of the origin of the characteristic Röntgen radiation has been proposed by Sir J. J. Thomson*.

Without any special assumption in regard to the constitution of the radiation, we can from this view determine the minimum velocity of the cathode rays necessary to produce the characteristic Röntgen radiation of a special type by

* Comp. J. J. Thomson, Phil. Mag. xxiii. p. 456 (1912).

On the Constitution of Atoms and Molecules.
By N. Bohr, Dr. phil., Copenhagen

Part III.—Systems containing Several Nuclei

§ 1. *Preliminary.*

ACCORDING to Rutherford's theory of the structure of atoms, the difference between an atom of an element and a molecule of a chemical combination is that the first consists of a cluster of electrons surrounding a single positive nucleus of exceedingly small dimensions and of a mass great in comparison with that of the electrons, while the latter contains at least two nuclei at distances from each other comparable with the distances apart of the electrons in the surrounding cluster.

The leading idea used in the former papers was that the atoms were formed through the successive binding by the nucleus of a number of electrons initially nearly at rest. Such a conception, however, cannot be utilized in considering the formation of a system containing more than a single nucleus ; for in the latter case there will be nothing to keep the nuclei together during the binding of the electrons. In this connexion it may be noticed that while a single nucleus carrying a large positive charge is able to bind a small number of electrons, on the contrary, two nuclei highly charged obviously cannot be kept together by the help of a few electrons. We must therefore assume that configurations containing several nuclei are formed by the interaction of systems—each containing a single nucleus—which already have bound a number of electrons.

§ 2 deals with the configuration and stability of a system already formed. We shall consider only the simple case of a system consisting of two nuclei and of a ring of electrons rotating round the line connecting them ; the result of the calculation, however, gives indication of what configurations are to be expected in more complicated cases. As in the former papers, we shall assume that the conditions of equilibrium can be deduced by help of the ordinary mechanics. In determining the absolute dimensions and the stability of the systems, however, we shall use the main hypothesis of Part I. According to this, the angular momentum of every electron round the centre of its orbit is equal to a universal value $\dfrac{h}{2\pi}$, where h is Planck's constant ; further, the

calculating the energy necessary to remove one of the electrons from the different rings. • • Let us consider a simple system consisting of a bound electron rotating in a circular orbit round a positive nucleus of charge Ne. From the expressions (1) on p. 478 we get for the velocity of the electron, putting $F = N$,

$$v = \frac{2\pi e^2}{h} N = 2 \cdot 1 \cdot 10^8 N.$$

The total energy to be transferred to the system in order to remove the electron to an infinite distance from the nucleus is equal to the kinetic energy of the bound electron. If, therefore, the electron is removed to a great distance from the nucleus by impact of another rapidly moving electron, the smallest kinetic energy possessed by the latter when at a great distance from the nucleus must necessarily be equal to the kinetic energy of the bound electron before the collision. The velocity of the free electron therefore must be at least equal to v.

According to Whiddington's experiments * the velocity of cathode rays just able to produce the characteristic Röntgen radiation of the so-called K-type—the hardest type of radiation observed—from an element of atomic weight A is for elements from Al to Se approximately equal to $A \cdot 10^8$ cm./sec. As seen this is equal to the above calculated value for v, if we put $N = \dfrac{A}{2}$.

Since we have obtained approximate agreement with experiment by ascribing the characteristic Röntgen radiation of the K-type to the innermost ring, it is to be expected that no harder type of characteristic radiation will exist. This is strongly indicated by observations of the penetrating power of γ rays[†].

It is worthy of remark that the theory gives not only nearly the right value for the energy required to remove an electron from the outer ring, but also the energy required to remove an electron from the innermost ring. The approximate agreement between the calculated and experimental values is all the more striking when it is recalled that the energies required in the two cases for an element of atomic weight 70 differ by a ratio of 1000.

---

* R. Whiddington, Proc. Roy. Soc. A. lxxxv. p. 323 (1911).
† Comp. E. Rutherford, Phil. Mag. xxiv. p. 453 (1912).

stability is determined by the condition that the total energy of the system is less than in any neighbouring configuration satisfying the same condition of the angular momentum of the electrons.

In § 3 the configuration to be expected for a hydrogen molecule is discussed in some detail.

§ 4 deals with the mode of formation of the systems. A simple method of procedure is indicated, by which it is possible to follow, step by step, the combination of two atoms to form a molecule. The configuration obtained will be shown to satisfy the conditions used in § 2. The part played in the considerations by the angular momentum of the electrons strongly supports the validity of the main hypothesis.

§ 5 contains a few indications of the configurations to be expected for systems containing a greater number of electrons.

∴

Concluding remarks.

In the present paper an attempt has been made to develop a theory of the constitution of atoms and molecules on the basis of the ideas introduced by Planck in order to account for the radiation from a black body, and the theory of the structure of atoms proposed by Rutherford in order to explain the scattering of α-particles by matter.

Planck's theory deals with the emission and absorption of radiation from an atomic vibrator of a constant frequency, independent of the amount of energy possessed by the system in the moment considered. The assumption of such vibrators, however, involves the assumption of quasi-elastic forces and is inconsistent with Rutherford's theory, according to which all the forces between the particles of an atomic system vary inversely as the square of the distance apart. In order to apply the main results obtained by Planck it is therefore necessary to introduce new assumptions as to the emission and absorption of radiation by an atomic system.

The main assumptions used in the present paper are :—

1. That energy radiation is not emitted (or absorbed) in the continuous way assumed in the ordinary electrodynamics, but only during the passing of the systems between different "stationary" states.

2. That the dynamical equilibrium of the systems in the stationary states is governed by the ordinary laws of mechanics, while these laws do not hold for the passing of the systems between the different stationary states.

3. That the radiation emitted during the transition of a system between two stationary states is homogeneous, and that the relation between the frequency ν and the total amount of energy emitted E is given by E $= h\nu$, where h is Planck's constant.

4. That the different stationary states of a simple system consisting of an electron rotating round a positive nucleus are determined by the condition that the ratio between the total energy, emitted during the formation of the configuration, and the frequency of revolution of the electron is an entire multiple of $\frac{h}{2}$. Assuming that the orbit of the electron is circular, this assumption is equivalent with the assumption that the angular momentum of the electron round the nucleus is equal to an entire multiple of $\frac{h}{2\pi}$.

5. That the "permanent" state of any atomic system—i. e., the state in which the energy emitted is maximum—is determined by the condition that the angular momentum of every electron round the centre of its orbit is equal to $\frac{h}{2\pi}$.

It is shown that, applying these assumptions to Rutherford's atom model, it is possible to account for the laws of Balmer and Rydberg connecting the frequency of the different lines in the line-spectrum of an element. Further, outlines are given of a theory of the constitution of the atoms of the elements and of the formation of molecules of chemical combinations, which on several points is shown to be in approximate agreement with experiments.

The intimate connexion between the present theory and modern theories of the radiation from a black body and of specific heat is evident ; again, since on the ordinary electrodynamics the magnetic moment due to an electron rotating in a circular orbit is proportional to the angular momentum, we shall expect a close relation to the theory of magnetons proposed by Weiss. The development of a detailed theory of heat radiation and of magnetism on the basis of the present theory claims, however, the introduction of additional assumptions about the behaviour of bound electrons in an electromagnetic field. The writer hopes to return to these questions later.

Nobel Prize Lecture:
The Structure of the Atom

Niels Bohr delivered this lecture in Stockholm on 11 December 1922. The main body of the address was quite extensive, consisting of a masterly and comprehensive exposition of the whole of atomic physics as then known, interpreted on the basis of the fundamental principles that Bohr had developed. The topics treated included atomic stability and electrodynamic theory, the origin of the quantum theory, the quantum theory of atomic constitution, the hydrogen spectrum, relationships between the elements, absorption and excitation of spectral lines, the quantum theory of multiply periodic systems, the correspondence principle, the natural system of the elements, and X-ray spectra and atomic constitution.

We reproduce here two excerpts: the opening section, in which Bohr summarizes current knowledge about the atom, and the main part of the section on the correspondence principle.[1]

A.P.F.

Ladies and Gentlemen:

Today, as a consequence of the great honor the Swedish Academy of Sciences has done me in awarding me this year's Nobel Prize for Physics for my work on the structure of the atom, it is my duty to give an account of the results of this work, and I think that I shall be acting in accordance with the traditions of the Nobel Foundation if I give this report in the form of a survey of the development which has taken place in the last few years within the field of physics to which this work belongs.

THE GENERAL PICTURE OF THE ATOM

The present state of atomic theory is characterized by the fact that we not only believe the existence of atoms to be proved beyond a doubt, but also we even believe that we have an intimate knowledge of the constituents of the individual atoms. I cannot on this occasion give a survey of the scientific developments that have led to this result; I will only recall the discovery of the electron toward the close of the last century, which furnished the direct verification and led to a conclusive formulation of the conception of the atomic nature of electricity which had evolved since the discovery by Faraday of the fundamental laws of electrolysis and Berzelius's electrochemical theory, and had its greatest triumph in the electrolytic dissociation theory of Arrhenius. This discovery of the electron and elucidation of its properties was the result of the work of a large number of investigators, among whom Lenard and J. J. Thomson may be particularly mentioned. The latter especially has made very important contributions to our subject by his ingenious attempts to develop ideas about atomic constitution on the basis of the electron theory. The present state of our knowledge of the elements of atomic structure was reached, however, by the discovery of the atomic nucleus, which we owe to Rutherford, whose work on the radioactive substances discovered toward the close of the last century has much enriched physical and chemical science.

According to our present conceptions, an atom of an element is built up of a nucleus that has a positive electrical charge and is the seat of by far the greatest part of the atomic mass, together with a number of electrons, all having the same negative charge and mass, which move at distances from the nucleus that are very great compared to the dimensions of the nucleus or of the electrons themselves. In this picture we at once see a striking resemblance to a planetary system, such as we have in our own solar system. Just as the simplicity of the laws that govern the motions of the solar system is intimately connected with the circumstance that the dimensions of the moving bodies are small in relation to the orbits, so the corresponding relations in atomic structure provide us with an explanation of an essential feature of natural phenomena insofar as these depend on the properties of the elements. It makes clear at once that these properties can be divided into two sharply distinguished classes.

To the first class belong most of the ordinary physical and chemical properties of substances, such as their state of aggregation, color, and chemical reactivity. These properties depend on the motion of the electron system and the way in which this motion changes under the influence of different external actions. On account of the large mass of the nucleus relative to that of the electrons and its smallness in comparison to the electron orbits, the electronic motion will depend only to a very small extent on the nuclear mass, and will be determined to a close approximation solely by the total electrical charge of the nucleus. Especially the inner

What is not always recognized is that a model being drawn from a field of knowledge other than that to which it is to be applied carries a certain amount of preexisting understanding of its own properties. The Rutherford-Bohr picture of the atom as a planetary system of electrons in orbit around a nucleus owes its strength not only to the basic principles of classical physics, but also to our familiarity with just such systems in astronomy.

J. M. Ziman,
Reliable Knowledge, 1979

structure of the nucleus and the way in which the charges and masses are distributed among its separate particles will have a vanishingly small influence on the motion of the electron system surrounding the nucleus. On the other hand, the structure of the nucleus will be responsible for the second class of properties that are shown in the radioactivity of substances. In the radioactive processes, we meet with an explosion of the nucleus whereby positive or negative particles, the so-called α- and β-particles, are expelled with very great velocities.

Our conceptions of atomic structure afford us, therefore, an immediate explanation of the complete lack of interdependence between the two classes of properties, which is most strikingly shown in the existence of substances which have to an extraordinarily close approximation the same ordinary physical and chemical properties, even though the atomic weights are not the same, and the radioactive properties are completely different. Such substances, of the existence of which the first evidence was found in the work of Soddy and other investigators on the chemical properties of the radioactive elements, are called isotopes, with reference to the classification of the elements according to ordinary physical and chemical properties. It is not necessary for me to state here how it has been shown in recent years that isotopes are found not only among the radioactive elements, but also among ordinary stable elements; in fact, a large number of the latter that were previously supposed simple have been shown by Aston's well-known investigations to consist of a mixture of isotopes with different atomic weights.

The question of the inner structure of the nucleus is still but little understood, although a method of attack is afforded by Rutherford's experiments on the disintegration of atomic nuclei by bombardment with α-particles. Indeed, these experiments may be said to open up a new epoch in natural philosophy in that for the first time the artificial transformation of one element into another has been accomplished. In what follows, however, we shall confine ourselves to a consideration of the ordinary physical and chemical properties of the elements and the attempts which have been made to explain them on the basis of the concepts just outlined.

The wonderful progress of our knowledge of the atomic constitution of matter and of the methods by which such knowledge can be acquired and interrelated has indeed carried us far beyond the scope of the deterministic pictorial description brought to such perfection by Newton and Maxwell. Following this development at close hand, I have often had occasion to think of the dominating influence of Rutherford's original discovery of the atomic nucleus, which at every stage presented us with so forceful a challenge.

Niels Bohr,
Rutherford Memorial Lecture, 1958

ဗၑ

THE CORRESPONDENCE PRINCIPLE

While this development of the theory of spectra was based on the working out of formal methods for the fixation of stationary states, the present lecturer succeeded shortly afterward in throwing light on the theory from a new viewpoint, by pursuing further the characteristic connection between the quantum theory and classical electrodynamics already traced out in the hydrogen spectrum. In connection with the important work of Ehrenfest and Einstein these efforts led to the formulation of the so-called *correspon-*

dence principle, according to which the occurrence of transitions between the stationary states accompanied by emission of radiation is traced back to the harmonic components into which the motion of the atom may be resolved and which, according to the classical theory, determine the properties of the radiation to which the motion of the particles gives rise.

According to the correspondence principle, it is assumed that every transition process between two stationary states can be coordinated with a corresponding harmonic vibration component in such a way that the probability of the occurrence of the transition is dependent on the amplitude of the vibration. The state of polarization of the radiation emitted during the transition depends on the further characteristics of the vibration, in a manner analogous to that in which, on the classical theory, the intensity and state of polarization in the wave system emitted by the atom as a consequence of the presence of this vibration component would be determined respectively by the amplitude and further characteristics of the vibration.

With the aid of the correspondence principle it has been possible to confirm and to extend the above-mentioned results. Thus, it was possible to develop a complete quantum-theory explanation of the *Zeeman effect* for the hydrogen lines, which, in spite of the essentially different character of the assumptions that underlie the two theories, is very similar throughout to Lorentz's original explanation based on the classical theory. In the case of the *Stark effect,* where, on the other hand, the classical theory was completely at a loss, the quantum-theory explanation could be so extended with the help of the correspondence principle as to account for the polarization of the different components into which the lines are split, and also for the characteristic intensity distribution exhibited by the components. This last question has been more closely investigated by Kramers, and the accompanying figure will give some impression of how completely it is possible to account for the phenomenon under consideration.

Figure 6 reproduces one of Stark's well-known photographs of the splitting up of the hydrogen lines. The picture displays very well the varied nature of the phenomenon, and shows in how peculiar a fashion the intensity varies from component to component. The components below are polarized perpendicular to the field, while those above are polarized parallel to the field.

Figure 7 gives a diagrammatic representation of the experimental and theoretical results for the line H_y, the frequency of which is given by the Balmer formula with $n'' = 2$ and $n' = 5$. The vertical lines denote the components into which the line is split up, of which the picture on the right gives the components which are polarized parallel to the field and that on the left those that are polarized perpendicular to it. The experimental results are represented in the upper half of the diagram, the distances from the dotted line representing the measured displacements of the components, and the lengths of the lines being proportional to the relative inten-

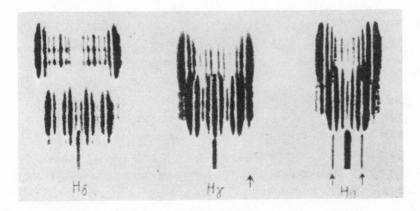

Figures 6 and 7 from the Nobel lecture, showing experimental and theoretical patterns for the Stark effect in Balmer lines of atomic hydrogen.

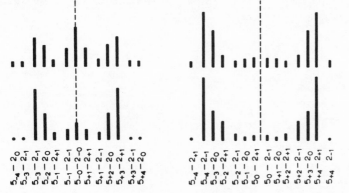

sity as estimated by Stark from the blackening of the photographic plate. In the lower half is given for comparison a representation of the theoretical results from a drawing in Kramers's paper.

The symbol $(n'_{s'} - n'_{s''})$ attached to the lines gives the transitions between the stationary states of the atom in the electric field by which the components are emitted. Besides the principal quantum integer n, the stationary states are further characterized by a subordinate quantum integer s, which can be negative as well as positive and has a meaning quite different from that of the quantum number k occurring in the relativity theory of the fine structure of the hydrogen lines, which fixed the form of the electron orbit in the undisturbed atom. Under the influence of the electric field both the form of the orbit and its position undergo large changes, but certain properties of the orbit remain unchanged, and the subordinate quantum number s is connected with these. In Figure 7 the position of the components corresponds to the frequencies calculated for the different transitions, and the lengths of the lines are proportional to the probabilities as calculated on the basis of the correspondence principle, by which also the polarization of the radiation is determined. It is seen that the theory reproduces completely the

95

Letter from Einstein to Bohr, congratulating him on the award of the Nobel Prize. Written 11 January 1923 aboard the Japanese ship *Haruna Maru*, near Singapore. The letter reads:

"Dear (or rather beloved) Bohr! Your cordial letter reached me shortly before my departure from Japan. I can say without exaggeration that it gave me as much pleasure as the Nobel Prize. I find especially charming your concern lest you might have received the prize before I did—that is truly 'Bohrish.' Your new investigations on the atom have accompanied me on my travels and have further increased my regard for your intellect. I believe that I have finally understood the connection between electricity and gravitation. Eddington has come closer to the truth than Weyl.

The trip is splendid. I am charmed by Japan and the Japanese and am sure that you would be too. Moreover, a sea voyage like this is a delightful existence for a dreamer—it is like a cloister. Add to this the caressing warmth near the equator—warm rain drips lazily down from the sky, engendering peace and a plant-like state of semiconsciousness. This little letter attests to it.

Hearty greetings. I look forward to seeing you again, at the latest in Stockholm. Yours in admiration, A. Einstein."

main features of the experimental results, and in the light of the correspondence principle we can say that the Stark effect reflects down to the smallest details the action of the electric field on the orbit of the electron in the hydrogen atom, even though in this case the reflection is so distorted that, in contrast with the case of the Zeeman effect, it would scarcely be possible directly to recognize the motion on the basis of the classical ideas of the origin of electromagnetic radiation.

Results of interest were also obtained for the spectra of elements of higher atomic number, the explanation of which in the meantime had made important progress through the work of Sommerfeld, who introduced several quantum numbers for the description of the electron orbits. Indeed, it was possible, with the aid of the correspondence principle, to account completely for the characteristic rules which govern the seemingly capricious occurrence of combination lines, and it is not too much to say that the quantum theory has not only provided a simple interpretation of the combination principle, but has further contributed materially to the clearing up of the mystery that has long rested over the application of this principle.

III. THE BIRTH AND GROWTH OF QUANTUM MECHANICS

Bohr on the Foundations
of Quantum Theory

Edward MacKinnon

Throughout his career Niels Bohr manifested an abiding concern with the foundational problems generated by atomic physics and quantum theory. This concern gradually led him from a consideration of particular theories and concepts to the more general question of the nature of concepts and the way they function to make descriptions and explanations possible. His mature position, unfortunately, was usually presented in popular lectures that seemed to trivialize the significance of the position he had developed. To physicists, these appeared to be one more summary of past achievements; to philosophers, the few who read Bohr's works, these lectures seemed to exhibit the sort of amateur philosophizing routinely presented by aging scientists on ceremonial occasions. This bothered Bohr. Shortly before his death he complained that no professional philosopher had ever understood his doctrine of complementarity.

My purpose here is to trace the development of Bohr's thought on foundational issues. His protracted struggle to fashion a coherent interpretation of quantum theory contributed much more to his ultimate philosophical position than did any reading of philosophical sources. Bohr's consequent position on the nature and goals of scientific explanation also underlay his continuing controversies with Albert Einstein on the interpretation of quantum theory. I will try to bring out the aspects of this debate that Bohr thought crucial.

EARLY ROUTINE REALISM

Bohr's master's thesis (or big question) was a survey of the results achieved by the contemporary electron theories of metals. His doctoral dissertation focused on a problem this survey had uncovered. Thomson, Langevin,

Edward MacKinnon has made a special study of basic philosophical and conceptual problems in modern physics. He is Professor of Philosophy at California State University, Hayward, California.

Lorentz, and others had explained, with some success, such properties of metals as paramagnetism and diamagnetism by introducing some ad hoc assumptions into an essentially classical framework. Bohr set himself the task of determining which conclusions concerning the properties of metals flowed from the general principles of classical physics without the introduction of special assumptions. His results were deflationary. Classical physics, by itself, was unable to explain either paramagnetism or diamagnetism. It also led to other conclusions at variance with experimental results. In a lecture he gave shortly after his arrival in England in late 1911, Bohr pinpointed the basic reason for the failure of classical physics to yield an adequate electron theory of metals and to account for blackbody radiation: Maxwell's equations do not supply a basis for describing the small-scale motion of individual electrons.

By June 1912, Niels was writing to his brother, Harald, that he had made a breakthrough which, unlike the impossible Thomson model of the atom, might capture a bit of reality. This work and the theory that emerged from it have been summarized in Part II. The essential point here is that Bohr's 1913 theory did supply a descriptive account of electronic motion. Electrons orbit the nucleus in circular coplanar paths and can jump from one allowed orbit to another. This work also seemed to sharpen the limits of the validity to be accorded Maxwell's equations. These equations cannot explain how the emission and absorption of radiation is connected with electronic transitions. However, they do supply an adequate basis for describing the propagation of radiation in free space. This evaluation supported Bohr's continued opposition to Einstein's light-quantum hypothesis.

At this stage of Bohr's development there was nothing particularly novel about his way of interpreting scientific theories. The core of a good theory should be a descriptive account of objective reality. Bohr's atomic theory had clearly not yet achieved this goal, but it had taken some significant steps in the right direction. Bohr's early work on the scattering of particles from atoms fitted into the same general explanatory perspective. Ideally, one would begin with a descriptive account of the structure of the atom, the forces on the electrons, and the orbital motion of each electron. Since this proved impossible, Bohr made simplifications and approximations that provided a sufficient starting point. Within this perspective, experimental results could be interpreted as supplying additional information about the atom. The scattering of particles was interpreted as giving information about the static behavior of atoms, while the absorption of particles gave information about the dynamic behavior of atoms. This was not dogmatic realism. It was simply normally functioning physics.

Even at the earliest stage, however, there was a significant difference in style between Einstein's and Bohr's ways of doing physics. The problems that most intrigued Einstein were those he later dubbed "the secrets of the Old One." He searched for the beautiful simplicity that, he was convinced, must lie hidden behind the tangled phenomena of experience. Einstein repeatedly devised ingenious thought experiments that idealized and clarified the hidden core of a paradoxical situation. Only after he had worked out the mathematical implications of some new position would Einstein search for experimental data that might test its experimental validity. The finer and often grubbier details of development were left for later and lesser men.

Bohr never manifested Einstein's genius for intuitive creation. His talent lay in his ability to bring order out of chaos, to adjust models and equations, to adapt theories and interpret data until he had the best overall fit. Whereas Einstein tried to grasp a hidden essence by disregarding anything he thought irrelevant, Bohr insisted that nothing be left out. In a letter to George de Hevesy, written 7 February 1913, Bohr listed the topics he was then working on and which he felt that an adequate theory of the atom must explain: atomic volume and its variation with valence, the periodicity of the system of elements, the conditions of atomic combination, excitation energies of characteristic X rays, dispersion, magnetism, and radioactivity. Within a few weeks he added the topics of spectral series, ionization energies, and the mechanism of emitting and absorbing radiation.

Bohr is the most profound of the four [Planck, Rutherford, Einstein, Bohr] and probably the one whose influence has been largest. He is always questioning, never certain of his answers.

D. ter Haar, 1959

RETREAT TO FORMALISM

Bohr accepted and expanded Sommerfeld's redevelopment of his theory, a redevelopment that hinged more on mathematical methods than on visualizable models. This led to Bohr's "second atomic theory" (the theory of the periodic table of the elements). In about 1920 Bohr began to make a sharp distinction between the descriptive elements of the theory and its formal principles. The truth of the descriptive elements depends on the correspondence between the description given and the reality described. The significance of the formal principles comes from the role they play in theories.

The stress on this distinction came from Bohr's increasing reliance on his own correspondence principle. This was not so much a principle of atomic physics as it was a method of extending classical laws to quantum domains. H. A. Lorentz's account of the production of electromagnetic radiation could not, in Bohr's opinion, be considered descriptively correct. Yet this theory had a completeness that the quantum theory of radiation lacked. It predicted polarizations, relative intensities, and forbidden transitions. Bohr adapted the correspondence principle into a tool for extending these classical formulas to the quantum theory of radiation. Such an adaptation had no theoretical justification, for classical and quantum physics rested on quite different presuppositions. Yet it worked surprisingly well. Accordingly, Bohr concluded, the classical theory of radiation, though descriptively incorrect, must have some sort of formal correctness.

Bohr's explanation of the periodic table of elements represented the crowning achievement of the new Bohr-Sommerfeld theory. By the time

103

Our understanding of atomic physics, of what we call the quantum theory of atomic systems, had its origins at the turn of the century and its great synthesis and resolutions in the 1920s. It was a heroic time. It was not the doing of any one man. It involved the collaboration of scores of scientists from many different lands, though from first to last the deeply creative and subtle and critical spirit of Niels Bohr guided, restrained, deepened, and finally transmuted the enterprise.

J. Robert Oppenheimer, quoted in Robert Jungk, *Brighter Than a Thousand Suns,* 1956

Bohr had completed this in 1922, his account rested on three formal principles and three descriptive props. The three formal principles were Bohr's own correspondence principle; Ehrenfest's *adiabatic principle,* used to determine which orbits are possible; and the principle of the permanence of quantum numbers, the pivotal assumption that Bohr had used in explaining the building up of the periodic table of elements. Three descriptive props were accorded a foundational role. First, atomic electrons orbit the nucleus in elliptic paths. Second, the azimuthal quantum number k ($=l+1$, in contemporary notation) characterizes an orbit's ellipticity. Third, the emission and absorption of radiation is due to orbital transitions of electrons.

The Bohr-Sommerfeld theory was encountering formidable difficulties when it was extended to such outstanding problems as the splitting of spectral lines in weak magnetic or strong electric fields, the scattering of light from atoms, the helium atom, the hydrogen molecule, and some interaction effects later explained through the Pauli exclusion principle and electron spin. To accommodate such intractable problems, Bohr, Landé, Pauli, Kramers, and especially the iconoclastic young Heisenberg introduced ad hoc assumptions and new models incompatible with the Bohr-Sommerfeld theory. The theory had effectively become a thing of rags and patches. By the summer of 1923 a consensus was beginning to emerge, especially among Max Born's associates, that something radically new was needed, a quantum mechanics to replace the patched up mixture of classical and quantum ingredients.

Bohr, already the elder statesman of atomic physics, now left the mathematics to others. He concentrated on analyzing the nature of the explanations atomic physics could supply. The three descriptive props he had thought essential had either been undermined by new developments or simply disregarded in solving particular problems. Bohr began to think that atomic physics should be interpreted as giving a purely formal, rather than a descriptive, account of atomic processes and structures. Though this formalistic method of interpreting theories bore resemblances to ideas developed by Pierre Duhem, Ernst Mach, and Henri Poincaré a generation earlier, Bohr gave little evidence of being influenced by their work.

To implement his new approach, Bohr turned to a surprising ally: Albert Einstein. By 1907 Einstein had come to accept the validity of the Planck radiation formula. Yet he never accepted Planck's justification of this law. In 1916 and 1917 Einstein had written three papers deriving Planck's law from the general principles of the Bohr theory: the assumption that molecules have only discrete energy states, and that radiation frequencies are determined by differences between energy states. In the third paper, Einstein introduced an assumption that was to haunt the remaining years of his life. His theory included both spontaneous and induced emission of radiation. Spontaneous emission seemed to be a pure chance event not determined by causal laws. Einstein thought that this, as well as the probability coefficients he introduced, would eventually be given a deterministic foundation.

In Einstein's 1917 radiation paper, Bohr saw an approach to atomic physics that used the general principles of the Bohr theory but that did not rely on any of the descriptive props that Bohr now found suspect. There remained, however, one major point of divergence. Einstein's paper defended and expanded the light-quantum hypothesis. In spite of the then current revival of interest in this hypothesis and the reinforcement it had recently received from its role in explaining the *Compton effect,* Bohr still rejected it. He discovered a possible way of using Einstein's approach, without also using the suspect light-quantum hypothesis which had a basic significance in Einstein's paper. John Slater, then a young postdoctoral fellow, had introduced the idea of a virtual radiation field through which atoms communicate with each other and induce transitions. Slater thought of this as a way of fusing the wave and particle models of light. In 1924 he came to Copenhagen and collaborated with Bohr and Kramers on a paper (the so-called B-K-S paper), which adapted this assumption but dropped the idea of light quanta.[1]

Bohr actually wrote the B-K-S paper himself. It anticipated the basic elements later given greater coherence in the "Copenhagen interpretation" of quantum mechanics. The guiding principle of the paper was the formal (in Bohr's special sense) nature of atomic explanation. The paper was almost exclusively concerned with clarifying the type of information quantum theory gives, rather than with the mathematical formalism. Space-time descriptions were seen as excluding the simultaneous use of causal accounts. The most distinctive conclusion of the paper, one that Slater did not like, was the reinterpretation of the principles of energy and momentum conservation as *statistical* principles. The paper postulated that these principles cannot be strictly applied to individual interactions.

Though this paper built on Einstein's work, it used statistical reasoning in a way quite different from Einstein's. Einstein, who had published twenty-four papers concerned with statistics, was clearly the master of statistical reasoning. His guiding principle had always been that atomic events are governed by laws of objective validity, though not necessarily by already known laws. Statistical methods are introduced either when the underlying laws are not adequately understood, or when the large numbers involved render precise calculation impossible. The B-K-S paper accorded the notion of probability a foundational rather than a derivative status, and did so without even examining the problems of deterministic foundations that perplexed Einstein. The paper also rejected the light-quantum hypothesis that Einstein had so long and staunchly defended. Einstein let it be known that he did not accept the B-K-S paper.

The physics community, even the popular press, soon became aware of this clash of the titans. Walther Bothe and Hans Geiger in Berlin, and Arthur H. Compton and Alfred W. Simon in the United States, began some ingenious and very difficult tests to determine whether energy and momentum are strictly conserved in atomic interactions. When the experimental results finally came in, they showed that energy and momentum are indeed conserved. Einstein had decisively won the first round in the Bohr-Einstein debates.

MUDDLED CONCEPTS AND A NEW MECHANICS

Bohr did not interpret these experimental results as a straightforward refutation of his statistical interpretation. The basic problem, as he then saw it, was a fundamental breakdown in the way science uses space-time pictures to describe natural phenomena. The purely formal, or hypothetical-deductive, interpretation of scientific theories championed in the B-K-S paper did not accord any descriptive significance to general principles or the concepts they embodied. This now seemed too extreme a renunciation. Yet the descriptive realism implicit in much of Bohr's earlier work seemed less tenable than ever. Bohr's reaction to this dilemma was to sharpen his correspondence principle. Earlier he had adapted it into a tool for extending classical laws to quantum domains. The technique that had worked for classical laws should also work for classical *concepts*.

Bohr had been using 'classical' in his own special sense since 1912. By 'classical concepts' he meant ordinary language terms, such as 'wave' and 'particle,' which are given specialized meanings in classical physics. The same terms were transferred into quantum physics, though quantum physics seemed to require a decisive break with some of the presuppositions underlying classical physics. Bohr's feeling — at this stage it was more a feeling than a developed doctrine — was that the correspondence principle approach should clarify the way in which a term, whose meaning is determined in one conceptual framework, can function in a quite different conceptual framework. This method of interpreting scientific concepts seemed promising, provided there was something to interpret. Still lacking was an atomic theory that was coherent and empirically adequate. This lacuna was soon filled by the development of *matrix mechanics, wave mechanics,* the Pauli *exclusion principle,* and the discovery of electron *spin*. Here I will focus on only one aspect of this crucial breakthrough in quantum theory. That aspect is the manner in which the new mathematical formalisms were given a physical interpretation.

Heisenberg's development of quantum mechanics was, in many respects, an extension of Bohr's own work. The development hinged on a skillful use of, and a skillful concealment of, the virtual oscillator model of the atom that had been introduced in the B-K-S paper. By the time quantum mechanics was transformed into matrix mechanics, it seemed to rest on the interpretative principle that Heisenberg had adapted from Pauli and Born: quantum theory should be interpreted exclusively in terms of relations between observable quantities. This was the type of purely formal, or

Men like Einstein or Niels Bohr grope their way in the dark toward their conceptions of general relativity or atomic structure by a type of experience and imagination different from that of the mathematician, although no doubt mathematics is an essential ingredient.

Hermann Weyl, "David Hilbert and His Mathematical Work," 1944 positivist, interpretation that Bohr himself had backed into—and was critically reexamining in the aftermath of the B-K-S paper.

Heisenberg and Pauli were closely associated with Bohr and with the mainstream of atomic physics, a stream that flowed through Copenhagen, Cambridge, Göttingen, and Munich. Wave mechanics was more an extension of Einstein's work than of Bohr's. Louis de Broglie accepted the wave-particle duality which Einstein had introduced into light theory and which had received an independent development at the hands of physicists struggling with the puzzling properties of X rays. De Broglie's treatment of the relation between the wave and particle aspects of matter was based exclusively on the special theory of relativity, rather than on the complex collection of spectroscopic data that was the center of concern of the mainstream atomic physicists.

As studies by Hanle and Wessels have shown,[2] Schrödinger's path to wave mechanics had ideal gas theory as its point of departure. In developing Bose-Einstein statistics, Einstein had retained the familiar view of a gas as a collection of molecules, but changed the way states are counted. Schrödinger reinterpreted this. He retained Boltzmann's method of counting energy states, but accommodated the new statistics by reinterpreting the bearers of these states. Thus, a gas was treated as a unit. Molecules lost much of their individuality and were treated as something analogous to wave crests. Then, following a suggestion of Einstein's, Schrödinger studied de Broglie's thesis. After attempting and failing to develop a wave mechanics based on the special theory of relativity, Schrödinger developed his nonrelativistic wave mechanics.

Schrödinger's wave mechanics presented a mathematical formalism, apparently quite different from matrix mechanics, loosely coupled to something of a wave interpretation. Schrödinger was well aware of difficulties with dispersion of wave packets and of the differences between de Broglie's wave interpretation, based on relativistic transformations in real space, and his own approach, based on a nonrelativistic formulation in phase space. The first two papers in his series on wave mechanics stressed the point that the new mathematical formulation led to the natural emergence of integral quantum numbers. The papers merely suggested that some sort of wave interpretation seemed appropriate.

Then Schrödinger proved that wave mechanics and matrix mechanics are mathematically equivalent. This immediately enlarged both approaches. Wave mechanics adopted the *operator* interpretation of variables, while matrix mechanics could now use the simpler and more familiar mathematical methods of wave mechanics, based on differential equations. This intertranslatability of formalisms also presented something of a priority problem. As both Schrödinger and Heisenberg realized, if matrix and wave mechanics are mathematically equivalent, then the only really significant difference between them is the physical interpretation to be accorded the mathematical formalism. The ensuing competition tended to harden each man in his own approach. Schrödinger insisted that a wave interpretation has an intuitive significance that Heisenberg's purely formal methods of interpretation lacked. Heisenberg countered that the intuitive appeal of the wave picture actually misrepresented the facts of atomic physics. Each sought problems that would establish the correctness of his own view.

The competition between Schrödinger and Heisenberg reflected a deeper challenge. The unprecedented success of Schrödinger's brilliant series of papers on wave mechanics came at a time when, as Mara Beller has shown, matrix mechanics was encountering formidable difficulties.[3] Schrödinger seemed to be beating the mainstream spectroscopists at their own game, and doing it by relying on methods and concepts familiar to physicists, rather than on the unfamiliar mathematics of matrix mechanics and its unpalatable rejection of space-time pictures of the atom. This avalanche of new developments urgently demanded an interpretation that was consistent and comprehensible.

During this period, Heisenberg was spending much of his time at Bohr's institute. Bohr invited Schrödinger to join them in the hope that the three of them might work out an acceptable interpretation of the new developments. Schrödinger came to Copenhagen in October 1926. He and Bohr never reached any accord. Schrödinger, ever the loner, left after two weeks to work out his own interpretation in the privacy he needed for creative work. Heisenberg stayed on. He had a position as Bohr's assistant for that academic year. For months he and Bohr had protracted, often heated, discussions concerning idealized experiments, physical interpretations, and the ground of meaning of the concepts involved in these interpretations.

Their dialogue eventually broke down because of two points on which they could reach no accord. The first was Schrödinger's wave interpretation. Though Heisenberg was, by this time, willing to use Schrödinger's mathematical methods, he completely rejected any wave interpretation of electrons. He insisted that one should begin with the mathematical formalism, work out its consequences, and then relate these to experiments, both actual and thought experiments. Bohr insisted that a consistent physical interpretation is more basic than the mathematical formalism that builds on it. Though Schrödinger's wave interpretation could not be considered literally correct, it must, Bohr argued, be accorded a significant role in the overall physical interpretation.

The second point of disagreement concerned the way concepts acquire meaning. Heisenberg insisted that the meaning of such crucial terms as 'position' and 'momentum" should be set by the operations through which these quantities are measured. Bohr insisted that such terms are not *assigned* meanings. Their meanings are already determined by their usage in classical physics. The crucial conceptual problem was one of extending concepts, whose meanings are set in one conceptual framework, to a new conceptual framework resting on seemingly incompatible foundations. This problem was to be solved not by assigning meanings but by extending the correspondence principle from formulas to concepts.

By February 1927 their dialogue had so decayed that cooperative effort proved extremely difficult. Bohr went to Norway for a skiing trip and there thought through the doctrine he later presented as complementarity. Heisenberg remained in Copenhagen and wrote the first draft of a paper demolishing Schrödinger's appeal to intuition. This was the origin of the indeterminacy (or uncertainty) principle. Heisenberg wrote the draft in the form of a long letter to Pauli. He was assuming, as in his earlier development of quantum mechanics, that if Pauli, the master of criticism, would accept his argument, he was on solid ground.

Pauli accepted the argument. Yet the indeterminacy paper initially served to deepen the rift between Bohr and Heisenberg. When Bohr returned from Norway he read the paper and thought that it should be treated the way initial drafts of his own papers were. It should serve as a basis for discussion and be written and rewritten until every detail was correct. Heisenberg ignored all such suggestions and sent the hastily written paper, with all its imperfections, to the *Zeitschrift für Physik*. The indeterminacy principle decisively undercut Schrödinger's wave picture, which in principle assumed precise specifications of both position and momentum. Heisenberg wanted his paper published as soon as possible.

The intervention of Oskar Klein, Bohr's new assistant, served to exacerbate the difficulties between Bohr and Heisenberg. Finally Pauli inter-

Bohr talking with Werner Heisenberg (center) and Wolfgang Pauli (right) in the lunchroom of the Niels Bohr Institute.

vened. In early June 1927, he went to Copenhagen and mediated the rift between Bohr and Heisenberg. Both accepted his face-saving solution that their disagreement concerned not fundamental concepts, but merely the order of precedence of these concepts. Heisenberg added a conciliatory postscript to the draft copy of the indeterminacy paper, indicating that the wave and particle pictures could be considered complementary.

From this time on, Bohr, Heisenberg, and Pauli shared something of a common doctrine, which came to be known as the Copenhagen (or 'orthodox') interpretation of quantum mechanics. Heisenberg's indeterminacy paper and Bohr's paper on complementarity were the two pillars of the Copenhagen interpretation. This interpretation was summarized in, and transmitted through, Pauli's article in the 1933 *Handbuch der Physik*. The Copenhagen interpretation includes the uncertainty principle; the idea that photons, electrons, and other "particles" exhibit both wave and particle properties: the probabilistic interpretation of the *wave function;* the correspondence between *eigenvalues,* given by the theory, and measured values; the complementary relationship between the Heisenberg and Schrödinger representations; and the correspondence principle conclusion that quantum mechanics merges with classical mechanics in the limits of large quantum numbers. These ideas are now an established part of functioning quantum theory. In 1927 most of them seemed novel and more than a bit bizarre.

THE CONSOLIDATION OF BOHR'S POSITION

It is important to distinguish the Copenhagen interpretation, which was well understood and widely accepted, from Bohr's epistemology, which was poorly understood and rarely accepted. From 1927, when the Copenhagen interpretation went public, until 1937, when Bohr's epistemological position was fixed, Bohr kept reworking his position through papers, addresses, and interminable discussions. For the purposes of this discussion, I will divide his development during this ten-year period into two phases: reformulations of his interpretation of quantum theory, a phase lasting till about 1930; and attempts to extend his position in a way that would give functioning physics an overall conceptual consistency.

A brief historical scenario may help relate the aspects I will consider to the developments I will merely skim. Bohr presented his paper on complementarity at a meeting in Como, Italy, which Einstein did not attend, and in October 1927 gave essentially the same paper at the Solvay conference, which Einstein did attend. Before this conference Bohr sent Einstein a letter outlining his position and also, at Heisenberg's request, included a copy of Heisenberg's indeterminacy paper. Einstein did not reply to the letter, did not give a paper at this conference (or at any Solvay conference after the first one in 1911), and did not comment on the paper Bohr gave. However, he made his views known by commenting on a paper given by Max Born, and in informal discussions.

Bohr walking with Einstein in Brussels during one of the Solvay Conferences (probably 1930). Photo by Paul Ehrenfest.

The development of Bohr's position can best be seen by focusing on two principles which came to play an organizing role: the first was complementarity, and the second was the epistemological irreducibility of atomic experiments. When Bohr introduced the term 'complementarity' in 1927, he used it to refer primarily to the complementary relationship obtaining between spatio-temporal descriptions and the application of causal principles. The causal principles basic to physics rested on the classical idealiza-

111

tion of motion by free particles and of radiation in empty space. Space-time descriptions, however, require an interaction between the system observed and the agency of measurement, an interaction which precludes any possibility of an isolated system. This primary sense of 'complementarity' generates a derivative complementarity between the definition of concepts and the possibility of observation. Thus, 'stationary state' implies precise energy. Such a precise specification is possible only when a system is not disturbed by measurement.

The original formulation of complementarity had some unfortunate implications, of which Bohr only gradually became aware. First, it implicitly presupposed a distinction between atomic systems as they exist and the same systems as known, without explaining how anything meaningful can be said about these systems apart from the conditions of their knowability. Second, it presupposed that a system is in a determined state prior and subsequent to the measurement that disturbs this state, though these prior and subsequent states were, by Bohr's own principles, unknowable. Finally, Bohr's original formulation implied unlimited accuracy in retrodicting the position and momentum that a particle had prior to the act of measurement.

Bohr's second basic principle, the epistemological irreducibility of atomic experiments, is well illustrated by the standard example: a double-slit electron interference experiment. Someone performing such an experiment with both slits open cannot meaningfully ask which slit the electron went through. The system being observed, and the apparatus used to observe it, form an epistemologically irreducible unit. The only meaningful way to determine which slit the electron went through is to close one slit. If the electron hits the plate then it went through the open slit. The idea was fairly clear; the terminology used to express it was not. The classical terminology used to describe trajectories implicitly presupposed a classical determinacy, something incompatible with the quantum indeterminacy.

When Bohr came to grips with this difficulty around 1929, he temporarily resorted to a locution Dirac had introduced, and spoke about the need to reckon with a free choice on the part of nature about the state a system enters. He also substituted 'reciprocity' for 'complementarity.' Neither 'free choices on the part of nature' nor 'reciprocity' was used after 1929. When Bohr returned to 'complementarity' he came to stress, as its primary application, the complementary relation between the wave and particle representations. This brought a new set of difficulties. Unlike space-time descriptions and causal principles, the wave and particle concepts are mutually exclusive in classical physics. Such considerations intensified Bohr's concern with the question of how concepts acquire meaning. His earliest writings reflect the traditional empiricist position that the meanings of key concepts linking knowledge to reality are basically determined by perception. However, the classical concepts that had emerged as problematic were, as Bohr clearly realized, idealizations. The meanings of such

The notion of complementarity does not imply any renunciation of detailed analysis limiting the scope of our inquiry, but simply stresses the character of objective description, independent of subjective judgment, in any field of experience where unambiguous communication essentially involves regard to the circumstances in which evidence is obtained. In logical respect, such a situation is well known from discussions about psychological and social problems where many words have been used in a complementary manner since the very origin of language.

Niels Bohr,
Rutherford Memorial Lecture, 1958

112

idealized concepts depend on the conceptual framework — for example, classical physics — in which they function. This growing tendency to develop a general doctrine of meaning was temporarily circumvented by some novel developments in physics that induced Bohr to consider a few particular concepts in a more searching way.

The final act in this first phase was another confrontation with Einstein. Though neither Bohr nor Einstein gave a paper at the 1930 Solvay conference, they had extensive semipublic discussions at the Hotel Metropole. There, during breakfast one day, Einstein proposed a now famous thought experiment. A box that contains trapped radiation and a clockwork mechanism controlling a shutter emits one photon at a precisely determined time. The energy may be determined by a precise weighing of the box before and after the emission using Einstein's equation $E = mc^2$. This experiment seems to prove that it is possible, at least in principle, to violate the indeterminacy relation between time and energy. At breakfast the next morning Bohr showed that when Einstein's own principle of equivalence is included in the account, one obtains exactly the results predicted by the indeterminacy principle.

By 1930 quantum physics had once more reached a crisis stage. There were three problem areas that seemed interrelated. Quantum field theory, introduced by Dirac when he was studying at Bohr's institute, seemed to do what Slater's virtual radiation field had failed to do. It supplied an interpretation of the phenomena Einstein had explained through his light quantum hypothesis, and did so in a way that clearly fitted the principle of complementarity. Bohr accepted this enthusiastically. By 1930, however, the problem of infinite self-energy emerged when Oppenheimer calculated the interaction of an electron with its own field.

Relativistic quantum mechanics, another contribution of Dirac's, was an instant success, in spite of the peculiar negative-energy solutions it allowed. At first, Dirac simply disregarded them. However, Klein showed that a sufficiently strong potential over a sufficiently small region could induce an electron to jump into a negative energy state. These states, accordingly, could not simply be ignored.

Nuclear physics presented a family of outstanding puzzles, stemming mainly from the question of whether electrons, known to be emitted in beta decay, must, for this reason, be assumed to be nuclear constituents. Electrons, so confined, would have such high kinetic energies that it was hard to see how they could remain confined. Second, the assumption that a nucleus is composed of protons and electrons led to conclusions concerning nuclear spin and statistics that conflicted with experimental results.

As the presiding figure in the atomic physics community, Bohr was deeply immersed in all of these problems. Yet his manner of tackling them was unique. He left the mathematical formulations, detailed theories, and experimental research to younger people. His task, as he saw it, was to articulate a network of fundamental physical concepts that was meaningful, coherent, and adequate to the data. This network involved such physical concepts as 'particle,' 'wave,' 'position,' 'momentum,' and 'spin.' It also included such epistemological and ontological concepts as 'observe,' 'report,' 'describe,' 'measure,' 'objective,' and 'real.' To exemplify what this entailed, I will consider Bohr's analysis of one key concept: 'particle.'

On 24 November 1929, Bohr wrote Dirac a letter explaining how he thought the difficulties generated by Dirac's relativistic wave equation should be handled. The classical theory of the electron, based on the classical concept of a particle, has a limited validity. This limit is characterized by the 'classical radius' of the electron, $e^2/mc^2 = 2.8 \times 10^{-13}$ cm. It may be, Bohr suggested, that the concept of a particle has no validity below this limit. If the particle concept goes, energy conservation may go with it.

Dirac answered immediately. Energy conservation must be preserved at all costs. To indicate how this might be done Dirac presented his first account of holes in a negative-energy sea appearing as positive particles, presumably protons. Bohr rejected this and replied: "In the difficulties of your old theory I still feel inclined to see a limit of the fundamental concepts on which atomic theory hitherto rests."

Bohr kept reworking this idea of finding the limits of applicability of the concept 'particle' as predicated for electrons, at that time the only particles that posed a real problem in this regard. His reasoning involved a unique fusion of semantics and physics. The key idea was that one could speak of the electron as a particle only when dealing with phenomena on a scale much larger than any size meaningfully related to electrons. Bohr took this size to be the Compton wavelength h/mc ($=3 \times 10^{-10}$ cm). As one dealt with phenomena on a scale smaller than this, it became progressively less meaningful to speak of the electron as a particle. For phenomena characterized by the classical electron radius, the concept 'particle' had no applicability at all.

By October 1931, when he gave an address to an Italian physics convention, Bohr had come to see this analysis of 'particle' as a means of handling *all* the outstanding difficulties previously noted. His idea may be summarized as follows. The radius of a nucleus is of the same order of magnitude as the classical radius of the electron. The concept of an electron as an individualized particle has, accordingly, no valid application *within* the nucleus. The charges of the absorbed electrons are conserved; their individual identities and mechanical properties are not. *Beta decay,* now interpreted as the creation of a mechanical particle, need not obey the conservation laws proper to already existing particles. This approach seemed to dispose of the difficulties of nuclear statistics, since electrons in the nucleus would not be counted as particles. It also seemed to banish the problems of infinite self-energy and the Klein paradox. Both required intense electric fields over very small regions. The particles required to produce such fields could not be localized over so small a region. Paradox lost.

Bohr's solution to these problems was an outright failure. The next year witnessed the discovery of the neutron and the discovery of the positron, the latter quickly interpreted as Dirac's antielectron. The following year, Fermi presented his theory of beta decay — a theory that made essential use of both the *neutrino* hypothesis, which Bohr had rejected, and the principles of energy and momentum conservation.

Bohr's reaction to this failure was similar to his earlier reaction to the failure of the B-K-S paper. In both cases he was convinced that his approach had an essential validity. The underlying difficulty he faced in the early thirties was an inherent fuzziness in the way he was attempting to relate concepts to objects. He realized that his unique exploratory type of conceptual analysis required supplementation by something more precise. He attempted once again to resharpen his old correspondence principle into a tool for examining the peculiar problems of extending concepts into new domains.

This led to Bohr's two greatest scientific achievements of the thirties. The first, done with the collaboration of Léon Rosenfeld, was to establish the consistency of quantum field theory against the objections brought by Lev Landau and Rudolf Peierls. The key point in the Bohr-Rosenfeld analysis is the idea that any assignment of sharp values to all points in a field is an idealization resting on the classical concept of a test particle and the correspondence principle.

Bohr's redevelopment of nuclear physics was a more spectacular success. After the discovery of the neutron, Heisenberg, Ettore Majorana, and D. Ivanenko independently developed the idea that the nucleus is composed exclusively of protons and neutrons. When Bohr finally accepted this he realized that it changed the status of nuclear physics. Thanks to their large masses, protons and neutrons within the nucleus have kinetic energies that are much smaller than their rest-mass energies. The classical concept of a particle, accordingly, should have an approximate validity in describing their activities. A principle implicit in Bohr's epistemology is that visualizable models are valid and useful within the limits in which classical concepts can be used to give descriptive accounts. Nuclear models were thus now in order. Bohr proceeded to develop both his collective model of nuclear dynamics and later the liquid-drop model (as discussed in detail in the chapter "Niels Bohr and Nuclear Physics").

In his earlier conceptual analyses Bohr had focused on the proper way of extending classical concepts to, and restricting their usage in, quantum domains. He had accepted the classical meanings of these concepts as essentially unproblematic. His later work forced him to come to grips with the problem of how *any* concept has meaning. Though he never developed a systematic theory, he anticipated some of the key features later developed in Wittgenstein's *Philosophical Investigations*. The meaning of a word is determined by its usage in language, not by the objects it can or may denote. The particular form that Bohr's analysis took may best be seen by considering two basic classical concepts: 'particle' and 'wave.'

These are ordinary language concepts given a specialized usage in classical physics. To call anything a 'particle' is to imply that its internal structure, if such there be, is irrelevant in the context. A particle has a space-time trajectory. It can impinge on a target, penetrate it, recoil from it, or be deflected by it. These interrelations are not determined by what electrons really are. They simply represent an unpacking of the cluster of conceptual entailments implicit in the ordinary usage of 'particle' by physicists. Similarly, the classical concept 'wave' is at the center of a different conceptual cluster. A wave has neither a precise location nor a proper trajectory. It propagates via vibrations in a medium. Waves do not collide, penetrate, or recoil. Waves are reflected, refracted, or absorbed.

The experimental sources of information about atomic systems can be roughly divided into two types. The first is collision experiments. A target is bombarded with atomic projectiles and the resulting fragments are examined. The second source of information is the radiation that is emitted, absorbed, or modified by atomic or subatomic processes. In describing any type of experiment and reporting experimental results, it is necessary to rely on either the 'particle' or the 'wave' conceptual cluster. One of the two can be adapted to any experimental situation. Neither can be adapted to all situations. Which one is appropriate depends on the questions that the researcher puts to nature.

Bohr extended this approach to such higher-order terms as 'observe,' 'objective,' 'real,' and 'exist.' Each clarification is rooted in Bohr's analysis of the conditions of the possibility of unambiguous communication of information. Thus, 'objective' is properly used in contrast with 'subjective.' To interpret atomic physics objectively does not mean explaining the meaning and truth of statements in terms of the objects these statements describe. It means, rather, showing how the subject-object distinction, implicit but indispensable in ordinary language usage, is preserved in the extension of this usage to scientific contexts.

IS QUANTUM MECHANICS COMPLETE?

After the 1930 Solvay conference, Einstein accepted both the indeterminacy principle and the probabilistic interpretation of the wave function. Yet he continued to insist that quantum mechanics, despite its unparalleled success, should be regarded as a temporary expedient, not as the definitive theory. Bohr and Einstein had little personal or professional contact, except through their mutual confidant, Paul Ehrenfest. This contact ceased with Ehrenfest's suicide in 1933. Yet even with this lack of extended contact, each seems to have carried on an almost incessant inner private dialogue with the other's position. In 1935 Einstein, in collaboration with Boris Podolsky and Nathan Rosen, wrote a highly influential article (the so-called E-P-R paper) pinpointing what he took to be the chief shortcoming of quantum mechanics. I would like to consider here not the details (which

Our experiments, to use Bohr's favorite phrase, are questions that we put to Nature; and in our theories we try to state what we have learned from her in a language ensuring unambiguous and objective communication. Hence the paramount importance of establishing a rigorous terminology, sufficiently general to make due allowance for the peculiar conditions under which we observe atomic systems.
    Léon Rosenfeld, "Niels Bohr's Contribution to Epistemology," 1963

are discussed in the essay by David Mermin), but the conflicting ideals of scientific explanation manifested in the positions of Einstein and Bohr.

The goal of Einstein's striving, from the beginning to the end of his long career, was to come to know physical reality as it exists objectively. For him 'objective knowledge' meant knowing things as they exist independent of the human perceiver. Quantum mechanics, Einstein argued in the 1935 paper, is incomplete because there are elements of physical reality not included in the descriptions that quantum mechanics supplies.

The complication here is that one cannot "look" at physical reality and compare it with the predictions of quantum theory. One can "look" at the physical reality treated by quantum theory only through the lenses that the theory itself supplies. One must use quantum theory to establish the incompleteness of quantum theory. The E-P-R paper attempted this by a thought experiment that soon became as famous as Newton's rotating bucket and Maxwell's demon. Consider two systems, such as two particles, that interact for a time and then separate. Since the separate systems are still described by a single wave function, they are correlated. One could measure the position of particle A and then use the wave function to infer the position of particle B; or one could measure the momentum of particle A and then use the wave function to infer the momentum of particle B. Since one can infer either the position or the momentum of particle B without in any way disturbing the particle, both its position and its momentum must be considered objectively real properties. Quantum mechanics does not allow a simultaneous specification of both a particle's position and its momentum. Hence, the quantum-mechanical description of physical reality must be considered incomplete. There are objectively real properties of bodies not included in the theory.

Bohr's answering article was not concerned with clarifying the relation between quantum theory and physical reality as it exists independent of any human perceiver. At issue, rather, were the conditions of the possibility of the unambiguous communication of information about atomic systems through propositions using the concepts 'position' and 'momentum.' Bohr showed, through an analysis of possible experiments, that the conditions which allow an unambiguous use of 'position' render 'momentum' essentially ambiguous, regardless of which particle's momentum is in question. Similarly, any experiment that makes possible the unambiguous use of 'momentum' renders 'position' inapplicable to either particle. Accordingly, Bohr concluded, the E-P-R criterion of reality must itself be considered ambiguous. Quantum theory is complete inasmuch as it embodies a rational utilization of all the sources of information concerning atomic systems that admit of simultaneous specification.

The only really adequate answer to such objections, Einstein realized, was to replace quantum mechanics by a more basic theory that does describe physical reality as it exists objectively. This Einstein never succeeded in doing, despite some thirty years of protracted effort. Yet the goal he

Einstein, along with certain physicists of his generation or of preceding generations (Langevin, Planck), never granted, it seems, the new ideas of Bohr and Heisenberg. At the Solvay Conference of October 1927, he was already raising serious objections. A few years later, in a paper written with Podolsky and Rosen, he expounded the difficulties which the actual interpretation of quantum mechanics appeared to him to raise. To Einstein's objections various physicists, and notably Mr. Bohr, composed acute replies, and it rather seems today that the physicists of the younger generation are almost unanimously in favor of granting the Bohr-Heisenberg interpretation, which seems to be the only one compatible with the totality of the known facts. Nevertheless, Einstein's objections, which bear the imprint of his profound mind, were surely useful because they forced the champions of the new conceptions to clarify subtle points. Even if one considers it possible to circumvent them, it is useful to have studied and reflected upon them at length.

Louis de Broglie, "A General Survey of the Scientific Work of Albert Einstein," 1949 strove toward reveals his notions of what a scientific explanation should be and do. He aspired to achieve a complete logical unification of relativity theory and quantum theory based on only one foundational concept: 'field.' Though he had to appeal to an almost mystic intuition to justify his conviction that what is basic in thought is also basic in reality, he clearly held for such a correspondence.

A foundationalist approach to human knowledge constitutes an abiding temptation for systematic thinkers. In the rationalist tradition, the foundation was generally thought to be some sort of indubitable or self-evident principles. The empirical tradition rejected such purported foundations and sought instead to ground human knowledge in such empirical reality as sense data or (the logical empiricist variation) some form of sense data reports. In spite of this reversal, these competing traditions shared a common ideal. Human knowledge, or at least the reliable aspects of human knowledge, could be properly redeveloped as a deductive system, one in which the truth of the foundation and the correctness of the inferences justified the truth of derived conclusions. Even the skeptical critics of such systematizations often reflected the same ideal, when they based their skepticism on doubts about the truth or certainty of the foundation or the indubitability of the chain of inferences.

The real strength of this ideal came not from the sharply limited success of the philosophical systems that added flesh to these formal bones, but from the progress of science. Physics, since the pioneering work of Newton, had grown through ever more inclusive deductive unifications. Its natural, almost inevitable goal seemed to be a comprehensive unification of fundamental physics on the basis of the fewest and most abstract concepts coupled to the most powerful deductive machinery. Einstein clearly manifested this ideal in his protracted and ultimately frustrated search for a grand unified field theory.

In the course of the intellectual odyssey sketched here, Bohr came to criticize, modify, and ultimately reject such a foundationalist conception of human knowledge. In its place he would substitute a dialectical process involving an initial antithesis and later synthesis of conceptual frameworks. The only shared conceptual framework is the one embodied in and transmitted through ordinary language. When new experiences entailing different types of information cannot be accommodated within the prevailing framework, it becomes necessary to break with this and develop something new. Eventually the new framework must so incorporate the old one that it can be interpreted as a rational generalization of what preceded it. This usually requires reinterpreting both the old and the new framework. A necessary, but not sufficient, condition for success is that the expanded framework preserve the conditions of the possibility of unambiguous communication of information concerning experimental results.

Thus, Bohr interpreted the physics stemming from Galileo and Newton as the first significant step beyond the representation of reality implicit in

118

ordinary language. The next major change, in Bohr's view, was the development of field theory by Faraday, Maxwell, and Hertz. This he saw as sharpening the relationship between causal determinism and space-time descriptions. It represented, in this sense, the completion of classical physics. Relativity, in Bohr's view, was a very significant advance, but one that did not essentially change the way physics is interpreted.

Quantum physics constitutes a decisive break with classical physics precisely because it manifests the inadequacy of the representation of reality provided by ordinary language plus classical physics. It can be seen as a rational generalization of what preceded it only by reinterpreting the significance of classical physics, especially in the way scientific descriptions are interpreted. From 1937 on, Bohr insisted that the term 'phenomenon' be used in quantum contexts to refer to observations obtained under circumstances that included a specification of the experimental apparatus. This supplied the final formulation of the two principles considered earlier — namely, complementarity and epistemological irreducibility. Each atomic phenomenon is an epistemologically irreducible unit. It is, accordingly, misleading to speak of disturbing the phenomena by experiment, creating physical attributes through the process of measurement, or reducing the wave packet. Such locutions imply a conceptual division of an epistemologically irreducible phenomenon.

Complementarity extends the framework of ordinary language and classical physics in the sense that the standard language used to report observations suffices for quantum experiments. The technical terms used to supplement ordinary language do not present deep conceptual problems. This way of interpreting complementarity is the justification for Bohr's highly controversial and widely misunderstood claim that complementarity must remain a part of all future advances in physics. He was not trying to freeze the progress of physics, but to specify a necessary condition for true progress. Complementarity, properly interpreted, represents the only epistemologically consistent way of giving an objective description of atomic and nuclear phenomena. Any further advances must ultimately be interpretable as rational generalizations of present physics. They must ultimately, after the ferment of further conceptual revolutions subsides, be formulated in such a way as to preserve the possibility of the unambiguous communication of information concerning experimental results and also incorporate the quantum of action as irreducible. Only complementarity meets these conditions.

Somewhat surprisingly, Bohr's philosophical position has a much greater affinity to the mainstream of ordinary language analysis than it does to almost any established position in the philosophy of science. The philosophical point of departure for both Bohr and the analysts is that humans are immersed in language. When this parallel is properly recognized and exploited, Bohr's epistemology can make a contribution to contemporary philosophy. I believe that it can supply an initial guide for those wishing to

For the clearest analysis of the conceptual principles of quantum mechanics we are indebted to Bohr, who, in particular, applied the concept of complementarity to interpret the validity of the quantum-mechanical laws. The uncertainty relations alone afford an instance of how in quantum mechanics the exact knowledge of one variable can exclude the exact knowledge of another. This complementary relationship between different aspects of one and the same physical process is indeed characteristic for the whole structure of quantum mechanics.

Werner Heisenberg,
Nobel Prize lecture, 1932

extend the methods and insights of ordinary language analysis to an analysis of the development and functioning of scientific theories, considered as extensions of human knowledge rather than as conceptually isolated units. Further advances would depend on searching criticism and on a more systematic presentation, decoupled from the particular problems that Bohr encountered. Both criticism and systematic redevelopment presuppose something that has rarely obtained: an understanding of Bohr's position on the part of philosophers.[4]

The Bohr-Einstein Dialogue

For the occasion of Einstein's seventieth birthday in 1949, a number of distinguished scientists were invited to contribute to a celebratory volume, *Albert Einstein: Philosopher-Scientist*.[1] Niels Bohr chose to use the opportunity to remember and record "the many occasions through the years on which I had the privilege to discuss with Einstein epistemological problems raised by the development of atomic physics." The essay that resulted must be regarded as a classic of scientific literature and is of abiding interest to those who are truly interested in physics and in the way it develops. Under the title "Discussion with Einstein on Epistemological Problems in Atomic Physics," Bohr opened with a review of the foundation work in quantum mechanics and of the contributions by Einstein, himself, and many others. The years that followed that initial work, between the Como conference of 1927 and the "clash of giants" over the E-P-R paper, were the most intense period of the debate. The description of this period forms the core of the Bohr-Einstein dialogue, and it is the part that we have chosen to reproduce here.[2]

In the excerpt, Bohr refers to two equations introduced earlier in his article. The first of these connects the energy E and momentum P of a light quantum (photon) with its frequency v and wavelength λ:

$$E = hv; \quad P = h\sigma, \tag{1}$$

where h is Planck's constant and $\sigma \; (= 1/\lambda)$ is the wave number.

The second equation gives the relationship between two classical conjugate variables, p and q, when expressed in the noncommutative algebra of early quantum mechanics:

$$qp - pq = \sqrt{-1}\, \frac{h}{2\pi}. \tag{2}$$

121

The necessary consequence of this relationship for the limitation imposed on the allowed simultaneous precision of measurements of the two conjugate quantities was first recognized by Werner Heisenberg, as Bohr proceeds to recall in the narrative below.

P.J.K.

୫๑

This phase of the development was, as is well known, initiated in 1927 by Heisenberg,[3] who pointed out that the knowledge obtainable of the state of an atomic system will always involve a peculiar "indeterminacy." Thus, any measurement of the position of an electron by means of some device, like a microscope, making use of high-frequency radiation, will, according to the fundamental relations (1), be connected with a momentum exchange between the electron and the measuring agency, which is the greater the more accurate a position measurement is attempted. In comparing such considerations with the exigencies of the quantum-mechanical formalism, Heisenberg called attention to the fact that the commutation rule (2) imposes a reciprocal limitation on the fixation of two conjugate variables, q and p, expressed by the relation

$$\Delta q \cdot \Delta p \approx h, \tag{3}$$

where Δq and Δp are suitably defined latitudes in the determination of these variables. In pointing to the intimate connection between the statistical description in quantum mechanics and the actual possibilities of measurement, this so-called indeterminacy relation is, as Heisenberg showed, most important for the elucidation of the paradoxes involved in the attempts of analyzing quantum effects with reference to customary physical pictures.

The new progress in atomic physics was commented upon from various sides at the International Physical Congress held in September 1927, at Como, in commemoration of Volta. In a lecture on that occasion,[4] I advocated a point of view conveniently termed "complementarity," suited to embrace the characteristic features of individuality of quantum phenomena, and at the same time to clarify the peculiar aspects of the observational problem in this field of experience. For this purpose, it is decisive to recognize that, *however far the phenomena transcend the scope of classical physical explanation, the account of all evidence must be expressed in classical terms.* The argument is simply that by the word "experiment" we refer to a situation where we can tell others what we have done and what we have learned and that, therefore, the account of the experimental arrangement and of the results of the observations must be expressed in unambiguous language with suitable application of the terminology of classical physics.

This crucial point, which was to become a main theme of the discussions reported in the following, implies the *impossibility of any sharp separation between the behavior of atomic objects and the interaction with the measuring instruments which serve to define the conditions under which the phenomena appear.* In fact, the individuality of the typical quantum effects finds its proper expression in the circumstance that any attempt of subdividing the phenomena will demand a change in the experimental arrangement introducing new possibilities of interaction between objects and measuring instruments which in principle cannot be controlled. Consequently, evidence obtained under different experimental conditions cannot be comprehended within a single picture, but must be regarded as *complementary* in the sense that only the totality of the phenomena exhausts the possible information about the objects.

Under these circumstances an essential element of ambiguity is involved in ascribing conventional physical attributes to atomic objects, as is at once evident in the dilemma regarding the corpuscular and wave properties of electrons and photons, where we have to do with contrasting pictures, each referring to an essential aspect of empirical evidence. An illustrative example, of how the apparent paradoxes are removed by an examination of the experimental conditions under which the complementary phenomena appear, is also given by the Compton effect, the consistent description of which at first had presented us with such acute difficulties. Thus, any arrangement suited to study the exchange of energy and momentum between the electron and the photon must involve a latitude in the space-time description of the interaction sufficient for the definition of wave-number and frequency which enter into the relation (1). Conversely, any attempt of locating the collision between the photon and the electron more accurately would, on account of the unavoidable interaction with the fixed scales and clocks defining the space-time reference frame, exclude all closer account as regards the balance of momentum and energy.

As stressed in the lecture, an adequate tool for a complementary way of description is offered precisely by the quantum-mechanical formalism, which represents a purely symbolic scheme permitting only predictions, on lines of the correspondence principle, as to results obtainable under conditions specified by means of classical concepts. It must here be remembered that even in the indeterminacy relation (3) we are dealing with an implication of the formalism which defies unambiguous expression in words suited to describe classical physical pictures. Thus, a sentence like "We cannot know both the momentum and the position of an atomic object" raises at once questions as to the physical reality of two such attributes of the object, which can be answered only by referring to the conditions for the unambiguous use of space-time concepts, on the one hand, and dynamical conservation laws, on the other hand. While the combination of these concepts into a single picture of a causal chain of events is the essence of classical mechanics, room for regularities beyond the grasp of such a description is just afforded by the circumstance that the study of the complementary phenomena demands mutually exclusive experimental arrangements.

The necessity, in atomic physics, of a renewed examination of the foundation for the unambiguous use of elementary physical ideas recalls in some way the situation that led Einstein to his original revision on the basis of all application of space-time concepts which, by its emphasis on the primordial importance of the observational problem, has lent such unity to our world picture. Notwithstanding all novelty of approach, causal description is upheld in relativity theory within any given frame of reference, but in quantum theory the uncontrollable interaction between the objects and the measuring instruments forces us to a renunciation even in such respect. This recognition, however, in no way points to any limitation of the scope of the quantum-mechanical description, and the trend of the whole argumentation presented in the Como lecture was to show that the viewpoint of complementarity may be regarded as a rational generalization of the very ideal of causality.

At the general discussion in Como, we all missed the presence of Einstein, but soon after, in October 1927, I had the opportunity to meet him in Brussels at the Fifth Physical Conference of the Solvay Institute, which was devoted to the theme "Electrons and Photons." At the Solvay meetings, Einstein had from their beginning been a most prominent figure, and several of us came to the conference with great anticipations to learn his reaction to the latest stage of the development which, to our view, went far in clarifying the problems which he had himself from the outset elicited so ingeniously. During the discussions, where the whole subject was reviewed by contributions from many sides and where also the arguments mentioned in the preceding pages were again presented, Einstein expressed, however, a deep concern over the extent to which causal account in space and time was abandoned in quantum mechanics.

To illustrate his attitude, Einstein referred at one of the sessions [5] to the simple example, illustrated by Figure 1, of a particle (electron or photon) penetrating through a hole or a narrow slit in a diaphragm placed at some distance before a photographic plate. On account of the diffraction of the wave connected with the motion of the particle and indicated in the figure by the thin lines, it is under such conditions not possible to predict with certainty at what point the electron will arrive at the photographic plate, but only to calculate the probability that, in an experiment, the electron will be found within any given region of the plate. The apparent difficulty, in this description, which Einstein felt so acutely, is the fact that, if in the experiment the electron is recorded at one point (A) of the plate, then it is out of the question of ever observing an effect of this electron at another point (B), although the laws of ordinary wave propagation offer no room for a correlation between two such events.

Einstein's attitude gave rise to ardent discussions within a small circle, in which Ehrenfest, who through the years had been a close friend of us both, took part in a most active and helpful way. Surely we all recognized that, in the above example, the situation presents no analogue to the application of

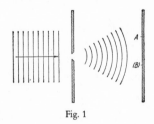

Fig. 1

Bohr and Einstein deep in discussion. The photographs were taken by Paul Ehrenfest in about 1927.

Bohr was here, and I am just as keen on him as you are. He is a very sensitive lad and goes about this world as if hypnotized.

Albert Einstein to Paul Ehrenfest, 1920

statistics in dealing with complicated mechanical systems, but rather recalled the background for Einstein's own early conclusions about the unidirection of individual radiation effects which contrasts so strongly with a simple wave picture. The discussions, however, centered on the question of whether the quantum-mechanical description exhausted the possibilities of accounting for observable phenomena or, as Einstein maintained, the analysis could be carried further and, especially, of whether a fuller description of the phenomena could be obtained by bringing into consideration the detailed balance of energy and momentum in individual processes.

To explain the trend of Einstein's arguments, it may be illustrative here to consider some simple features of the momentum and energy balance in connection with the location of a particle in space and time. For this

125

purpose, we shall examine the simple case of a particle penetrating through a hole in a diaphragm without or with a shutter to open and close the hole, as indicated in Figures 2a and 2b, respectively. The equidistant parallel lines to the left in the figures indicate the train of plane waves corresponding to the state of motion of a particle which, before reaching the diaphragm, has a momentum P related to the wave-number σ by the second of equations (1). In accordance with the diffraction of the waves when passing through the hole, the state of motion of the particle to the right of the diaphragm is represented by a spherical wave train with a suitably defined angular aperture θ and, in case of Figure 2b, also with a limited radial extension. Consequently, the description of this state involves a certain latitude Δp in the momentum component of the particle parallel to the diaphragm and, in the case of a diaphragm with a shutter, an additional latitude ΔE of the kinetic energy.

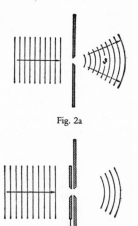

Fig. 2a

Fig. 2b

Since a measure for the latitude Δq in location of the particle in the plane of the diaphragm is given by the radius a of the hole, and since $\theta \approx (1/\sigma a)$, we get, using (1), just $\Delta p \approx \theta P \approx (h/\Delta q)$, in accordance with the indeterminacy relation (3). This result could, of course, also be obtained directly by noticing that, due to the limited extension of the wave-field at the place of the slit, the component of the wave-number parallel to the plane of the diaphragm will involve a latitude $\Delta \sigma \approx (1/a) \approx (1/\Delta q)$. Similarly, the spread of the frequencies of the harmonic components in the limited wave-train in Figure 2b is evidently $\Delta \nu \approx (1/\Delta t)$, where Δt is the time interval during which the shutter leaves the hole open and, thus, represents the latitude in time of the passage of the particle through the diaphragm. From (1), we therefore get

$$\Delta E \cdot \Delta t \approx h, \tag{4}$$

again in accordance with the relation (3) for the two conjugated variables E and t.

From the point of view of the laws of conservation, the origin of such latitudes entering into the description of the state of the particle after passing through the hole may be traced to the possibilities of momentum and energy exchange with the diaphragm or the shutter. In the reference system considered in Figures 2a and 2b, the velocity of the diaphragm may be disregarded and only a change of momentum Δp between the particle and the diaphragm needs to be taken into consideration. The shutter, however, which leaves the hole opened during the time Δt, moves with a considerable velocity $v \approx (a/\Delta t)$, and a momentum transfer Δp involves therefore an energy exchange with the particle, amounting to $v\Delta p \approx (1/\Delta t)\Delta q\Delta p \approx (h/\Delta t)$, being just of the same order of magnitude as the latitude ΔE given by (4) and, thus, allowing for momentum and energy balance.

The problem raised by Einstein was now to what extent a control of the momentum and energy transfer, involved in a location of the particle in space and time, can be used for a further specification of the state of the particle after passing through the hole. Here, it must be taken into consideration that the position and the motion of the diaphragm and the shutter have so far been assumed to be accurately coordinated with the space-time reference frame. This assumption implies, in the description of the state of these bodies, an essential latitude as to their momentum and energy which need not, of course, noticeably affect the velocities, if the diaphragm and the shutter are sufficiently heavy. However, as soon as we want to know the momentum and energy of these parts of the measuring arrangement with an accuracy sufficient to control the momentum and energy exchange with the particle under investigation, we shall, in accordance with the general indeterminacy relations, lose the possibility of their accurate location in space and time. We have, therefore, to examine how far this circumstance will affect the intended use of the whole arrangement and, as we shall see, this crucial point clearly brings out the complementary character of the phenomena.

Returning for a moment to the case of the simple arrangement indicated in Figure 1, it has so far not been specified to what use it is intended. In fact, it is only on the assumption that the diaphragm and the plate have well-defined positions in space that it is impossible, within the frame of the quantum-mechanical formalism, to make more detailed predictions as to the point of the photographic plate where the particle will be recorded. If, however, we admit a sufficiently large latitude in the knowledge of the position of the diaphragm it should, in principle, be possible to control the momentum transfer to the diaphragm and, thus, to make more detailed predictions as to the direction of the electron path from the hole to the recording point. As regards the quantum-mechanical description, we have to deal here with a two-body system consisting of the diaphragm as well as of the particle, and it is just with an explicit application of conservation laws to such a system that we are concerned in the Compton effect where, for instance, the observation of the recoil of the electron by means of a cloud chamber allows us to predict in what direction the scattered photon will eventually be observed.

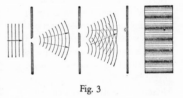

Fig. 3

The importance of considerations of this kind was, in the course of the discussions, most interestingly illuminated by the examination of an arrangement where between the diaphragm with the slit and the photographic plate is inserted another diaphragm with two parallel slits, as is shown in Figure 3. If a parallel beam of electrons (or photons) falls from the left on the first diaphragm, we shall, under usual conditions, observe on the plate an interference pattern indicated by the shading of the photographic plate shown in front view to the right of the figure. With intense beams, this pattern is built up by the accumulation of a large number of individual processes, each giving rise to a small spot on the photographic plate, and the distribution of these spots follows a simple law derivable from the wave analysis. The same distribution should also be found in the statistical

127

account of many experiments performed with beams so faint that in a single exposure only one electron (or photon) will arrive at the photographic plate at some spot, shown in the figure as a small star. Since, now, as indicated by the broken arrows, the momentum transferred to the first diaphragm ought to be different if the electron was assumed to pass through the upper or the lower slit in the second diaphragm, Einstein suggested that a control of the momentum transfer would permit a closer analysis of the phenomenon and, in particular, a conclusion as to which of the two slits the electron had passed before arriving at the plate.

A closer examination showed, however, that the suggested control of the momentum transfer would involve a latitude in the knowledge of the position of the diaphragm which would exclude the appearance of the interference phenomena in question. In fact, if ω is the small angle between the conjectured paths of a particle passing through the upper or the lower slit, the difference of momentum transfer in these two cases will, according to (1), be equal to $h\sigma\omega$, and any control of the momentum of the diaphragm with an accuracy sufficient to measure this difference will, due to the indeterminacy relation, involve a minimum latitude of the position of the diaphragm, comparable with $1/\sigma\omega$. If, as in the figure, the diaphragm with the two slits is placed in the middle between the first diaphragm and the photographic plate, it will be seen that the number of fringes per unit length will be just equal to $\sigma\omega$ and, since an uncertainty in the position of the first diaphragm of the amount of $1/\sigma\omega$ will cause an equal uncertainty in the positions of the fringes, it follows that no interference effect can appear. The same result is easily shown to hold for any other placing of the second diaphragm between the first diaphragm and the plate, and would also be obtained if, instead of the first diaphragm, another of these three bodies were used for the control, for the purpose suggested, of the momentum transfer.

This point is of great logical consequence, since it is only the circumstance that we are presented with a choice of *either* tracing the path of a particle *or* observing interference effects, which allows us to escape from the paradoxical necessity of concluding that the behavior of an electron or a photon should depend on the presence of a slit in the diaphragm through which it could be proved not to pass. We have here to do with a typical example of how the complementary phenomena appear under mutually exclusive experimental arrangements and are just faced with the impossibility, in the analysis of quantum effects, of drawing any sharp separation between an independent behavior of atomic objects and their interaction with the measuring instruments which serve to define the conditions under which the phenomena occur.

Our talks about the attitude to be taken in the face of a novel situation as regards analysis and synthesis of experience touched naturally on many aspects of philosophical thinking, but, in spite of all divergencies of approach and opinion, a most humorous spirit animated the discussions. On his side, Einstein mockingly asked us whether we could really believe that

To understand correctly the views of the two disputants it should be recalled that for Bohr these thought-experiments were not the reason but the necessary consequence of a much more profound truth underlying the quantum-mechanical description and, in particular, the uncertainty relations. Bohr consequently had the advantage that, from his point of view, he was justified in extending the chain of reasoning until he could appropriately resort to the indeterminacy relations to support his thesis. Einstein, on the other hand, had the advantage that if he could disprove the Heisenberg relations by a closer analysis of the mechanics of one single thought-experiment, Bohr's contention of the incompatibility of a simultaneous causal and space-time description of phenomena and with it his whole theory would be refuted.

Max Jammer, *The Philosophy of Quantum Mechanics*, 1974

the providential authorities took recourse to dice-playing *("ob der liebe Gott würfelt"),* to which I replied by pointing at the great caution, already called for by ancient thinkers, in ascribing attributes to Providence in everyday language. I remember also how at the peak of the discussion Ehrenfest, in his affectionate manner of teasing his friends, jokingly hinted at the apparent similarity between Einstein's attitude and that of the opponents of relativity theory; but instantly Ehrenfest added that he would not be able to find relief in his own mind before concord with Einstein was reached.

Einstein's concern and criticism provided a most valuable incentive for us all to reexamine the various aspects of the situation as regards the description of atomic phenomena. To me it was a welcome stimulus to clarify still further the role played by the measuring instruments and, in order to bring into strong relief the mutually exclusive character of the experimental conditions under which the complementary phenomena appear, I tried in those days to sketch various apparatus in a pseudorealistic style of which the following figures are examples. Thus, for the study of an interference phenomenon of the type indicated in Figure 3, it suggests itself to use an experimental arrangement like that shown in Figure 4, where the solid parts of the apparatus, serving as diaphragms and plateholder, are firmly bolted to a common support. In such an arrangement, where the knowledge of the relative positions of the diaphragms and the photographic plate is secured by a rigid connection, it is obviously impossible to control the momentum exchanged between the particle and the separate parts of the apparatus. The only way in which, in such an arrangement, we could ensure that the particle passed through one of the slits in the second diaphragm is to cover the other slit by a lid, as indicated in the figure; but if the slit is covered, there is of course no question of any interference phenomenon, and on the plate we shall simply observe a continuous distribution, as in the case of the single fixed diaphragm in Figure 1.

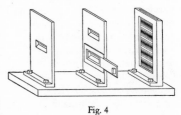

Fig. 4

In the study of phenomena in the account of which we are dealing with detailed momentum balance, certain parts of the whole device must naturally be given the freedom to move independently of others. Such an apparatus is sketched in Figure 5, where a diaphragm with a slit is suspended by weak springs from a solid yoke bolted to the support on which also other immobile parts of the arrangement are to be fastened. The scale on the diaphragm together with the pointer on the bearings of the yoke refer to such study of the motion of the diaphragm as may be required for an estimate of the momentum transferred to it, permitting one to draw conclusions as to the deflection suffered by the particle in passing through the slit. Since, however, any reading of the scale, in whatever way performed, will involve an uncontrollable change in the momentum of the diaphragm, there will always be, in conformity with the indeterminacy principle, a reciprocal relationship between our knowledge of the position of the slit and the accuracy of the momentum control.

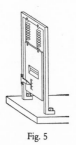

Fig. 5

In the same semiserious style, Figure 6 represents part of an arrangement

129

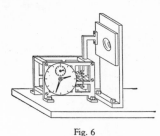

Fig. 6

I cannot believe that the deep philosophical enquiry into the relation between subject and object and into the true meaning of the distinction between them depends on the quantitative results of physical and chemical measurements with weighing scales, spectroscopes, microscopes, telescopes, with Geiger-Müller counters, Wilson chambers, photographic plates, arrangements for measuring the radioactive decay, and whatnot. It is not very easy to say *why* I do not believe it. I feel a certain incongruity between the applied means and the problem to be solved.

Erwin Schrödinger,
Science and Humanism, 1951

suited for the study of phenomena which, in contrast to those just discussed, involve time coordination explicitly. It consists of a shutter rigidly connected with a robust clock resting on the support which carries a diaphragm and on which further parts of similar character, regulated by the same clockwork or by other clocks standardized relatively to it, are also to be fixed. The special aim of the figure is to underline that a clock is a piece of machinery, the working of which can completely be accounted for by ordinary mechanics and will be affected neither by reading of the position of its hands nor by the interaction between its accessories and an atomic particle. In securing the opening of the hole at a definite moment, an apparatus of this type might, for instance, be used for an accurate measurement of the time an electron or a photon takes to come from the diaphragm to some other place, but evidently it would leave no possibility of controlling the energy transfer to the shutter with the aim of drawing conclusions as to the energy of the particle which has passed through the diaphragm. If we are interested in such conclusions we must, of course, use an arrangement whereby the shutter devices can no longer serve as accurate clocks, but whereby the knowledge of the moment when the hole in the diaphragm is open involves a latitude connected with the accuracy of the energy measurement by the general relation (4).

The contemplation of such more or less practical arrangements and their more or less fictitious use proved most instructive in directing attention to essential features of the problems. The main point here is the distinction between the *objects* under investigation and the *measuring instruments* which serve to define, in classical terms, the conditions under which the phenomena appear. Incidentally, we may remark that, for the illustration of the preceding considerations, it is not relevant that experiments involving an accurate control of the momentum or energy transfer from atomic particles to heavy bodies like diaphragms and shutters would be very difficult to perform, if practicable at all. It is only decisive that, in contrast to the proper measuring instruments, these bodies together with the particles would in such a case constitute the system to which the quantum-mechanical formalism has to be applied. As regards the specification of the conditions for any well-defined application of the formalism, it is moreover essential that the *whole experimental arrangement* be taken into account. In fact, the introduction of any further piece of apparatus, such as a mirror, in the way of a particle might imply new interference effects essentially influencing the predictions as regards the results to be eventually recorded.

The extent to which renunciation of the visualization of atomic phenomena is imposed upon us by the impossibility of their subdivision is strikingly illustrated by the following example, to which Einstein very early called attention and often has reverted. If a semireflecting mirror is placed in the way of a photon, leaving two possibilities for its direction of propagation, either the photon may be recorded on one, and only one, of two photographic plates situated at great distances in the two directions in question, or else we may, by replacing the plates by mirrors, observe effects exhibiting an interference between the two reflected wave-trains. In any attempt of a pictorial representation of the behavior of the photon we would, thus, meet with the difficulty: to be obliged to say, on the one hand, that the photon always chooses *one* of the two ways and, on the other hand, that it behaves as if it had passed *both* ways.

It is just arguments of this kind which recall the impossibility of subdividing quantum phenomena and reveal the ambiguity in ascribing customary physical attributes to atomic objects. In particular, it must be realized that—besides in the account of the placing and timing of the instruments forming the experimental arrangement—all unambiguous use of space-time concepts in the description of atomic phenomena is confined to the recording of observations which refer to marks on a photographic plate or to similar practically irreversible amplification effects like the building of a water drop around an ion in a cloudchamber. Although, of course, the existence of the quantum of action is ultimately responsible for the properties of the materials of which the measuring instruments are built and on which the functioning of the recording devices depends, this circumstance is not relevant for the problems of the adequacy and completeness of the quantum-mechanical description in its aspects here discussed.

These problems were instructively commented upon from different sides at the Solvay meeting,[6] in the same session where Einstein raised his general objections. On that occasion an interesting discussion arose also about how to speak of the appearance of phenomena for which only predictions of a statistical character can be made. The question was whether, as to the occurrence of individual effects, we should adopt a terminology proposed by Dirac, that we were concerned with a choice on the part of "nature," or, as suggested by Heisenberg, we should say that we have to do with a choice on the part of the "observer" constructing the measuring instruments and reading their recording. Any such terminology would, however, appear dubious since, on the one hand, it is hardly reasonable to endow nature with volition in the ordinary sense, while, on the other hand, it is certainly not possible for the observer to influence the events which may appear under the conditions he has arranged. To my mind, there is no other alternative than to admit that, in this field of experience, we are dealing with individual phenomena and that our possibilities of handling the measuring instruments allow us only to make a choice between the different complementary types of phenomena we want to study.

The epistemological problems touched upon here were more explicitly dealt with in my contribution to the issue of *Naturwissenschaften* in celebration of Planck's seventieth birthday in 1929. In this article, a comparison was also made between the lesson derived from the discovery of the universal quantum of action and the development which has followed the discovery of the finite velocity of light and which, through Einstein's pioneer work, has so greatly clarified basic principles of natural philosophy. In

"No elementary phenomenon is a phenomenon until it is a registered phenomenon." This summary of the central lesson of the quantum takes its two key words from Bohr. "Registered" as Bohr uses it means "brought to a close by an irreversible act of amplification" and "communicable in plain language." This adjective, equivalent in most respects to "observed," has a special feature as compared to that more frequently seen word. It explicitly denies the view that quantum theory rests in any way whatsoever on "consciousness." The critical word, "phenomenon," Bohr found himself forced to introduce in his discussions with Einstein to stress how different "reality" is from Einstein's "any reasonable concept of reality."

John A. Wheeler, "Beyond the Black Hole," 1979

relativity theory, the emphasis on the dependence of all phenomena on the reference frame opened quite new ways of tracing general physical laws of unparalleled scope. In quantum theory, it was argued, the logical comprehension of hitherto unsuspected fundamental regularities governing atomic phenomena has demanded the recognition that no sharp separation can be made between an independent behavior of the objects and their interaction with the measuring instruments which define the reference frame.

In this respect, quantum theory presents us with a novel situation in physical science; but attention was called to the very close analogy with the situation as regards analysis and synthesis of experience, which we meet in many other fields of human knowledge and interest. As is well known, many of the difficulties in psychology originate in the different placing of the separation lines between object and subject in the analysis of various aspects of psychical experience. Actually, words like "thoughts" and "sentiments," equally indispensable to illustrate the variety and scope of conscious life, are used in a similar complementary way as are space-time coordination and dynamical conservation laws in atomic physics. A precise formulation of such analogies involves, of course, intricacies of terminology, and the writer's position is perhaps best indicated in a passage in the article, hinting at the mutually exclusive relationship which will always exist between the practical use of any word and attempts at its strict definition. The principal aim, however, of these considerations, which were not least inspired by the hope of influencing Einstein's attitude, was to point to perspectives of bringing general epistemological problems into relief by means of a lesson derived from the study of new but fundamentally simple physical experience.

At the next meeting with Einstein, at the Solvay conference in 1930, our discussions took quite a dramatic turn. As an objection to the view that a control of the interchange of momentum and energy between the objects and the measuring instruments was excluded if these instruments should serve their purpose of defining the space-time frame of the phenomena, Einstein brought forward the argument that such control should be possible when the exigencies of relativity theory were taken into consideration. In particular, the general relationship between energy and mass, expressed in Einstein's famous formula

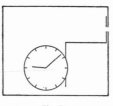

Fig. 7

$$E = mc^2 \tag{5}$$

should allow, by means of simple weighing, to measure the total energy of any system and, thus, in principle to control the energy transferred to it when it interacts with an atomic object.

As an arrangement suited for such purpose, Einstein proposed the device indicated in Figure 7, consisting of a box with a hole in its side, which could be opened or closed by a shutter moved by means of a clockwork within the box. If, in the beginning, the box contained a certain amount of radiation and the clock was set to open the shutter for a very short interval at a chosen time, it could be achieved that a single photon was released through the hole at a moment known with as great accuracy as desired. Moreover, it would apparently also be possible, by weighing the whole box before and after this event, to measure the energy of the photon with any accuracy wanted, in definite contradiction to the reciprocal indeterminacy of time and energy quantities in quantum mechanics.

This argument amounted to a serious challenge and gave rise to a thorough examination of the whole problem. At the outcome of the discussion, to which Einstein himself contributed effectively, it became clear, however, that this argument could not be upheld. In fact, in the consideration of the problem, it was found necessary to look more closely into the consequences of the identification of inertial and gravitational mass implied in the application of relation (5). Especially, it was essential to take into account the relationship between the rate of a clock and its position in a gravitational field — well known from the red-shift of the lines in the sun's spectrum — following from Einstein's principle of equivalence between gravity effects and the phenomena observed in accelerated reference frames.

Our discussion concentrated on the possible application of an apparatus incorporating Einstein's device — an apparatus drawn in Figure 8 in the same pseudorealistic style as some of the preceding figures. The box, of which a section is shown in order to exhibit its interior, is suspended in a spring-balance and is furnished with a pointer to read its position on a scale fixed to the balance support. The weighing of the box may thus be performed with any given accuracy Δm by adjusting the balance to its zero position by means of suitable loads. The essential point is now that any determination of this position with a given accuracy Δq will involve a minimum latitude Δp in the control of the momentum of the box connected with Δq by the relation (3). This latitude must obviously again be smaller than the total impulse which, during the whole interval T of the balancing procedure, can be given by the gravitational field of a body with a mass Δm, or

$$\Delta p \approx \frac{h}{\Delta q} < T \cdot g \cdot \Delta m, \qquad (6)$$

where g is the gravity constant. The greater the accuracy of the reading q of the pointer, the longer must, consequently, be the balancing interval T, if a given accuracy Δm of the weighing of the box with its content shall be obtained.

Now, according to general relativity theory, a clock, when displaced in the direction of the gravitational force by an amount Δq, will change its rate in such a way that its reading in the course of a time interval T will differ by an amount ΔT given by the relation

$$\frac{\Delta T}{T} = \frac{1}{c^2} g \Delta q. \qquad (7)$$

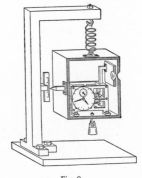

Fig. 8

George Gamow's make-believe apparatus for Einstein's *Gedanken-experiment* to beat the uncertainty relation on the product $\Delta E \cdot \Delta t$ by weighing a box before and after the escape of a photon.

By comparing (6) and (7) we see, therefore, that after the weighing procedure there will in our knowledge of the adjustment of the clock be a latitude

$$\Delta T > \frac{h}{c^2 \Delta m}.$$

Together with formula (5), this relation again leads to

$$\Delta T \cdot \Delta E > h,$$

in accordance with the indeterminacy principle. Consequently, a use of the apparatus as a means of accurately measuring the energy of the photon will prevent us from controlling the moment of its escape.

The discussion, so illustrative of the power and consistency of relativistic arguments, thus emphasized once more the necessity of distinguishing, in the study of atomic phenomena, between the proper measuring instruments which serve to define the reference frame and those parts which are to be regarded as objects under investigation and in the account of which quantum effects cannot be disregarded. Notwithstanding the most suggestive confirmation of the soundness and wide scope of the quantum-mechanical way of description, Einstein, in a following conversation with me, expressed a feeling of disquietude as regards the apparent lack of firmly laid-down principles for the explanation of nature, on which all could agree. From my viewpoint, however, I could only answer that, in dealing with the task of bringing order into an entirely new field of experience, we could hardly trust in any accustomed principles, however broad, apart from the demand of avoiding logical inconsistencies and, in this respect, the mathematical formalism of quantum mechanics should surely meet all requirements.

The Solvay meeting in 1930 was the last occasion where, in common discussions with Einstein, we could benefit from the stimulating and mediating influence of Ehrenfest; but shortly before his deeply deplored death in 1933 he told me that Einstein was far from satisfied and with his usual acuteness had discerned new aspects of the situation which strengthened his critical attitude. In fact, by further examining the possibilities for the application of a balance arrangement, Einstein had perceived alternative procedures which, even if they did not allow the use he originally intended, might seem to enhance the paradoxes beyond the possibilities of logical solution. Thus, Einstein had pointed out that, after a preliminary weighing of the box with the clock and the subsequent escape of the photon, one was still left with the choice of either repeating the weighing or opening the box and comparing the reading of the clock with the standard time scale. Consequently, we are at this stage still free to choose whether we want to draw conclusions either about the energy of the photon or about the moment when it left the box. Without in any way interfering with the photon between its escape and its later interaction with other suitable measuring instruments, we are thus able to make accurate predictions pertaining *either* to the moment of its arrival *or* to the amount of energy liberated by its absorption. Since, however, according to the quantum-mechanical formalism, the specification of the state of an isolated particle cannot involve both a well-defined connection with the time scale and an accurate fixation of the energy, it might thus appear as if this formalism did not offer the means for an adequate description.

Once more Einstein's searching spirit had elicited a peculiar aspect of the situation in quantum theory, which in a most striking manner illustrated how far we have here transcended customary explanation of natural phenomena. Still, I could not agree with the trend of his remarks as reported by Ehrenfest. In my opinion, there could be no other way to deem a logically consistent mathematical formalism as inadequate than by demonstrating the departure of its consequences from experience or by proving that its predictions did not exhaust the possibilities of observation, and Einstein's argumentation could be directed to neither of these ends. In fact, we must realize that in the problem in question we are not dealing with a *single* specified experimental arrangement, but are referring to *two* different, mutually exclusive arrangements. In the one, the balance, together with another piece of apparatus such as a spectrometer, is used for the study of the energy transfer by a photon; in the other, a shutter regulated by a standardized clock, together with another apparatus of similar kind, accurately timed relative to the clock, is used for the study of the time of propagation of a photon over a given distance. In both these cases, as also assumed by Einstein, the observable effects are expected to be in complete conformity with the predictions of the theory.

The problem again emphasizes the necessity of considering the *whole* experimental arrangement, the specification of which is imperative for any well-defined application of the quantum-mechanical formalism. Incidentally, it may be added that paradoxes of the kind contemplated by Einstein

135

are encountered also in such simple arrangements as sketched in Figure 5. In fact, after a preliminary measurement of the momentum of the diaphragm, we are in principle offered the choice, when an electron or photon has passed through the slit, either to repeat the momentum measurement or to control the position of the diaphragm and, thus, to make predictions pertaining to alternative subsequent observations. It may also be added that it obviously can make no difference, as regards observable effects obtainable by a definite experimental arrangement, whether our plans of constructing or handling the instruments are fixed beforehand or whether we prefer to postpone the completion of our planning until a later moment when the particle is already on its way from one instrument to another.

In the quantum-mechanical description, our freedom of constructing and handling the experimental arrangement finds its proper expression in the possibility of choosing the classically defined parameters entering in any proper application of the formalism. Indeed, in all such respects quantum mechanics exhibits a correspondence with the state of affairs familiar from classical physics — a correspondence which is as close as possible, considering the individuality inherent in the quantum phenomena. Just in helping to bring out this point so clearly, Einstein's concern had therefore again been a most welcome incitement to explore the essential aspects of the situation.

The next Solvay meeting, in 1933, was devoted to the problems of the structure and properties of atomic nuclei, in which field such great advances were made just in that period due to the experimental discoveries as well as to new fruitful applications of quantum mechanics. It need in this connection hardly be recalled that the evidence obtained by the study of artificial nuclear transformations gave a most direct test of Einstein's fundamental law regarding the equivalence of mass and energy, which was to prove an evermore important guide for researches in nuclear physics. It may also be mentioned how Einstein's intuitive recognition of the intimate relationship between the law of radioactive transformations and the probability rules governing individual radiation effects was confirmed by the quantum-mechanical explanation of spontaneous nuclear disintegrations. In fact, we are here dealing with a typical example of the statistical mode of description, and the complementary relationship between energy-momentum conservation and time-space coordination is most strikingly exhibited in the well-known paradox of particle penetration through potential barriers.

Einstein himself did not attend this meeting, which took place at a time darkened by the tragic developments in the political world which were to influence his fate so deeply and add so greatly to his burdens in the service of humanity. A few months earlier, on a visit to Princeton — where Einstein was then guest of the newly founded Institute for Advanced Study, to which he soon became permanently attached — I had, however, opportu-

When it came to attacking a real problem, a serious problem of physics, he was marvellous to watch. I always felt that he moved with the skill of a spider in apparently empty space, judging accurately how much weight each slender thread of argument could bear. When he had explored the field, his assurance grew and his speech became vigorous and full of vivid images. I remember an occasion when after a lengthy discussion on the fundamental problems of quantum theory a visitor said, "It makes me quite giddy to think about these problems." Bohr immediately rounded on him and said, "But, but, but . . . if anybody says he can think about quantum theory *without* getting giddy it merely shows that he hasn't understood the first thing about it!" He never trusted a purely formal or mathematical argument. "No, no," he would say, "You are not thinking; you are just being logical."

O. R. Frisch,
What Little I Remember, 1979

nity to talk with him again about the epistemological aspects of atomic physics; but the difference between our ways of approach and expression still presented obstacles to mutual understanding. While, so far, relatively few persons had taken part in the discussions reported in this article, Einstein's critical attitude toward the views on quantum theory adhered to by many physicists was soon after brought to public attention through a paper with the title "Can Quantum-Mechanical Description of Physical Reality Be Considered Complete?" which was published in 1935 by Einstein, Podolsky, and Rosen.[7]

The argumentation in this paper is based on a criterion which the authors express in the following sentence: "If, without in any way disturbing a system, we can predict with certainty (i.e., with probability equal to unity) the value of a physical quantity, then there exists an element of physical reality corresponding to this physical quantity." By an elegant exposition of the consequences of the quantum-mechanical formalism as regards the representation of a state of a system, consisting of two parts which have been in interaction for a limited time interval, it is next shown that different quantities, the fixation of which cannot be combined in the representation of one of the partial systems, can nevertheless be predicted by measurements pertaining to the other partial system. According to their criterion, the authors therefore conclude that quantum mechanics does not "provide a complete description of the physical reality," and they express their belief that it should be possible to develop a more adequate account of the phenomena.

Due to the lucidity and apparently incontestable character of the argument, the paper of Einstein, Podolsky, and Rosen created a stir among physicists and has played a large role in general philosophical discussion. Certainly the issue is of a very subtle character and suited to emphasize how far, in quantum theory, we are beyond the reach of pictorial visualization. It will be seen, however, that we are here dealing with problems of just the same kind as those raised by Einstein in previous discussions, and, in an article which appeared a few months later,[8] I tried to show that from the point of view of complementarity the apparent inconsistencies were completely removed. The trend of the argumentation was in substance the same as that exposed in the foregoing pages, but the aim of recalling the way in which the situation was discussed at that time may be an apology for citing certain passages from my article.

Thus, after referring to the conclusions derived by Einstein, Podolsky, and Rosen on the basis of their criterion, I wrote:

Such an argumentation, however, would hardly seem suited to affect the soundness of quantum-mechanical description, which is based on a coherent mathematical formalism covering automatically any procedure of measurement like that indicated. The apparent contradiction in fact discloses only an essential inadequacy of the customary viewpoint of

Dear Einstein: It seems to me that the concept of probability is terribly mishandled these days. Probability surely has as its substance a statement as to whether something *is* or *is not* the case—an uncertain statement, to be sure. But nevertheless it has meaning only if one is indeed convinced that the something in question quite definitely either *is* or *is not* the case. A probabilistic assertion presupposes the full reality of its subject. No reasonable person would express a conjecture as to whether Caesar rolled a five with his dice at the Rubicon. But the quantum-mechanics people sometimes act as if probabilistic statements were to be applied *just* to events whose reality is vague.

Erwin Schrödinger to Albert Einstein, 18 November 1950

natural philosophy for a rational account of physical phenomena of the type with which we are concerned in quantum mechanics. Indeed the *finite interaction between object and measuring agencies* conditioned by the very existence of the quantum of action entails — because of the impossibility of controlling the reaction of the object on the measuring instruments, if these are to serve their purpose — the necessity of a final renunciation of the classical ideal of causality and a radical revision of our attitude toward the problem of physical reality. In fact, as we shall see, a criterion of reality like that proposed by the named authors contains — however cautious its formulation may appear — an essential ambiguity when it is applied to the actual problems with which we are here concerned.

As regards the special problem treated by Einstein, Podolsky, and Rosen, it was next shown that the consequences of the formalism as regards the representation of the state of a system consisting of two interacting atomic objects correspond to the simple arguments mentioned in the preceding in connection with the discussion of the experimental arrangements suited for the study of complementary phenomena. In fact, although any pair, q and p, of conjugate space and momentum variables obeys the rule of non-commutative multiplication expressed by (2), and can thus only be fixed with reciprocal latitudes given by (3), the difference $q_1 - q_2$ between two space coordinates referring to the constituents of the system will commute with the sum $p_1 + p_2$ of the corresponding momentum components, as follows directly from the commutability of q_1 with p_2 and of q_2 with p_1. Both $q_1 - q_2$ and $p_1 + p_2$ can, therefore, be accurately fixed in a state of the complex system and, consequently, we can predict the values of either q_1 or p_1 if either q_2 or p_2, respectively, are determined by direct measurements. If, for the two parts of the system, we take a particle and a diaphragm, like that sketched in Figure 5, we see that the possibilities of specifying the state of the particle by measurements on the diaphragm just correspond to the situation described earlier in this essay, where it was mentioned that, after the particle has passed through the diaphragm, we have in principle the choice of measuring either the position of the diaphragm or its momentum and, in each case, must make predictions as to subsequent observations pertaining to the particle. As repeatedly stressed, the principal point is here that such measurements demand mutually exclusive experimental arrangements.

The argumentation of the article was summarized in the following passage:

From our point of view we now see that the wording of the above-mentioned criterion of physical reality proposed by Einstein, Podolsky, and Rosen contains an ambiguity as regards the meaning of the expression "without in any way disturbing a system." Of course there is, in a case like that just considered, no question of a mechanical disturbance of the system under investigation during the last critical stage of the measuring procedure. But even at this stage there is essentially the question of *an influence on the very conditions which define the possible types of predictions regarding the future behavior of the system.* Since these conditions constitute an inherent element of the description of any phenomenon to which the term "physical reality" can be properly attached, we see that the argumentation of the mentioned authors does not justify their conclusion that quantum-mechanical description is essentially incomplete. On the contrary, this description, as appears from the preceding discussion, may be characterized as a rational utilization of all possibilities of unambiguous interpretation of measurements, compatible with the finite and uncontrollable interaction between the objects and the measuring instruments in the field of quantum theory. In fact, it is only the mutual exclusion of any two experimental procedures, permitting the unambiguous definition of complementary physical quantities, which provides room for new physical laws, the coexistence of which might at first sight appear irreconcilable with the basic principles of science. It is just this entirely new situation as regards the description of physical phenomena that the notion of *complementarity* aims at characterizing.

Rereading these passages, I am deeply aware of the inefficiency of expression which must have made it very difficult to appreciate the trend of the argumentation aiming to bring out the essential ambiguity involved in a reference to physical attributes of objects when dealing with phenomena where no sharp distinction can be made between the behavior of the objects themselves and their interaction with the measuring instruments. I hope, however, that the present account of the discussions with Einstein in the foregoing years, which contributed so greatly to making us familiar with the situation in quantum physics, may give a clearer impression of the necessity of a radical revision of basic principles for physical explanation in order to restore logical order in this field of experience.

Einstein's own views at that time are presented in an article entitled "Physics and Reality," published in 1936 in the *Journal of the Franklin Institute*.[9] Starting from a most illuminating exposition of the gradual development of the fundamental principles in the theories of classical physics and their relation to the problem of physical reality, Einstein here argues that the quantum-mechanical description is to be considered merely as a means of accounting for the average behavior of a large number of atomic systems, and his attitude to the belief that it should offer an exhaustive description of the individual phenomena is expressed in the following words: "To believe this is logically possible without contradiction; but it is so very contrary to my scientific instinct that I cannot forgo the search for a more complete conception."

Even if such an attitude might seem well-balanced in itself, it nevertheless implies a rejection of the whole argumentation exposed in the preceding, aiming to show that, in quantum mechanics, we are not dealing with an arbitrary renunciation of a more detailed analysis of atomic phenomena, but with a recognition that such an analysis is *in principle* excluded. The peculiar individuality of the quantum effects presents us, as regards the comprehension of well-defined evidence, with a novel situation unforeseen in classical physics and irreconcilable with conventional ideas suited for our orientation and adjustment to ordinary experience. It is in this respect that quantum theory has called for a renewed revision of the foundation for the unambiguous use of elementary concepts, as a further step in the development which, since the advent of relativity theory, has been so characteristic of modern science.

[After a brief discussion of developments subsequent to 1936, Bohr concluded his account with the following remarks.]

The discussions with Einstein which have formed the theme of this article have extended over many years which have witnessed great progress in the field of atomic physics. Whether our actual meetings have been of short or long duration, they have always left a deep and lasting impression on my mind, and when writing this report I have, so to say, been arguing with Einstein all the time even when entering on topics apparently far removed from the special problems under debate at our meetings. As regards the account of the conversations, I am, of course, aware that I am relying only on my own memory, just as I am prepared for the possibility that many features of the development of quantum theory, in which Einstein has played so large a part, may appear to himself in a different light. I trust, however, that I have not failed in conveying a proper impression of how much it has meant to me to be able to benefit from the inspiration which we all derive from every contact with Einstein.

No more profound intellectual debate has ever been conducted — and, since they were both men of the loftiest spirit, it was conducted with noble feeling on both sides. If two men are going to disagree, on the subject of most ultimate concern to them both, then that is the way to do it.

C. P. Snow,
Variety of Men, 1967

A Bolt From the Blue:
The E-P-R Paradox

N. David Mermin

David Mermin is Professor of Physics at Cornell University. His special fields are theoretical solid-state and statistical physics.

Fifty years ago Einstein, Podolsky, and Rosen published a striking argument that quantum theory provides only an incomplete description of physical reality — that is, that some elements of physical reality fail to have a counterpart in quantum-theoretical description.[1] What made this a challenge to the quantum-theoretical *Weltanschauung* was the very mild character of the sufficient condition for the reality of a physical quantity on which their argument hinged: "If, without in any way disturbing a system, we can predict with certainty the value of a physical quantity, then there exists an element of physical reality corresponding to this physical quantity." The argument consists in pointing out that it is possible to construct a quantum-mechanical state ψ with the following properties:

1. The state ψ describes two noninteracting systems (I and II). In almost all discussions the two systems are taken to be two particles, and the absence of relevant interaction is built in by taking ψ to assign negligible probability to finding the particles closer together than some macroscopically large distance.

2. By measuring an *observable A* of system I, one can predict with certainty the result of a subsequent measurement of a corresponding observable P of system II.

3. If, instead, one chooses to measure a different observable B of system I, one can predict with certainty the result of a subsequent measurement of a corresponding observable Q of system II.

4. The observables P and Q are represented in quantum theory by noncommuting operators, which means that they cannot both have definite values. Einstein, Podolsky, and Rosen give an example in which P and Q are the position and momentum of particle II along a given direction, and it is in terms of this example that Bohr's reply is formulated. Subse-

141

quently, David Bohm gave a very simple and clearcut example in which P and Q are distinct components of the intrinsic angular momentum of a spinning particle.[2] In recent experiments by Alain Aspect and his colleagues, they are the linear polarization of a photon along distinct nonperpendicular directions.[3]

The route to the E-P-R conclusion is straightforward. Because the systems do not interact, the measurements on system I can be made "without in any way disturbing" system II. Therefore, the values of P and Q are both elements of reality. But the quantum-mechanical description of reality given by the wave function ψ does not assign definite values to P and Q. Therefore, the wave function does not provide a complete description of reality.

The E-P-R article appeared in the *Physical Review* on 15 May 1935. According to Léon Rosenfeld,

> this onslaught came down upon us as a bolt from the blue. Its effect on Bohr was remarkable . . . A new worry could not come at a less propitious time. Yet, as soon as Bohr had heard my report of Einstein's argument, everything else was abandoned . . . In great excitement, Bohr immediately started dictating to me the outline of . . . a reply. Very soon, however, he became hesitant: "No, this won't do, we must try all over again . . . we must make it quite clear . . ." So it went on for a while, with growing wonder at the unexpected subtlety of the argument . . . The next morning he at once took up the dictation again, and I was struck by a change in the tone of the sentences: there was no trace in them of the previous day's sharp expressions of dissent. As I pointed out to him that he seemed to take a milder view of the case, he smiled: "That's a sign," he said, "that we are beginning to understand the problem."[4]

Bohr's reply appeared in the *Physical Review* five months later, with exactly the same title as that used by Einstein, Podolsky, and Rosen: "Can Quantum-Mechanical Description of Physical Reality be Considered Complete?" He remarks that the E-P-R argument can

> hardly seem suited to affect the soundness of quantum-mechanical description, which is based on a coherent mathematical formalism covering automatically any procedure of measurement like that indicated. The apparent contradiction in fact discloses only an essential inadequacy of the customary viewpoint of natural philosophy for a rational account of physical phenomena of the type with which we are concerned in quantum mechanics . . . A criterion of reality like that proposed by the named authors contains — however cautious its formulation may appear — an essential ambiguity when it is applied to the actual problems with which we are here concerned . . .

[This ambiguity] regards the meaning of the expression "without in any way disturbing a system." Of course there is in a case like that just considered no question of a mechanical disturbance of the system under investigation during the last critical stage of the measuring procedure. But even at this stage there is essentially the question of *an influence on the very conditions which define the possible types of predictions regarding the future behavior of the system.* Since these conditions constitute an inherent element of the description of any phenomenon to which the term "physical reality" can be properly attached, we see that the argumentation of the mentioned authors does not justify their conclusion that quantum-mechanical description is essentially incomplete. On the contrary, this description . . . may be characterized as a rational utilization of all possibilities of unambiguous interpretation of measurements, compatible with the finite and uncontrollable interaction between the objects and the measuring instruments in the field of quantum theory. In fact, it is only the mutual exclusion of any two experimental procedures, permitting the unambiguous definition of complementary physical quantities, which provides room for new physical laws, the coexistence of which might at first sight appear irreconcilable with the basic principles of science.[5]

At the end of their article, Einstein, Podolsky, and Rosen anticipate some aspects of this reply:

One could object to this conclusion on the ground that our criterion of reality is not sufficiently restrictive. Indeed, one would not arrive at our conclusion if one insisted that two or more physical quantities can be regarded as simultaneous elements of reality *only when they can be simultaneously measured or predicted.* On this point of view, since either one or the other, but not both simultaneously, of the quantities P and Q can be predicted, they are not simultaneously real. This makes the reality of P and Q depend upon the process of measurement carried out on the first system, which does not disturb the second system in any way. No reasonable definition of reality could be expected to permit this.

One of the central points at issue is what it means to "disturb" something. I learned my quantum metaphysics primarily through the writings of Heisenberg. As I understood it, the unavoidable, uncontrollable disturbances accompanying a measurement were local, "mechanical," and not especially foreign to naive classical intuition (photons bumping into electrons in the course of a position measurement — that sort of thing). When I read the E-P-R paper (in the late 1950s), it gave me a shock. Bohr's casual extension of Heisenberg's straightforward view of a "disturbance" seemed to me radical and bold. That most physicists were not, apparently, shocked at the time — that Bohr was generally and immediately viewed as having once again set things straight — surprised and perplexed me. Until I learned about J. S. Bell's 1964 paper, "On the Einstein-Podolsky-Rosen Paradox," I must confess to having been on Einstein's side of the dispute.[6]

Dear Schrödinger: You are the only contemporary physicist, besides Laue, who sees that one cannot get around the assumption of reality — if only one is honest. Most of them simply do not see what sort of risky game they are playing with reality — reality as something independent of what is experimentally established. They somehow believe that the quantum theory provides a description of reality, and even a *complete* description.

Albert Einstein to Erwin Schrödinger, 22 December 1950

Current attitudes toward the E-P-R argument and Bell's simple but remarkable analysis vary widely. Physicists today are by and large immune to worries about the meaning of quantum mechanics, and quite oblivious, or, if it is brought forcefully to their attention, indifferent to the intense interest the subject still generates among a few. Thus, Abraham Pais, in his biography of Einstein, expresses the prevailing opinion by suggesting that the E-P-R paper will ultimately be of interest only for the insight it reveals into Einstein's state of mind (through the phrase "no reasonable definition of reality could be expected to permit this").[7] There is a rather different minority view, of which perhaps the most extreme example is Stapp's opinion that Bell's analysis of the E-P-R paper constitutes "the most profound discovery of science."[8]

Before Bell's paper, the grounds for dismissing the E-P-R argument were entirely metaphysical. Pauli, for example, tried to explain to Born in 1954 that the only reason for rejecting Einstein's views was that "one should no more rack one's brains about the problem of whether something one cannot know anything about exists all the same than about the ancient question of how many angels are able to sit on the point of a needle. But it seems to me that Einstein's questions are ultimately always of this kind."[9] Bell transformed the issue by showing that the E-P-R position was inconsistent with the quantitative numerical predictions of the quantum theory and therefore susceptible to direct experimental test.

How can one demonstrate that "something one cannot know anything about" cannot exist? It turns out to be quite simple. Add to the E-P-R observables A and B of system I a third observable, C, and let the corresponding observables for system II be P, Q, and R. Suppose a measurement of any one of these six observables can have only two outcomes. We can label the outcomes "yes" and "no" and regard the observables as questions. Suppose the pair of observables A and P invariably yield the same answers, as do the pair B and Q, and the pair C and R, so that by measuring the corresponding system I observable "one can predict with certainty the result of a subsequent measurement" of the corresponding system II observable, or vice versa. Einstein, Podolsky, and Rosen would then conclude that the answers to P, Q, R, A, B, and C are all elements of reality, even when it is physically impossible to measure more than a single one of the observables A, B, and C, or more than one of P, Q, and R.

But now suppose that whenever one asked the questions associated with any other pair of observables (that is, the pairs AQ, AR, BP, BR, CQ, CP), the answers invariably differed. The answer to P would then always be the same as the answer to A but always opposite to the answers to B and C; thus, B and C would always have the same answer, which would always be opposite to the answer to A. On the other hand, since the answer to Q would always be opposite to the answers to A and C, the answers to A and C would always have to be the same.

Thus, on the one hand A and C would always have to have different answers, but on the other hand they would always have to have the same answers. The only conclusion, contrary to the E-P-R conclusion, is that the answers to all six questions cannot all be elements of reality.

Bell's argument differs from this in only one respect. He considers a version (Bohm's) of the E-P-R experiment in which the state of affairs is exactly as described above except that (in one particularly simple case)[10] the six pairs of questions that fail to always agree do not disagree all the time, but only three times out of four. This is still enough to establish that all the answers cannot exist, though the discussion now acquires a statistical character. The answer to P remains always the same as the answer to A, but it is now opposite to the answer to B and C only three-fourths of the time (not necessarily the same three-fourths). Evidently A and C (or A and B) can then agree only one-fourth of the time. A little thought reveals that, in addition, B and C must have the same answers at least half the time. If this reasoning is repeated, starting with a consideration of the answer Q, however, the step requiring a little thought now leads to the conclusion that A and C must have the same answers at least half the time.

We have again arrived at a contradiction: on the one hand A and C can agree only one-fourth of the time, but on the other hand they must agree at least half the time. The inescapable conclusion is that all six questions cannot have answers.

Thus, if the quantum theory is correct in its quantitative predictions for this type of E-P-R experiment, then the E-P-R conclusion is untenable even for people who reject, or fail fully to grasp, the notion of complementarity. Even a neutralist position like Pauli's must be abandoned. That quantum-theoretical predictions are correct in precisely this context has been conclusively demonstrated in the elegant series of experiments by Aspect and his colleagues, mentioned above.

In the Aspect experiments, systems I and II are a pair of photons, emitted by a calcium atom in a radiative cascade after appropriate pumping by lasers. The observables A, B, C, . . . (and P, Q, R, . . .) are the linear polarization of photon I (and photon II) along various directions — that is, the question, for a given photon and a given direction, is whether the photon is polarized along ("yes") or perpendicular to ("no") that direction. The initial and final atomic states have spherical symmetry, as a consequence of which quantum theory predicts (and experiment confirms) that the photons will behave in the same way if their polarizations are measured along the same directions. But if the two polarizations are measured along directions 120 degrees apart, then quantum theory predicts (and experiment confirms) that the photons will behave in the opposite way three-fourths of the time. (Aspect et al. were interested in a somewhat modified version of Bell's argument in which the angles of greatest interest were multiples of 22.5 degrees, but they collected data for many different angles.)

This is a precise realization of the experiment used above to illustrate Bell's analysis, and the fact that the experiment confirmed the quantum-theoretical predictions to within a few percent establishes that the E-P-R reality criterion is not valid.

There are some remarkable features to these experiments. The two polarization analyzers were placed as far as thirteen meters apart without any noticeable change in the results, thereby effectively eliminating the possibility that the strange quantum correlations might somehow diminish as the distance between I and II grew to macroscopic dimensions. At such a distance it is hard to deny that the polarization measurement of photon I is made "without in any way disturbing" photon II, in the sense of Bohr's "mechanical disturbance of the system under investigation during the last critical stage of the measuring procedure." Indeed, at this large separation, a hypothetical disturbance originating when one photon passed through its analyzer could not reach the other analyzer in time to affect the outcome of the second polarization measurement, even if it traveled at the fastest possible speed (the speed of light).

In the third of the Aspect papers, the feasibility of bizarre conspiracy theories, designed to salvage the E-P-R reality criterion, is considerably diminished by the use of an ingenious mechanism for the extremely rapid switching of the directions along which the two photon polarizations are measured. The two switching rates are different, uncorrelated, and so high that several changes are made even during the very short time it would take a signal traveling at the speed of light to reach from one analyzer to the other. Thus, if one takes the view that the "reality" of the polarization of photon I along a given direction depends on the choice of directions along which the polarization of photon II is measured, then that "reality" must be transmitted at superluminal speeds. No reasonable definition of reality could be expected to permit this.

Einstein surely would have found these experiments shocking. In 1948 he observed that an entirely appropriate response to the E-P-R experiment by "those physicists who regard the descriptive methods of quantum mechanics as definitive in principle" would be to

> drop the requirement for the independent existence of physical reality present in different parts of space; they would be justified in pointing out that the quantum theory nowhere makes explicit use of this requirement.
>
> I admit this, but would point out: when I consider the physical phenomena known to me, and especially those which are being so successfully encompassed by quantum mechanics, I still cannot find any fact anywhere which would make it appear likely that [this] requirement will have to be abandoned.[11]

The Aspect experiments provide such facts. They would not have surprised Bohr, but although some physicists today might regard them as no more than an extremely complicated confirmation of Malus's classical law, they surely would have pleased and interested him. Combined with Bell's elementary analysis they provide a simple, direct, and compelling demonstration of complementarity in one of its most dramatic manifestations.

PHYSICAL REVIEW

MAY 15, 1935

VOLUME 47

Can Quantum-Mechanical Description of Physical Reality Be Considered Complete?

A. EINSTEIN, B. PODOLSKY AND N. ROSEN, *Institute for Advanced Study, Princeton, New Jersey*
(Received March 25, 1935)

In a complete theory there is an element corresponding to each element of reality. A sufficient condition for the reality of a physical quantity is the possibility of predicting it with, certainty, without disturbing the system. In quantum mechanics in the case of two physical quantities described by non-commuting operators, the knowledge of one precludes the knowledge of the other. Then either (1) the description of reality given by the wave function in

quantum mechanics is not complete or (2) these two quantities cannot have simultaneous reality. Consideration of the problem of making predictions concerning a system on the basis of measurements made on another system that had previously interacted with it leads to the result that if (1) is false then (2) is also false. One is thus led to conclude that the description of reality as given by a wave function is not complete.

OCTOBER 15, 1935

PHYSICAL REVIEW

VOLUME 48

Can Quantum-Mechanical Description of Physical Reality be Considered Complete?

N. BOHR, *Institute for Theoretical Physics, University, Copenhagen*
(Received July 13, 1935)

It is shown that a certain "criterion of physical reality" formulated in a recent article with the above title by A. Einstein, B. Podolsky and N. Rosen contains an essential ambiguity when it is applied to quantum phenomena. In this connection a viewpoint termed "complementarity" is explained from which quantum-mechanical description of physical phenomena would seem to fulfill, within its scope, all rational demands of completeness.

IN a recent article[1] under the above title A. Einstein, B. Podolsky and N. Rosen have presented arguments which lead them to answer the question at issue in the negative. The trend of their argumentation, however, does ... with which we ... shall

---

PHYSICAL REVIEW LETTERS

12 JULY 1982

VOLUME 49, NUMBER 2

Experimental Realization of Einstein-Podolsky-Rosen-Bohm *Gedankenexperiment*: A New Violation of Bell's Inequalities

Alain Aspect, Philippe Grangier, and Gérard Roger
*Institut d'Optique Théorique et Appliquée, Laboratoire associé au Centre National de la Recherche Scientif...
Université Paris-Sud, F-91406 Orsay, France*
(Received 30 December 1981)

The linear-polarization correlation of pairs of photons emitted in a radiative cascade of calcium has been measured. The new experimental scheme, using two-channel polarizers (i.e., optical analogs of Stern-Gerlach filters), is a straightforward transposition of Einstein-Podolsky-Rosen-Bohm *gedankenexperiment*. The present results, in excellent agreement with the quantum mechanical predictions, lead to the greatest violation of generalized Bell's inequalities ever achieved.

Experimental Test of Bell's Inequalities Using Time-Varying Analyzers

Alain Aspect, Jean Dalibard,[a] and Gérard Roger
Institut d'Optique Théorique et Appliquée, F-91406 Orsay Cédex, France
(Received 27 September 1982)

Correlations of linear polarizations of pairs of photons have been measured with time-varying analyzers. The analyzer in each leg of the apparatus is an acousto-optical switch followed by two linear polarizers. The switches operate at incommensurate frequencies near 50 MHz. Each analyzer amounts to a polarizer which jumps between two orientations in a time short compared with the photon transit time. The results are in good agreement with quantum mechanical predictions but violate Bell's inequalities by 5 standard deviations.

PACS numbers: 03.65.Bz, 35.80.+s

Bell's inequalities apply to any correlated measurement on two correlated systems. For instance, in the optical version of the Einstein-Podolsky-Rosen-Bohm *Gedankenexperiment*,[1] a source emits pairs of photons (Fig. 1). Measurements of the correlations of linear polarizations are performed on two photons belonging to the same pair. For pairs emitted in suitable states, the correlations are strong. To account for these correlations, Bell[2] considered theories which invoke common properties of both members of the

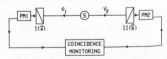

FIG. 1. Optical version of the Einstein-Podolsky-Rosen-Bohm *Gedankenexperiment*. The pair of photons ν_1 and ν_2 is analyzed by linear polarizers I and II (in orientations $\vec{a}$ and $\vec{b}$) and photomultipliers. The coincidence rate is monitored.

BERTLMANN'S SOCKS AND THE NATURE OF REALITY

J.S. Bell

CERN, CH-1211, Genève 23, Suisse

1. Introduction.- The philosopher in the street, who has not suffered a course in quantum mechanics, is quite unimpressed by Einstein-Podolsky-Rosen correlations /1/. He can point to many examples of similar correlations in everyday life. The case of Bertlmann's socks is often cited. Dr. Bertlmann likes to wear two socks of different colours. Which colour he will have on a given foot on a given day is quite unpredictable. But when you see (Fig. 1) that the first sock is pink you can be already sure that the second sock will not be pink. Observation of the first, and experience of Bertlmann, gives immediate information about the second. There is no accounting for tastes, but apart from that there is no mystery here. And is not the EPR business just the same ?

Les chaussettes
de M. Bertlmann
et la nature
de la réalité

Fondation Hugot
juin 17 1980

pink   not pink

ON THE EINSTEIN PODOLSKY ROSEN PARADOX*

J. S. BELL†
Department of Physics, University of Wisconsin, Madison, Wisconsin
(Received 4 November 1964)

I. Introduction

THE paradox of Einstein, Podolsky and Rosen [1] was advanced as an argument that quantum mechanics could not be a complete theory but should be supplemented by additional variables. These additional variables were to restore to the theory causality and locality [2]. In this note that idea will be formulated mathematically and shown to be incompatible with the statistical predictions of quantum mechanics. It is the requirement of locality, or more precisely that the result of a measurement on one system be unaffected by operations on a distant system with which it has interacted in the past, that creates the essential difficulty. There have been attempts [3] to show that even without such a separability or locality requirement no "hidden variable" interpretation of quantum mechanics is possible. These attempts have been examined elsewhere [4] and found wanting. Moreover, a hidden variable interpretation of elementary quantum theory [5] has been explicitly constructed. That particular interpretation has indeed a grossly non-local structure. This is characteristic, according to the result to be proved here, of any such theory which reproduces exactly the quantum mechanical predictions.

VIEW LETTERS

12 JULY 1982

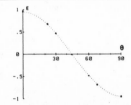

FIG. 3. Correlation of polarizations as a function of the relative angle of the polarimeters. The indicated errors are ±2 standard deviations. The dotted curve is not a fit to the data, but quantum mechanical predictions for the actual experiment. For ideal polarizers, the curve would reach the values ±1.

AL REVIEW LETTERS

20 DECEMBER 1982

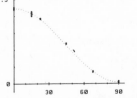

FIG. 4. Average normalized coincidence rate as a function of the relative orientation of the polarizers. Indicated errors are ±1 standard deviation. The dashed curve is not a fit to the data but the predictions by quantum mechanics for the actual experiment.

A pictorial history of the Einstein-Podolsky-Rosen paradox and its sequel in terms of real experiments.

Delayed-Choice Experiments

P. J. Kennedy

P. J. Kennedy, author of the brief biography that appears at the beginning of this volume, here describes some remarkable recent tests of Bohr's basic ideas in quantum theory.

The underlying meaning to be attributed to quantum theory has been the subject of discussion — and, indeed, of controversy — ever since the theory was first formulated. These discussions, largely initiated by Bohr and Einstein in the 1920s and 1930s (see "The Bohr-Einstein Dialogue"), have remained lively ever since. In recent times the controversy has received new impetus from a set of exciting experiments, or proposed experiments, which have been made possible by advances in measurement techniques in the 1980s and which will facilitate experimental investigation of some of the apparent paradoxes that have featured so prominently in the long argument.

In Bohr's account of his dialogue with Einstein, there are two passages of special interest concerning the two-slit interference experiment, as well as a particular variant suggested by Einstein:

> The extent to which renunciation of the visualization of atomic phenomena is imposed upon us by the impossibility of their subdivision is strikingly illustrated by the following example, to which Einstein very early called attention and often has reverted. If a semireflecting mirror is placed in the way of a photon, leaving two possibilities for its direction of propagation, the photon may either be recorded on one, and only one, of two photographic plates situated at great distances in the two directions in question, or else we may, by replacing the plates by mirrors, observe effects exhibiting an interference between the two reflected wave-trains. In any attempt of a pictorial representation of the behavior of the photon we would, thus, meet with the difficulty: to be obliged to say, on the one hand, that the photon always chooses *one* of the two ways and, on the other hand, that it behaves as if it had passed *both* ways.

It is just arguments of this kind which recall the impossibility of subdividing quantum phenomena and reveal the ambiguity in ascribing customary physical attributes to atomic objects.[1]

This quotation, which is concerned with the complementary attributes of a quantum-mechanical phenomenon, emphasizes the element of choice which exists in deciding on the final irreversible registration of the phenomenon in a particular apparatus. Einstein had suggested a "prototype" experiment in which a dilemma appears to arise in choosing between observation of direction (one path) and phase (both paths).

In the same essay, Bohr made a deeply perceptive comment:

It may also be added that it obviously can make no difference as regards observable effects obtainable by a definite experimental arrangement, whether our plans of constructing or handling the instruments are fixed beforehand or whether we prefer to postpone the completion of our planning until a later moment when the particle is already on its way from one instrument to another.

Evidently Bohr discerned, very early in the dialogue, that this choice between complementary modes of registration need not be made until the transit of the particle through an apparatus is almost complete; or, in the language of Einstein's experiment, until after the photon appears to have chosen to travel one way or both ways.

Delayed-choice experiments, as they are now known, have been discussed in several articles by J. A. Wheeler.[2] Such experiments are intended to exemplify the apparent paradox of the Bohr-Einstein dialogue and, even if they served only to illustrate or dramatize the epistemological dilemma there revealed, they would be of great interest. In fact, some of them can offer a quantitative and measurable distinction between the predictions of quantum theory and those of other local (hidden) variable theories, and so are of even greater philosophical importance.

One such experiment, in which two photons, created in a single atomic event, are allowed to separate spatially before registration, has been described in the previous chapter by David Mermin. In that experiment, devised by Aspect and his colleagues, the delayed choice of the setting of the registration apparatus was made during the time of flight of the two photons. In later experiments, it is arranged that the setting of one register cannot be causally connected to the second distant register (other than by a superluminal or other exceptional mechanism) before that register is activated by the arriving photon. Mermin has described how such an experiment may be interpreted in terms of Bell's theorem and that it gives results in close quantitative agreement with the predictions of quantum mechanics. The prototype single-particle, two-mirror experiment, as envisaged by Einstein, forms the basis of several current investigations (see the simplified diagrams). The diagrams could illustrate the working of an inter-

Basic arrangements for delayed-choice experiments: above, using movable detectors; below, using a removable half-silvered mirror (M_2).

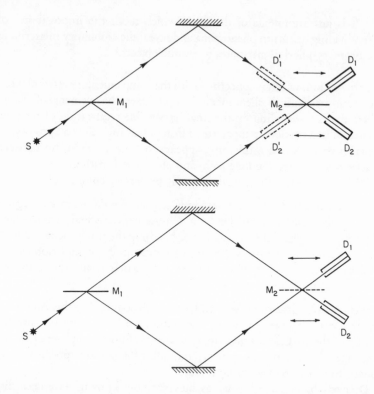

ferometer in a normal way, with a strong source S; the beam is split at the half-silvered mirror M_1 and recombined at the half-silvered mirror M_2. A choice between registering direction (which path) or phase interference (both paths) can be made by moving the detectors D_1' and D_2' to positions D_1 and D_2 in the upper drawing, or by removing or inserting M_2 in the lower drawing. When the intensity of the source is so reduced that only one photon may be considered to be in transit in the apparatus at a given moment, we meet the well known difficulty of description that Einstein remarked upon. The photon appears to travel on one path or on both paths, according to the type of registration chosen. If, in addition, this choice is delayed until the photon is between M_1 and M_2, and if the choice is made randomly, then the "pictorial representation of the behavior of the photon" is even more difficult.

This apparently simple prescription is not easy to achieve technically, but there are two possible approaches. In one, photons with transit times of a few nanoseconds (1 nanosecond $= 10^{-9}$ second) are used, and the random choice of registration is made by very fast switching. In the other method, ultracold neutrons with transit times of the order of tenths of a second are used, and the time problems of the delayed random choice are thereby reduced.

The first type of experiment may be exemplified by the work of C. O. Alley and his colleagues.[3] Here 100-picosecond pulses (1 picosecond $=$

10^{-12} second) from a laser enter the apparatus through an attenuator, such that some 10^6 quanta arrive singly over a period of twenty-four hours. The path between the beam splitter M_1 and the beam recombiner M_2 is about 4.5 meters, which is long when compared with the quantum wave packet. If the upper path of the interferometer is blocked, the information recorded after M_2 will be concerned with the path of the photon; but if *both paths* are open, the information gained will concern the phase of arrival. The experiment is thus close to that proposed by Einstein. The random delayed decision to block or unblock the upper path is made using a fast-acting *Pockels cell* (≈ 10 nanoseconds) during the period that the photon is between M_1 and M_2. As an additional experimental variable, a second Pockels cell, working in a different mode, introduces a controlled phase shift into the lower path. The whole experiment is susceptible of analysis in terms of the count rates predictable by quantum mechanics. The first results of the experiment support Bohr's contentions that they should not depend on the time at which the type of registration is chosen, and that, between the time the photon enters the apparatus and the time the actual mode of registration of an amplified signal is activated, no specific space-time trajectory can be assigned to the photon — it behaves as a "smoky dragon."

The alternative approach, in which ultracold neutrons are used as the "elementary quantum phenomena," has been proposed by several groups.[4] With modern techniques, neutrons with velocities of about 10 meters per second and with de Broglie wave lengths of some 100 nanometers can be introduced into interferometers like those shown in the simplified diagrams. With very low-intensity sources the neutrons can be treated as crossing the apparatus, one at a time, in times of tenths of a second. After division of the wave packet at M_1, the long-lived quantum-mechanical

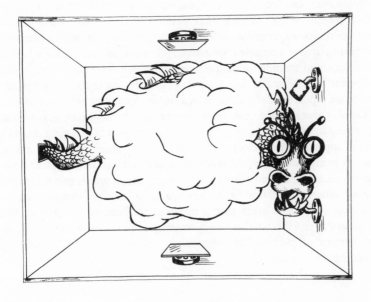

Wheeler's "smoky dragon," which symbolizes Bohr's "elementary quantum phenomenon." Drawing by Field Gilbert.

phenomenon is completed only when registration takes place near M_2. In this experiment various devices (for example, cadmium absorbers, copper reflectors, or spin flipping cells) can be used to block, unblock, or partially block either of the two paths, and this can be easily done randomly during the long transit times.

Although highly complex and difficult to perform, these experiments can still be discussed in the terms used by Einstein: that we can be led to believe that by one random choice we seem to determine the path the neutron or photon has taken, and with the other choice — the registration of interference terms — we seem to establish that the neutron or photon has traveled along both paths. If this random choice is delayed until after the transit through the beam splitter is complete, then this "later" choice of registration appears to determine the choice made by the quantum at some "earlier" time. It was these apparent paradoxes which led Bohr to insist that the use of such language is improper in describing quantum phenomena, and that such phenomena are known and discussable only in terms of the irreversible amplified registration in macroscopic counters. In John Wheeler's favorite way of putting it: "No elementary phenomenon is a phenomenon until it is a recorded phenomenon."[5] Greenberger and his colleagues have attempted to analyze the cold-neutron experiments in these terms, and their analysis leads to differences in the counting rates predicted by quantum mechanics and by other, hidden variable theories.[6] Such inequalities, similar to those deduced by Bell for the two-quantum experiments, lead to subtle tests of the completeness of quantum mechanics and of reality theories.

Apart from these quantitative tests, delayed choice experiments introduce philosophical and epistemological implications of great significance in that, although in the experiments described earlier the delay was a matter of only fractions of a second, the original statement by Bohr does not preclude much longer delays. Wheeler has pointed out that one could devise a delayed-choice experiment in which the "potential" quantum phenomenon is initiated at a remote cosmological distance from the place where the final choice of completion is made and in which this choice is delayed by many millions of years.[7] Thus, an act of present-day decision may seem to create an aspect of the past — or at least to determine what we may say about that aspect of the past. As Wheeler has said: "Registration equipment operating in the here and now has an undeniable part in bringing about that which appears to have happened. Useful as it is under every-day circumstances to say that the world exists 'out there' independent of us, that view can no longer be upheld. There is a strange sense in which this is a 'participatory universe.'"[8] How such ideas are to be incorporated in a view of ourselves and our relation to the universe has become a matter of lively speculation, and it is a tribute to Bohr that, in his centenary year, a topic at the frontier of philosophical discussion still owes much to his original thinking and criticism.[9]

Recent decades have taught us that physics is a magic window. It shows us the illusion that lies behind reality — and the reality that lies behind illusion. Its scope is immensely greater than we once realized. We are no longer satisfied with insights only into particles, or fields of force, or geometry, or even space and time. Today we demand of physics some understanding of existence itself.

John A. Wheeler, *Quantum Theory and Measurement*, 1983

152

On Bohr's Views Concerning the Quantum Theory

David Bohm

The views that I shall express in my main contribution to this colloquium later on are in certain ways rather similar to those of Bohr, while in other ways they are basically different. It will perhaps be useful here to go into Bohr's notions in some detail, in order to bring out what these similarities and differences are.[1]

First, it should be said of Bohr that his writings show an unusually strong emphasis on *coherence and consistency of language*. In this regard, one of his most important contributions is that he saw, at least implicitly, that the *form* of a coherent communication has to be in harmony with its *content*. For example, if someone shouts "Be quiet!" the noisy, angry, and turbulent form of such speech excites and stirs up the hearer; and this evidently interferes with the communication of the intended content (which is the need for a situation of peace and calmness). What has generally not been noticed is that the way in which physicists have usually discussed quantum theory has a rather similar disharmony of form and content, which tends also to lead to confusion. Bohr's writings are characterized by a highly implicit and carefully balanced mode of saying things, which makes reading his work rather arduous but which is in harmony with the very subtle content of quantum theory.

One of the major sources of disharmony, and therefore confusion, in most discussions of the content treated in quantum theory is that there is a strong tendency to continue to use a basic *form* of discourse that was appropriate only in classical physics. This form is that of describing the world as a union of disjoint elements. Thus, most physicists (especially those who follow along lines initiated by John von Neumann) continue to talk about a "quantum system" as if it were constituted of interacting components (for example, particles) which exist separately from each other

David Bohm is Professor of Physics, Emeritus, at Birkbeck College, University of London. He was born and received his professional education in the United States, and has taught there as well as in Brazil and Israel. For many years he has studied the fundamentals of quantum theory, particularly with respect to the possible existence of "hidden variables" in the formulation of the theory.

153

and from the instrument that is used in "observing the quantum state of the system." What is most characteristic of Bohr is that he does not use such a language form at all. Indeed, his discussion implies that this mode of description is irrelevant in the quantum context. What is relevant instead is the *wholeness* of the form of the experimental conditions and the content of the experimental results (from which it follows that it will not be consistent in this context to talk in terms of the classical notion of a union of disjoint elements).

In order to bring out Bohr's point of view in more detail it will be useful to discuss Heisenberg's well known microscope experiment, but in a rather different way from that which has generally been adopted. To do this, we can begin by asking in terms of *classical physics* what it means to make measurements of position and momentum. We start, however, not as is usually done, with a *light* microscope but, rather, with an *electron microscope*.

In Figure 1 there is in the target an "observed particle" at *O*, assumed to have initially a known momentum (for example, it may be at rest, with zero momentum). Electrons of known energy are incident on the target, and one of these is deflected from the particle at *O*. It goes into the electron microscope, following an orbit that leads it to the focus at *P*. From here, the particle leaves a track, *T*, in a certain direction, as it penetrates the photographic emulsion.

Now, the *directly observable results* of this experiment are the position *P* and the direction of the track, *T*. But, of course, these *in themselves* are of no interest. It is only by knowing the *experimental conditions* (the structure of the microscope, the target, the energy of the incident beam of electrons) that the experimental results can become relevant in the context of a physical inquiry. With the aid of an adequate description of these conditions, one can use the experimental results described above to make *inferences* about the position of the "observed particle" at *O*, and about the momentum transferred to it in the process of deflecting the incident electron. As a result, one "knows" both the position and the momentum of this particle at the time of deflection of the incident electron.

All this is quite straightforward in *classical* physics. Heisenberg's novel

From the beginning, the attitude toward the apparent paradoxes in quantum theory was characterized by the emphasis on the features of wholeness in the elementary processes, connected with the quantum of action. While so far it had been clear that energy content and other invariant quantities could be strictly defined only for isolated systems, Heisenberg's analysis revealed the extent to which the state of an atomic system is influenced during any observation by the unavoidable interaction with the measuring tools.

Niels Bohr, "The Genesis of Quantum Mechanics," 1962

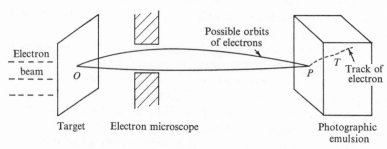

Fig. 1

step was to consider the implications of the "quantum" character of the electron that provides the "link" between the *experimental results* and *what is to be inferred from these results*. The electron can no longer be described as being just a classical particle. Rather, it also has to be described in terms of a "wave," as shown in Figure 2. Electron waves incident on the target are said to be diffracted by the atom at O, then pass through the microscope, where they are also diffracted, and are focused at the photographic emulsion. Here, they are said to determine only the "probability" that a track T begins at P and goes in a certain direction.

However, as was implicit in Heisenberg's discussion (and as was later brought out more explicitly by Bohr), this whole situation involves something radically new and not "coherent" with classical notions. Heisenberg tried to express this novelty by saying that both the position x and momentum p of the "electron link" between O and PT are "uncertain," the extent of this uncertainty being measured by the "uncertainty relationship" $\Delta x \Delta p \geq h$. But this involves a very significant kind of disharmony between the form of the language and the content to which Heisenberg implicitly intended to call attention. The form of the language implies that the "link" electron actually has a definite orbit that is, however, not precisely known to us. Bohr gave a thorough and consistent discussion of this whole situation, which made it clear that the orbit of the electron is not "uncertain" but, rather, that it is what he called *ambiguous*.[2] Unfortunately, even this word does not give a very clear notion of what is meant here. Perhaps one could say instead that both the notion of a particle following a well defined orbit (whether known or unknown) and the notion of a wave following a similarly well defined wave equation are not *relevant* to the "quantum" situation. What we have to deal with here is a radically new form of description that is incompatible with either of the old forms.

Now, because of the irrelevance in the "quantum" context of the description of the "electron link" in terms of well defined particle orbits or in terms of well defined wave motions, it followed that from the observed results of an experiment, one could no longer make inferences of unlimited precision about the observed object. But something more also followed, the

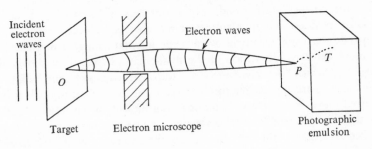

Fig. 2

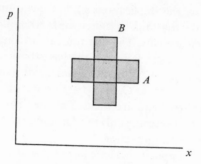

Fig. 3

very deep and far-reaching significance of which most physicists tended to overlook. To see what this is, note that from a particular set of experimental conditions (as determined by the structure of the microscope, for example), we could in some rough sense say that the limits of relevance of the classical description of the "observed object" are indicated by a certain cell in the *phase space* of this object, denoted by A in Figure 3. If, however, we had different experimental conditions (for example, a microscope of another aperture, or electrons of different energy), then these limits would be indicated by *another* cell in this phase space, indicated by B in the diagram. Both cells would have the same area, h, but their "shapes" would be different.

Now, in the corresponding discussion of the classical situation, it is possible to say that the experimental results do nothing more than permit inferences about an observed object, which exists separately and independently, in the sense that it can consistently be said to "have" these properties whether or not it interacts with anything else (such as an observing apparatus). A description of the experimental conditions is needed to permit these inferences to be carried out properly, but this description is in no way needed for saying what is meant by the properties of the observed object — that is, position and momentum.

However, in the "quantum" context the situation is very different. Here, certain relevant features of what is called the observed particle (the "shapes" of the cells in phase space) cannot properly be described except in conjunction with a description of the experimental conditions. Nor can one say that the "shapes" correspond only to our lack of knowledge about the precise position and momentum of the observed object, considered as separate and disjoint from the overall experimental arrangement. Indeed, as a more extensive discussion of the mathematical formalism shows, the region in which "the wave function of the observed object" (and its Fourier coefficients) are appreciable corresponds to the "shape" of the cell, as discussed above. But this wave function is coordinated, or "correlated," to that of the "link electron" in such a way that one has no meaning without the other. As a result, "the wave function of the observed object," which gives the fullest possible formal means of determining averages of physical properties, cannot be discussed relevantly apart from the experimental conditions, which provide a necessary context for a treatment of the wave function of the "link electron."

Thus, the mathematical formalism contains a reflection of the general situation with regard to "shapes" of cells in phase space that has been described here in informal terms. Therefore, the description of the experimental conditions does not drop out as a mere intermediary link of inference, but remains amalgamated with the description (both formal and informal) of what is called the observed object. This means that the "quantum" context calls for a new kind of description, which does not make use of the potential or actual separability of "observed object" and "observing apparatus." Instead, as has been indicated earlier, the form of the experimental conditions and the content of the experimental results now have to be one whole, in which analysis into disjoint elements is not relevant.

The irrelevance of such an analysis (for example, in terms of the notion that the "observed object is disturbed by the observing apparatus") was brought out in a very forceful way in the famous discussion between Einstein, Podolsky, and Rosen[3] and Bohr.[4] This showed that such wholeness of description is needed even when observations are carried out very far from each other in space and under conditions in which one would have said in terms of classical physics that no mechanical or dynamical contact or interaction between these observations could be possible.

In a context in which the detailed description of the "observed object" is not relevant, so that the "cells" in phase space could be replaced by points, however, the "shapes" of the cells would not matter, Thus, in such a context, all significant aspects of the "observed object" could meaningfully be described without bringing in a description of the experimental conditions. Thus, one could consistently use the traditional notion that the object can be discussed in terms of a "state" or of properties, which need not refer in any essential way to anything else at all. Therefore, the description in terms of a potential or actual disjunction between observed object and observing apparatus is a relevant simplification, in a context in which the fine details of the "quantum" description do not matter. But where the details are significant, this simplification cannot properly be carried out, and one has to return to a consideration of the wholeness of the total experimental situation.

What is meant here by "wholeness" could be indicated in a somewhat informal and metaphoric way by calling attention to a pattern (for example, the pattern in a carpet). Insofar as what is relevant *is* the pattern, it has no meaning to say that different parts of such a pattern (such as various flowers or trees that are to be seen in the carpet) are disjoint objects in interaction. Similarly, in the "quantum" situation, terms like "observed object," "observing instrument," "experimental conditions," and "experimental results" are just aspects of a single overall "pattern" that are, in effect, abstracted and "pointed out" or "made relevant" by our mode of discourse.

The belief in an external world independent of the perceiving subject is the basis of all natural science. Since, however, sense perception only gives information of this external world or of "physical reality" indirectly, we can only grasp the latter by speculative means. It follows from this that our notions of physical reality can never be final. We must always be ready to change these notions — that is to say, the axiomatic structure of physics — in order to do justice to perceived facts in the most logically perfect way.

Albert Einstein,
The World as I See It, 1934

157

Thus, it has no meaning to say, for example, that there is an "observed object" that interacts with the "observing instrument."

Thus far, Bohr's views are in general harmony with those adopted in my discussion of "hidden variables."[5] But now we come to an important difference between Bohr's views and my own. For Bohr went on to say that the terms of discussion of the experimental conditions and of the experimental results were *necessarily* those of "everyday language," suitably "refined" where necessary, so as to take the form of classical dynamics. It was apparently Bohr's belief that this was the only possible language for the *unambiguous communication* of the results of an experiment. He saw, moreover, that this kind of communication could be consistent with all that is meant by the term "quantum" only if it were further ruled that the experimental conditions needed for the precise definition of one of a pair of "complementary" variables (for example, position) are not compatible with those needed for the precise definition of the other member of the pair (for example, momentum). Since in classical physics *both* are required for the prediction of the future behavior of a "system," it follows that classical theory is no longer valid as a means of making *inferences* of this kind. Rather, it is now only a source of terms of *description of the experimental phenomena* ("position," "momentum," and so on).

Both the experimental conditions and the experimental results are, as in classical physics, to be described in terms of positions and momenta of various objects and parts of objects that make up the total experimental arrangement. But (as indicated in the discussion of the microscope experiment) the connections between these conditions and results and the inferences that are to be drawn from them must now be obtained from the quantum algorithm, with its purely statistical interpretation. So, in a certain sense, Bohr takes quantum theory to be a kind of "generalization" of classical theory. What is to be observed is always described in classical language (regarded here as a refinement of ordinary "everyday" language), but the generalization consists in replacing the classical algorithm (the differential equation applying to individual systems) with a quantum algorithm (matrix theories applying only to statistical ensembles).

Because the experimental facts actually did disclose a statistical fluctuation of results, in accordance with what was to be inferred from the quantum algorithm, Bohr concluded that his general mode of using language was indeed relevant to the facts that were available to him at the time. But he went further than this. Since the classical language was supposed to be the only possible means of unambiguous communication, and since the terms of this language could not consistently be defined together, he also concluded that, if the quantum algorithm is used for making statistical inferences in the usual way, it is impossible to find *any unambiguous language at all* that could treat the order of occurrence of these statistical fluctuations as relevant. Therefore it would necessarily be a source of confusion merely to entertain the notion of "hidden variables" in terms of which these contingent fluctuations would be revealed unambiguously in a new field of novel orders of necessity. In this way Bohr was led to the conclusion that the "quantum" implies absolute contingency — that is, the necessity for "complete randomness" in the results of what are called individual experiments.

On the other hand, there is a consistent way of doing just what Bohr would have termed meaningless and irrelevant, in a theoretical language form that has a certain similarity to that of Bohr. I have described this method in another paper.[6]

A key difference between these forms is, however, that in terms of the language of my paper, one can discuss a possible new kind of significance for the *order* of successive operations (that is, "measurements"). According to current quantum theory, this order has to be "random." Indeed, there is no room, either in the formalism or in the informal language of the theory, even to talk about a "nonrandom" order of this kind. Moreover, as one can see, the particular order determined by the contingent parameters (that is, "hidden variables") also could not be incorporated into any classical theory. Thus, if this order is significant, one will have to describe the *experimental results themselves* (and, more generally, the experimental conditions as well) in terms of a new language form that is neither "classical" nor "quantum."

What is called for, in my view, is therefore a movement in which physicists freely explore novel forms of language, which take into account Bohr's very significant insights but which do not remain fixed statically to Bohr's adherence to the need for classical language forms, limited by the quantum algorithm, in the description of experimental conditions and experimental results.

Some preliminary steps of this kind have been suggested elsewhere.[7] It seems to me that the work done thus far clearly reveals the existence of considerable scope for such exploratory experimentation with new language forms, and that such experimentation shows genuine possibilities for being fruitful.[8]

Waves and Particles: 1923–1924

John C. Slater

John C. Slater was born in 1900 and was deeply involved in the early development of modern quantum theory. His name is particularly associated with the application of quantum theory to molecular and solid-state physics. He was head of the Physics Department at the Massachusetts Institute of Technology from 1930 to 1951. He died in 1976.

The scientific event of 1923 was the discovery of the Compton effect by Arthur Compton, who was then at Washington University, St. Louis, and later at the University of Chicago[1] . . . Here we had what looked like definite experimental evidence of the correctness of the corpuscular theory of light, a theory which had had some followers ever since the days of Newton. But we had equally definite experimental evidence of the correctness of the wave theory, which dated back to Huygens: the wave theory had to be used to explain the interference and diffraction effects, whose reality was observed in X-ray diffraction just as much as in ordinary optics. Which theory was right? . . . The straightforward deduction from everything that was known was that both theories were to be used simultaneously, one for one type of phenomena, the other for others.

This was not as absurd a situation as it seemed at first. A number of scientists — W. F. G Swann, among others — had suggested that the purpose of the electromagnetic field was not to carry a continuously distributed density of energy, but to guide the photons in some manner. This was the point of view which appealed to me, and during my period at the Cavendish Laboratory in the fall of 1923, I elaborated it. In electromagnetic theory one finds an energy density, the electrical-energy density being proportional to the square of the electric field while the magnetic-energy density is proportional to the square of the magnetic field . . . It is perfectly possible to assume that this energy density serves to determine the probability of finding a photon at a given point in space: the probable number of photons per unit volume times the energy $h\nu$ of a photon must equal the continuous energy density as calculated by electromagnetic theory. I was willing, as were many other scientists, to accept this type of

160

relation between the continuous electromagnetic wave and the point carriers of radiant energy, the photons.

One had, then, to assume the existence of these electromagnetic waves, but there was a great difficulty associated with them. Bohr's theory of the hydrogen atom assumed the existence of electronic stationary states, and assumed that radiation was emitted when the atom jumped from an upper state of energy E_2 to a lower state of energy E_1. The emitted energy $E_2 - E_1$ was assumed to be the energy of the photon $h\nu$, which therefore determined the frequencies of the emitted radiation in terms of the energy levels E_2 and E_1. The Bohr-Sommerfeld quantum conditions, which had been known in their simplest form since 1913, and in the more sophisticated form suggested by Sommerfeld since 1916, predicted the energy levels, and the energy differences gave frequencies which agreed with the known spectrum of hydrogen.

All of this was simple enough, but the difficulty was with the assumption that the radiation was emitted in the form of a photon at the instant the atom jumped from the state E_2 to the state E_1. Any student of physics or of mathematics knows that a wave-train of finite length has a frequency spectrum which is not strictly monochromatic, but which instead has a frequency breadth $\Delta\nu$ which is of the order of magnitude of $1/T$, where T is the length of time during which the train is emitted. I had first learned this years earlier, through reading A. A. Michelson's fascinating book *Light Waves and Their Uses*, which my father had in his library. If this time is much longer than the period of oscillation, which is $1/\nu$, the breadth $\Delta\nu$ will be very small compared to the frequency ν. The observed sharpness of spectral lines shows that this must be the case. The experiments are consistent with emitted wave-trains which have perhaps the order of 10^5 or more waves in the train. How, I asked myself, could a physicist with the insight of Bohr have suggested that the radiation was emitted instantaneously? Surely it must have taken long enough for 10^5 waves to be emitted.

A good deal was known even then about the lifetimes of atoms in their stationary states. An atom is excited to a level above the ground state, stays in that state for a while, and then falls back to the ground state. The interesting fact is that the lifetime in the excited state is of the same order of magnitude as the time required to emit the 10^5 waves, more or less, which are needed to produce the observed sharpness of the spectral lines. The situation thus appeared to me to be perfectly clear. It must be that all the time an atom is in an excited state, it must be emitting electromagnetic waves of all the frequencies corresponding to transitions to lower states which would be allowed by Bohr's theory. The intensities of these waves would have to be determined by Bohr's correspondence principle, which had been worked out some years earlier, and which tied together the quantum probability of emission with corresponding amplitudes in the classical oscillation of the electron in the atom. One would not only have these spontaneous oscillators producing the waves, but also induced oscillations

Supposing the identity of X rays and light to be established, the supposition is this, I take it. The energy travels from point to point like a corpuscle: the disposition of the lines of travel is governed by a wave theory. Seems pretty hard to explain, but that is surely how it stands at the moment.

W. H. Bragg to Ernest Rutherford, 18 January 1912

produced by external radiation falling on the atom, which by interference with the external radiation would result in absorption of radiation according to the electromagnetic theory.

These electromagnetic waves, in my view, would not directly carry energy, as they would in the classical electromagnetic theory, but would be connected with the probability of finding photons at the given point . . .

This represents substantially the view I had when I went to Copenhagen at Christmastime in 1923. While at the Cavendish Laboratory in the fall of that year, I had talked over the ideas with Ralph Fowler, who had liked them, and I had put them on paper not only in memoranda of my own but in a letter to Edwin Kemble. As soon as I discussed the ideas with Bohr and Kramers, I found them enthusiastic about the idea of the electromagnetic waves emitted by oscillators during the stationary states — they at once coined the name "virtual oscillators" for these. But to my consternation, I found that they completely refused to admit the real existence of photons. It had never occurred to me that they would object to what seemed like so obvious a deduction from many types of experiments. The result was that they insisted on our writing a joint paper in which the electromagnetic field was described as having a continuously distributed energy density whose intensity determined the probability of transition of an atom from one stationary state to another. One had then to assume only a statistical conservation of energy between the continuously distributed energy density in the electromagnetic field and the quantized energy of the atoms. They grudgingly allowed me to send a note to *Nature* indicating that my original idea had included the real existence of the photons, but that I had given up the notion at their instigation. This conflict, in which I acquiesced to their point of view but by no means was convinced by any arguments they tried to bring up, led to a great coolness between me and Bohr, which was never completely removed . . .

It was astonishing to see how rapidly my point of view was vindicated. Various experimentalists pointed out that if the intensity of the electromagnetic field determined only the probability of transition, there was no good reason to expect [in the Compton effect] that the recoil electron, whose recoil was given by probability arguments, would come off at the same time as the scattered photon, which also came off by probability arguments. In 1925 Bothe and Geiger, and slightly later Compton and Simon, set up ingenious experiments with counters to measure simultaneously the recoil electron and the scattered photon, and they showed clearly that the electron and photon actually appeared simultaneously. This settled the argument, and I believe that no one continued to doubt the real existence of photons, conservation of energy between photons and atoms, and the probability relation between the motion of photons and the intensity of electromagnetic waves.

An attempt to encompass individual atomic reactions within the framework of classical radiation theory had been made in collaboration with Kramers and Slater. Although at first we encountered difficulties regarding the strict conservation of energy and momentum, these investigations led to further development of the notion of virtual oscillators as the connecting link between atoms and radiation fields.

Niels Bohr, "The Genesis of Quantum Mechanics," 1962

Reminiscences from 1926 and 1927

Werner Heisenberg

During the first few months of 1926, Göttingen first became familiar with the work of the Viennese physicist Erwin Schrödinger, who was approaching atomic theory from an entirely fresh side.[1] The year before, Louis de Broglie in France had drawn attention to the fact that the strange wave-particle dualism which, at the time, seemed to prevent a rational explanation of light phenomena might be equally involved in the behavior of matter — for instance, of electrons. Schrödinger developed this idea further and, by means of a new wave equation, formulated the law governing the propagation of material waves under the influence of an electromagnetic field. In Schrödinger's model, the stationary states of an atomic shell are compared with the stationary vibrations of a system — for instance, a vibrating string — except that all the magnitudes normally considered as energies of the stationary states are treated as frequencies of the stationary vibrations. The results Schrödinger obtained in this way fitted in very well with the new quantum mechanics, and Schrödinger quickly succeeded in proving that his own wave mechanics was mathematically equivalent to quantum mechanics; in other words, that the two were but different mathematical formulations of the same structures. Needless to say, we were delighted by this new development, for it greatly strengthened our confidence in the correctness of the new mathematical formulation. Moreover, Schrödinger's procedure lent itself readily to the simplification of calculations that had severely strained the powers of quantum mechanics.

Unfortunately, however, the physical interpretation of the mathematical scheme presented us with grave problems. Schrödinger believed that, by associating particles with material waves, he had found a way of clearing the obstacles that had so long blocked the path of quantum theory. Accord-

Werner Heisenberg was born in 1901 and began a teaching career at the University of Göttingen in 1924. Although he went to Copenhagen in 1926 and during that time became very close to Bohr, most of his working life was spent in Germany, primarily at Leipzig and Göttingen. He received the Nobel Prize for physics in 1932 for his central role in the creation of quantum mechanics. He died in 1976.

ing to him, these material waves are fully comparable to such processes in space and time as electromagnetic or sound waves. Such obscure ideas as quantum jumps would completely disappear . . . Thus, when an atom passes from one stationary state to the next, we can no longer say that it changes its energy suddenly and radiates the difference in the form of an Einsteinian light quantum. Radiation is the result of quite a different process — namely, of the simultaneous excitation of two stationary material vibrations whose interference gives rise to the emission of electromagnetic waves (for example, light). This hypothesis seemed to me too good to be true, and I mustered what arguments I could to show that discontinuities are a fact of life, however inconvenient . . .

Toward the end of the 1926 summer term, Sommerfeld invited Schrödinger to address the Munich seminar . . . During the subsequent discussion, I raised a number of objections [to Schrödinger's own interpretation of quantum mechanics] and, in particular, pointed out that Schrödinger's conception would not even help explain Planck's radiation law. For this I was taken to task by Wilhelm Wien, who told me rather sharply that while he understood my regrets that quantum mechanics was finished, and with it all such nonsense as quantum jumps, the difficulties I had mentioned would undoubtedly be solved by Schrödinger in the very near future . . .

And so I went home rather sadly. It must have been that same evening that I wrote to Niels Bohr about the unhappy outcome of the discussion. Perhaps it was as a result of this letter that he invited Schrödinger to spend part of September in Copenhagen. Schrödinger agreed, and I, too, sped back to Denmark.

Bohr's discussions with Schrödinger began at the railway station and were continued daily from early morning until late at night. Schrödinger stayed at Bohr's house so that nothing would interrupt the conversations. And although Bohr was normally most considerate and friendly in his dealings with people, he now struck me as an almost remorseless fanatic, one who was not prepared to make the least concession or grant that he could ever be mistaken. It is hardly possible to convey just how passionate the discussions were, how deeply rooted the convictions of each man — a fact that marked their every utterance. All I can hope to do here is to produce a very pale copy of conversations in which two men were fighting for their particular interpretations of the new mathematical scheme with all the powers at their command.

Schrödinger: "Surely you realize that the whole idea of quantum jumps is bound to end in nonsense. You claim first of all that if an atom is in a stationary state, the electron revolves periodically but does not emit light, when, according to Maxwell's theory, it must. Next, the electron is said to jump from one orbit to the next and to emit radiation. Is this jump supposed to be gradual or sudden? If it is gradual, the orbital frequency and energy of the electron must change gradually as well. But in that case, how

You discuss the question of the explanation of radiation by means of beats or by means of difference tones in a very penetrating way that is also very instructive for me. The frequency discrepancy in the Bohr model, on the other hand, seems to me (and has indeed seemed to me since 1914) to be something so monstrous, that I should like to characterize the excitation of light in this way as really almost *inconceivable*.
Erwin Schrödinger to H. A. Lorentz, 6 June 1926

do you explain the persistence of sharp spectral lines? On the other hand, if the jump is sudden, Einstein's idea of light quanta will admittedly lead us to the right wave number, but then we must ask ourselves how precisely the electron behaves during the jump. Why does it not emit a continuous spectrum, as electromagnetic theory demands? And what laws govern its motion during the jump? In other words, the whole idea of quantum jumps is sheer fantasy."

Bohr: "What you say is absolutely correct. But it does not prove that there are no quantum jumps. It proves only that we cannot imagine them, that the representational concepts with which we describe events in daily life and experiments in classical physics are inadequate when it comes to describing quantum jumps. Nor should we be surprised to find it so, seeing that the processes involved are not the objects of direct experience."

Schrödinger: "I don't wish to enter into long arguments about the formation of concepts; I prefer to leave that to the philosophers. I wish only to know what happens inside an atom. I don't really mind what language you choose for discussing it. If there are electrons in the atom, and if these are particles — as all of us believe — then they must surely move in some way. I am not concerned with a precise description of this motion, but it ought to be possible to determine in principle how they behave in the stationary state or during the transition from one state to the next. But from the mathematical form of wave mechanics or quantum mechanics alone, it is clear that we cannot expect reasonable answers to these questions. The moment, however, that we change the picture and say that there are no discrete electrons, only electron waves or waves of matter, then everything looks quite different. We no longer wonder about the sharp lines. The emission of light is as easily explained as the transmission of radio waves through the aerial of the transmitter, and what seemed to be insoluble contradictions have suddenly disappeared."

Bohr: "I beg to disagree. The contradictions do not disappear; they are simply pushed to one side. You speak of the emission of light by the atom or more generally of the interaction between the atom and the surrounding radiation field, and you think that all the problems are solved once we assume that there are material waves but no quantum jumps. But just take the case of thermodynamic equilibrium between the atom and the radiation field — remember, for instance, the Einsteinian derivation of Planck's radiation law. This derivation demands that the energy of the atom should assume discrete values and change discontinuously from time to time; discrete values for the frequencies cannot help us here. You can't seriously be trying to cast doubt on the whole basis of quantum theory!"

Schrödinger: "I don't for a moment claim that all these relationships have been fully explained. But then you, too, have so far failed to discover a satisfactory physical interpretation of quantum mechanics. There is no reason why the application of thermodynamics to the theory of material waves should not yield a satisfactory explanation of Planck's formula as

Schrödinger's visit to Copenhagen in the fall of 1926 afforded a special opportunity for lively exchange of views. On this occasion, Heisenberg and I tried to convince him that his beautiful treatment of dispersion phenomena could not be brought into conformity with Planck's law of black-body radiation without expressly taking into account the individual character of the absorption and emission processes.

Niels Bohr, "The Genesis of Quantum Mechanics," 1962

The Heisenberg-Bohr tranquilizing philosophy — or religion? — is so delicately contrived that, for the time being, it provides a gentle pillow for the true believer from which he cannot very easily be aroused. So let him lie there. But this religion has so damned little effect on me that, in spite of everything, I say

    not: E and v,

    but rather: E or v.

Albert Einstein to Erwin Schrödinger, 31 May 1928

well — an explanation that will admittedly look somewhat different from all previous ones."

Bohr: "No, there is no hope of that at all. We have known what Planck's formula means for the past twenty-five years. And quite apart from that, we can see the inconstancies, the sudden jumps in atomic phenomena, quite directly — for instance, when we watch sudden flashes of light on a scintillation screen or the sudden rush of an electron through a cloud chamber. You cannot simply ignore these observations and behave as if they did not exist at all."

Schrödinger: "If all this damned quantum jumping were really here to stay, I should be sorry I ever got involved with quantum theory."

Bohr: "But the rest of us are extremely grateful that you did; your wave mechanics has contributed so much to mathematical clarity and simplicity that it represents a gigantic advance over all previous forms of quantum mechanics."

And so the discussions continued day and night. After a few days Schrödinger fell ill, perhaps as a result of his enormous effort; in any case, he was forced to keep to his bed, with a feverish cold. While Mrs. Bohr nursed him and brought in tea and cake, Niels Bohr kept sitting on the edge of the bed talking at Schrödinger: "But you must surely admit that . . ." No real understanding could be expected since, at the time, neither side was able to offer a complete and coherent interpretation of quantum mechanics. For all that, we in Copenhagen felt convinced toward the end of Schrödinger's visit that we were on the right track, though we fully realized how difficult it would be to persuade even leading physicists that they must abandon all attempts to construct perceptual models of atomic processes.

During the last few months of 1926 the physical interpretation of quantum mechanics was the central theme of all conversations between Bohr and myself. I was then living on the top floor of the institute, in a cozy little attic flat with slanting walls, and windows overlooking the trees at the entrance to Faelled Park. Bohr would often come into my attic late at night, and we would construct all sorts of imaginary experiments to see whether we had really grasped the theory. In doing so, we discovered that the two of us were trying to resolve the difficulties in rather different ways. Bohr was trying to allow for the simultaneous existence of both particle and wave concepts, holding that, though the two were mutually exclusive, both together were needed for a complete description of atomic processes. I disliked this approach. I wanted to start from the fact that quantum mechanics as we then knew it already imposed a unique physical interpretation of some magnitudes occurring in it — for instance, the time averages of energy, momentum, fluctuations, and so forth — so that it looked very much as if we no longer had any freedom with respect to that interpretation. Instead, we would have to try to derive the correct general interpretation by strict logic from the ready-to-hand, more special interpretation.

For that reason, I was — certainly quite wrongly — rather unhappy

I shall take care above all that you remain master of your own actions to the greatest possible extent, and especially that, at those times over and above the "official" periods dedicated to the Physical Society, you have the opportunity to withdraw and to occupy yourself as you see fit. I know from experience how pleasant it often is to have a possibility of this kind.

Max Planck to Erwin Schrödinger, 4 June 1926

about a brilliant piece of work Max Born had done in Göttingen. In it, he had treated collisions by Schrödinger's method and assumed that the square of the Schrödinger wave function measures, in each point of space and at every instant, the probability of finding an electron in this point at that instant. I fully agreed with Born's thesis as such, but disliked the fact that it looked as if we still had some freedom of interpretation; I was firmly convinced that Born's thesis itself was the necessary consequence of the fixed interpretation of special magnitudes in quantum mechanics. This conviction was strengthened further by two highly informative mathematical studies by Dirac and Jordan.

Luckily, at the end of our talks, Bohr and I would generally come to the same conclusions about particular physical experiments, so that there was good reason to think that our divergent efforts might yet lead to the same result. On the other hand, neither of us could tell how so simple a phenomenon as the trajectory of an electron in a cloud chamber could be reconciled with the mathematical formulations of quantum or wave mechanics. Such concepts as trajectories or orbits did not figure in quantum mechanics, and wave mechanics could be reconciled with the existence of a densely packed beam of matter only if the beam spread over areas much larger than the diameter of an electron . . .

The Niels Bohr Institute in the 1920s.

167

The Birth and Growth of Quantum Mechanics

Niels Bohr and Werner Heisenberg in conversation ("Ja, ja, Heisenberg, aber . . .") at the Copenhagen Conference, 1934.

Bohr decided in February 1927 to go skiing in Norway, and I was quite glad to be left behind in Copenhagen, where I could think about these hopelessly complicated problems undisturbed. I now concentrated all my efforts on the mathematical representation of the electron path in the cloud chamber, and when I realized fairly soon that the obstacles before me were quite insurmountable, I began to wonder whether we might not have been asking the wrong sort of question all along. But where had we gone wrong? The path of the electron through the cloud chamber obviously existed; one could easily observe it. The mathematical framework of quantum mechanics existed as well, and was much too convincing to allow for any changes. Hence, it ought to be possible to establish a connection between the two, hard though it appeared to be.

It must have been one evening after midnight when I suddenly remembered my conversation with Einstein, and particularly his statement "It is the theory which decides what we can observe." I was immediately convinced that the key to the gate that had been closed for so long must be sought right here. I decided to go on a nocturnal walk through Faelled Park and to think further about the matter. We had always said so glibly that the path of the electron in the cloud chamber could be observed. But perhaps what we really observed was something much less. Perhaps we merely saw a series of discrete and ill-defined spots through which the electron had passed. In fact, all we do see in the cloud chamber are individual water droplets which must certainly be much larger than the electron. The right question should therefore be: Can quantum mechanics represent the fact that an electron finds itself approximately in a given place and that it moves approximately with a given velocity, and can we make these approximations so close that they do not cause experimental difficulties?

A brief calculation after my return to the institute showed that one could indeed represent such situations mathematically, and that the approximations are governed by what would later be called the uncertainty principle of quantum mechanics: the product of the uncertainties in the measured values of the position and momentum cannot be smaller than Planck's constant. This formulation, I felt, established the much-needed bridge between the cloud chamber observations and the mathematics of quantum mechanics. True, it had still to be proved that any experiment whatsoever was bound to set up situations satisfying the uncertainty principle, but this struck me as plausible a priori, since the processes involved in the experiment or the observation had necessarily to satisfy the laws of quantum mechanics. On this presupposition, experiments are unlikely to produce situations that do not accord with quantum mechanics. "It is the theory which decides what we can observe." I resolved to prove this by calculations based on simple experiments during the next few days.

Here, too, I was helped by the memory of a conversation I had once had with Burkhard Drude, a fellow student at Göttingen. When discussing the difficulties involved in the concept of electron orbits, he had said that it ought to be possible, in principle, to construct a microscope of extraordinarily high resolving power in which one could see or photograph the electron paths inside the atom. Such a microscope would not, of course, work with ordinary light rays, but perhaps with gamma rays. Now this ran counter to my hypothesis, according to which not even the best microscope could cross the limits set by the uncertainty principle. Hence I had to demonstrate that the principle was obeyed even in this case. This I managed to do, and the proof strengthened my confidence in the consistency of the new interpretation . . .

Then Niels Bohr returned from his skiing holiday, and we had a fresh round of difficult discussions. For Bohr, too, had pursued his own ideas on wave-corpuscle dualism. Central to his thought was the concept of complementarity, which he had just introduced to describe a situation in which it is possible to grasp one and the same event by two distinct modes of interpretation. These two modes are mutually exclusive, but they also complement each other, and it is only through their juxtaposition that the perceptual content of a phenomenon is fully brought out. At first, Bohr raised a number of objections against the uncertainty principle, which he probably considered too special a case of the general rule of complementarity. But he soon afterward realized — manfully assisted by the Swedish physicist Oskar Klein, who was also working in Copenhagen — that there was no serious difference between the two interpretations, and that all that mattered now was to represent the facts in such a way that despite their novelty they could be grasped and accepted by all physicists.

The matter was thrashed out in the autumn of 1927 at two physics conferences: the General Physics Congress in Como, at which Bohr gave a comprehensive account of the new situation, and the Solvay Congress in Brussels. In accordance with the wishes of the Solvay Foundation, the latter was attended by a small group of specialists eager to discuss the problems of quantum theory in detail. We all stayed at the same hotel, and the keenest arguments took place not in the conference hall but during the hotel meals. Bohr and Einstein were in the thick of it all. Einstein was quite unwilling to accept the fundamentally statistical character of the new quantum theory. Needless to say, he had no objections against probability statements whenever a particular system was not known in every last detail — after all, the old statistical mechanics and thermodynamics had been based on just such statements. However, Einstein would not admit that it was impossible, even in principle, to discover all the partial facts needed for the complete description of a physical process. "God does not throw dice" was a phrase we often heard from his lips in these discussions. And so he refused point-blank to accept the uncertainty principle, and tried to think up cases in which the principle would not hold.

The discussion usually started at breakfast, with Einstein serving us up another imaginary experiment by which he thought he had definitely refuted the uncertainty principle. We would at once examine his fresh offering, and on the way to the conference hall, to which I generally accompanied Bohr and Einstein, we would clarify some of the points and discuss their relevance. Then, in the course of the day, we would have further discussions on the matter, and, as a rule, by suppertime we would have reached the point where Niels Bohr could prove to Einstein that even his latest experiment failed to shake the uncertainty principle. Einstein would look a bit worried, but by next morning he was ready with a new imaginary experiment more complicated than the last, and this time, so he avowed, bound to invalidate the uncertainty principle. This attempt would fare no better by evening, and after the same game had been continued for a few days Einstein's friend Paul Ehrenfest, a physicist from Leyden in Holland, said: "Einstein, I am ashamed of you; you are arguing against the new quantum theory just as your opponents argue about relativity theory." But even this friendly admonition went unheard.

Once again it was driven home to me how terribly difficult it is to give up an attitude on which one's entire scientific approach and career have been based. Einstein had devoted his life to probing into that objective world of physical processes which runs its course in space and time, independent of us, according to firm laws. The mathematical symbols of theoretical physics were also symbols of this objective world and as such enabled physicists to make statements about its future behavior. And now it was being asserted that, on the atomic scale, this objective world of time and space did not even exist and that the mathematical symbols of theoretical physics referred to possibilities rather than to facts. Einstein was not prepared to let us do what, to him, amounted to pulling the ground from under his feet. Later in life, also, when quantum theory had long since become an integral part of modern physics, Einstein was unable to change his attitude — at best, he was prepared to accept the existence of quantum theory as a temporary expedient. "God does not throw dice" was his unshakable principle, one that he would not allow anybody to challenge. To which Bohr could only counter, "Nor is it our business to prescribe to God how he should run the world."[2]

At the Niels Bohr Institute in 1929

Nevill Mott

Sir Nevill Mott, Professor Emeritus at Cambridge University, was a pioneer in the application of quantum theory to solid-state physics. He also (with H. S. W. Massey) wrote a classic book on the theory of atomic collisions. He was Director of the Cavendish Laboratory, Cambridge, from 1954 to 1971, and shared the Nobel Prize for physics in 1977.

My parents had both studied physics at Cambridge, and ever since I first heard about it at school and at home I had wanted to do research on quantum theory, and quantum theory for me was the theory of Bohr. But when I had finished my exams and started research in the autumn of 1926, the new quantum mechanics had just burst upon the world and the first thing to do was to learn about it. Apart from Dirac, I don't think anyone in Cambridge understood it very well; there were no lectures on it, and so the only thing to do was to learn German and read the original papers, particularly those of Schrödinger and Born's *Wellenmechanik der Stossvorgänge* (Wave mechanics of collision processes). After applying Born's theory to some collision problems, I was ready in 1928 to go abroad, and thought first of Göttingen; but my supervisor R. H. Fowler advised Copenhagen, and how right he was! I remember the long train journey there from a holiday in the Alps, the train ferries, and the arrival at Bohr's institute in September 1929, where I stayed until Christmas.

Life at the institute in Blegdamsvej was quite unlike anything in Cambridge. In Cambridge in those days (not now!) the idea was that one worked in one's college rooms. In Copenhagen one worked at the institute; we were in and out of each others' rooms all day, as was Bohr too. Douglas Hartree was there; he kept his windows open, and Bohr did not like very much to talk in Hartree's "Nordpol." In Cambridge we stood in a queue to see Fowler when we needed him; in Copenhagen "How are you getting on?" was an almost daily question from Bohr as one met him casually in the building. And—I do not remember how often—Bohr would take his small group for Sunday walks in the forests north of the city, talking about everything under the sun. The group was a small one; people came in and

George Gamow (left) with Wolfgang Pauli on a Swiss lake steamer.

out, including the great ones, Heisenberg and Pauli; but Hartree and George Gamow were the only ones I remember as having been there all the time. Gamow was fresh from his triumph in explaining alpha particle decay, and I was rather jealous, having nothing to my credit but electron scattering according to Dirac's relativistic equation. "Ah — Motti — you must make a theory of the alpha particle," Gamow used to say after he had borrowed twenty-five øre from me to buy cigarettes. But I couldn't do that, and went on with scattering problems.

The first was the scattering of Dirac electrons with spin. The theory made it clear that the scattered wave would be partially polarized, and that this could be detected by a double scattering experiment. This intrigued

173

Bohr very much; I believe that he first wondered whether a fully polarized beam was possible. He showed me that a direct determination of the spin of a free electron could not be made; nor could a Stern-Gerlach experiment with free electrons — it would be frustrated by the uncertainty principle.[1] Following Bohr's thought on this was a wonderful education; it left me, I believe, with an instinctive faith, which no philosopher could shake, in the probability interpretation of quantum mechanics and with it the uncertainty principle, almost as if I'd learned it in my cradle.

My next venture in Copenhagen was the use of antisymmetrical wave functions in collision problems. The idea was in a paper of J. R. Oppenheimer's, on collisions of electrons with atoms, an extension of Bohr's theory. I went further on this by pointing out that the collisions between two free electrons would also be modified, and that, using an unpolarized target, the scattering through forty-five degrees of an unpolarized electron beam should give half as many scattered electrons as predicted by the Rutherford formula. It was Fowler who pointed out to me that alpha particles, having zero spin, obeyed *Bose statistics*, in contrast to the *Fermi-Dirac* law for electrons; and so for alpha particles scattered at forty-five degrees there should be *twice* as many as the Rutherford formula would predict. It was when I got back to Cambridge that Chadwick slowed down alpha particles to the point at which nuclear non-Coulomb forces became unimportant, and found that for the scattering of alpha particles by helium it was indeed so. For the electron case, I remember again how intrigued and at first skeptical Bohr was, and how we went round and round the subject till he was convinced.

After my months in Copenhagen, I went to Göttingen, which was disappointing; Born was unwell, most of the founders of quantum mechanics had left, and there, in contrast to Bohr's institute, one paddled one's own canoe. With my indifferent German at that time, it was hard to make contacts. Then I had a year at Manchester with W. L. Bragg and three years at the Cavendish Laboratory, living through the excitement of the discoveries of 1932. I do not remember any contacts with Bohr at that time. In 1933 I was appointed to a chair at Bristol, and turned to what we now call solid-state physics — not deliberately, but because experimental work in progress there was very susceptible to treatment by the quantum-mechanical theories of Bloch, Peierls, and Wilson, so admirably set out already in Bethe's article in the *Handbuch der Physik*. Apart from the war years, I have remained in solid-state science ever since. I remember a visit to Copenhagen — I think soon after the war, but it may have been before — during which Bohr asked me eagerly to tell him what great things were happening in solids. But we did not get around to this discussion.

Everything I've done in solid-state physics has resulted from the confidence Bohr gave me that simple quantum mechanics is usable, beautiful, and intuitively right.

Niels Bohr and the Physics of Simple Phenomena

H. B. G. Casimir

Niels Bohr was a daring and imaginative innovator, but many physicists of my generation see him first of all as a wise philosopher, pondering the very foundations of physical theory. They remember him either as tirelessly expounding his ideas, trying all the time — not always successfully — to reconcile clarity and truth, or, alternatively, as being lost in thought and hardly aware of the outside world. But during my stay in Copenhagen in 1929 and 1930, I got to know an entirely different Niels Bohr, a man who could thoroughly enjoy simple things: a long walk in the country, a bicycle ride, a sailing trip with friends, a skiing holiday, playing ball with his sons, felling trees in the pine grove near his country house.

He was fascinated by traditional crafts. For instance, he took me along to have a look at the digging of a deep well by a time-honored method. The well is lined with brick strengthened by wooden hoops. A ring is made above ground level, soil is excavated, and the ring goes down. A second ring is made on top of the first one; a man down in the pit excavates further; the sand is hoisted up and the two rings together go down. A third ring is added, and so on. It must be a somewhat scary experience to work deep down in the pit and to watch how the whole cylinder, which may finally be thirty meters or more in height, comes sliding down. I remember the face of the digger as he was hoisted up, laughing and shaking the sand from his hair, while Bohr asked him a number of questions. (I hope, but I am not certain, that I remember the technique correctly after all these years.) On another occasion, Bohr explained to me his admiration for the skill of the cooper. He was a pretty good handyman himself, but mainly on mechanical lines. As far as I know, he never tinkered with radio or other types of electronics.

He also liked simple physics, of which I shall give a few examples. They

Hendrik Casimir, a Dutch theoretical physicist, has spent most of his career with Philips, Eindhoven, where he was Director of Research for many years. He was President of the European Physical Society from 1972 to 1975 and is now retired. His book *Haphazard Reality* contains many fascinating memoirs of the development of modern physics, and in particular of Copenhagen in the 1930s.

175

may seem fairly trivial, but I think there existed all the same a connection between his enjoying them and his gift for elucidating profound problems by considering simple — though subtle — examples.

Close to Bohr's institute there is a body of water about three kilometers long and some 150 or 200 meters wide: the Sortedamsø. The footpath along its shore provided a modest but welcome opportunity for stretching one's legs. One rather windy evening Bohr drew my attention to the fact that the reflections of lanterns on the other side of the lake were lengthened to long, luminous ribbons. Why weren't they broadened as well? Of course, the answer is simple in principle, but I had some difficulty working out the details of the geometry involved without pen and paper. Sortedamsø is crossed by several bridges. One day when Bohr and I were passing over one of them he said, "Look, I'll show you a curious resonance phenomenon." The parapet of the bridge was built in the following way. Stone pillars, about four feet high and ten feet apart, were linked near their tops by stout iron bars (or, rather more likely, tubes) let into the stone. Halfway between each two pillars an iron ring was anchored in the stonework of the bridge and two heavy chains, one on each side, were suspended between shackles welded to the top bar close to the stone pillars and the ring. Bohr grasped one chain near the top bar and set it swinging, and to my surprise the chain at the other side of the top bar began to swing too. "A remarkable example of resonance," Bohr repeated. I was much impressed, but suddenly Bohr began to laugh. Of course resonance was quite out of the question; the coupling forces were extremely small and the oscillations were strongly damped. What happened was that Bohr, when moving the chain, was rotating the top bar, which was let into, but not fastened to, the stone pillars, and in that way he had moved the two chains simultaneously. I was crestfallen that I had shown so little practical sense, but Bohr consoled me, saying that Heisenberg had also been taken in; he had even given a whole lecture on resonance. That bridge became known in Bohr's institute as the resonance bridge. But the bridge has since been broadened, and the parapet is no longer there. This little story is instructive. It illustrates how Bohr, the depth of his thought notwithstanding, kept close to physical reality and that he had a keen appreciation of orders of magnitude. I shall return to this point in a moment.

Bohr's first piece of research, on the determination of surface tension by the study of jets of liquids, was an experimental investigation — Bohr had made most of the equipment with his own hands — but it contained a good deal of theory. Was it because of this early work that he enjoyed showing how a table-tennis ball can float on top (or rather, almost on top) of a jet of water shooting up from a fountain? One more aspect of hydrodynamics: Bohr liked to skip flat pebbles on the surface of a lake or stream and to emphasize the importance of giving adequate angular momentum to the pebble so that it would maintain its orientation.

As to angular momentum, many years later Bohr amused himself with

Wolfgang Pauli (left) and Niels Bohr studying the behavior of a Tippetop.

the so-called Tippetop, a little top of special design which you spin with your fingers and which—provided it is spinning sufficiently fast—will always end up rotating around the axis corresponding to the smallest moment of inertia and with its heaviest part up. (The same trick, as I was told, can be done with a hardboiled egg: spin it rapidly about a short axis while its long axis is horizontal and it will rise on its point, on its "narrow end.") It so happens that the thesis with which I obtained my doctorate at Leiden in 1931, but on which I had already been working at Copenhagen, also dealt with spinning tops. I think the general theorems on groups and their representations that I deduced in that paper appealed less to Bohr than did the Tippetop in later years. "How is this rotation business coming along?" he would ask me from time to time, but out of kindness rather than out of real interest.

Once we (I have forgotten who else was present; perhaps Gamow and Landau) discussed the famous one-way optical system proposed by Rayleigh in 1885. The device works as follows: Two *Nicol prisms* (today one would use sheets of Polaroid) are placed at an angle of 45 degrees, and between them there is a substance in a longitudinal magnetic field which gives rise to a *Faraday rotation* of 45 degrees. The light traveling in one direction will go right through; light going in the opposite direction will be absorbed. In order to explain the geometry, books were placed on tables at the appropriate angles, and Bohr, using his fountain pen to indicate the polarization vector, walked to and fro between these "Nicols." Bohr greatly admired the work of Rayleigh, especially that on the resolving power of optical instruments. It certainly influenced his way of dealing with uncertainty relations.

A feeling for orders of magnitude is characteristic of inventors. The successful inventors I have known have always had a number of rules of thumb that enabled them to make rapid estimates and thus kept their imagination within reasonable bounds. Bohr as a physicist was in this respect closer to the inventive engineer than many, more formalistic theorists. Why was it a reasonable approach in Bohr's early treatment of the hydrogen atom (and also later in quantum mechanics) first to neglect the interaction with the radiation field? Because that interaction is small. Why is it small? Because the fine structure constant, $\alpha = e^2/\hbar c$, is small — approximately $\frac{1}{137}$. In classical theory the emission of radiation leads to a reactive force

$$F_R = \frac{e^2}{c^3}\dddot{x} = \frac{e^2}{c^3}a\omega^3 = \frac{e^2}{a^2}\left(\frac{a}{\lambdabar}\right)^3,$$

where a is the dimension of the atom, ω the angular frequency of the electron in its orbit, and $\lambdabar$ the corresponding wavelength divided by 2π. In the simplest form of Bohr's theory of the hydrogen atom we have $a/\lambdabar = \alpha$, and in more elaborate forms it is of the same order of magnitude. So the force of radiative reaction is of the order of α^3 times the Coulomb force — that is, smaller by a factor of the order of a million. That argument returns time and time again in Bohr's discussions.

Bohr also made me aware of the hierarchy of lengths in atomic physics. The formula e^2/mc^2 represents the classical radius of the electron. Multiply by $1/\alpha$ (that is, by 137) and you get $\hbar/mc$, the *Compton wavelength* divided by 2π. Multiply once more by $1/\alpha$ and you find $\hbar^2/me^2$, the so-called Bohr radius. One more factor $1/\alpha$ leads to $\hbar^3c/me^4$, the Rydberg wavelength divided by 4π. And — a result seldom mentioned — one more factor 137 gives you one micrometer to within 1 percent (0.9937μ to be exact).

Bohr liked to derive approximate formulas by very simple arguments. I remember the following: An electron is flying past a nucleus with charge Ze. What, according to classical mechanics, is the deflection? Of course, the basic solution (a hyperbolic orbit under a $1/r^2$ force law) was obtained by Newton. Applied to the electron deflection problem, it gives the result that if θ is the total deflection, v_0 the initial velocity of the electron, and q the distance of the nucleus from the unperturbed trajectory, then we have

$$\tan\frac{\theta}{2} = \frac{Ze^2}{mv_0{}^2q}.$$

But Bohr liked to argue as follows. The electron will strongly experience the field of the nucleus during a time q/v_0, and during that time the force perpendicular to the trajectory is about Ze^2/q^2. So the transverse momentum imparted to the electron is about Ze^2/qv_0 and the angle of deflection must be about Ze^2/qmv_0^2. As it happens, the result is exact.

If you are looking for striking applications of the simple theorem that the change of momentum is equal to force multiplied by time, you will find them in Bohr's treatment of measurements. Take, for instance, his famous refutation of Einstein's proposal to beat the relation $\delta E \cdot \delta t > \hbar$ by weighing a box from which a light quantum is escaping at a time determined by a clock that opens a shutter.

When I came to Copenhagen for the first time, in the spring of 1929, I was at once initiated in this art of discussing *"Gedankenexperimente."* Bohr, at the conference that marked the beginning of my stay, gave a talk in which he showed that the magnetic moment of a free electron cannot be determined by a *Stern-Gerlach* type of experiment, because of the uncertainties imposed by quantum mechanics.[1] Consider, for instance, the following arrangement (see figure). One pole of the magnet is a sharp wedge. Its symmetry plane is the z, y plane; its edge is along the y-axis. The other pole has the same symmetry but is much blunter. We may even assume that it is flat. The electron is supposed to enter the magnetic field in the symmetry

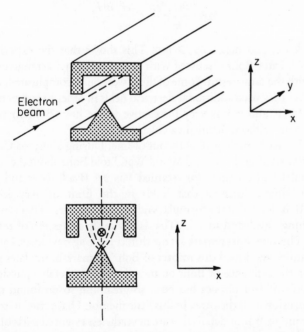

A Stern-Gerlach magnet arrangement for electrons. Actually, to prevent the electrons from being deviated sideways by the vertical magnetic field, a supplementary horizontal electric field or other counteracting agency would be required. In the lower diagram (end view) the cross in the circle indicates the position of the axis of the electron beam with respect to the magnet poles.

plane with a velocity parallel to the y-axis. The force on the magnetic moment will be given by

$$F_m = \frac{e\hbar}{2mc} \cdot \frac{\partial H_z}{\partial z}.$$

However, if there is a field component H_x, this will lead to a Lorentz force in the same direction:

$$F_l = \frac{ev}{c} H_x.$$

Now, according to *Maxwell's equations,*

$$\frac{\partial H_z}{\partial z} + \frac{\partial H_x}{\partial x} = 0.$$

Therefore,

$$\delta H_x = -\frac{\partial H_z}{\partial z} \cdot \delta x,$$

and, if we want to measure the spin, we must have

$$\delta x \cdot \frac{\partial H_z}{\partial z} \cdot \frac{ev}{c} < \frac{e\hbar}{mc} \frac{\partial H_z}{\partial z}$$

or $\delta x < \hbar/mv$, and hence $\delta p_x > mv$. This shows that the experiment is impossible. Later Bohr discussed with me several other arrangements but always with the same result. Some of them were quite complicated. As far as I remember they used additional electrical fields in order to compensate the Lorentz force, and it was very instructive to see how Bohr managed to analyze them without detailed calculations.

A final remark, perhaps slightly out of line. During a visit to Copenhagen a few years after the Second World War, I told Bohr about the results D. Polder and I had obtained for retarded *Van der Waals forces* and about a strikingly simple formula that holds in the limit of very long distances.[2] Bohr looked at the formula, said something like, "That is nice, that is something new," and added, "That is obviously a question of zero point energy." This one short remark set me thinking along new lines. I am afraid I have hardly mentioned this impact of Bohr's remark in my later publications. But the influence of Bohr on myself, on innumerable physicists, on the whole world of physics has been so great and so profound that one cannot mention it all the time. In this little memoir I have tried to tell about some seemingly minor details that are nevertheless essential parts of a much greater whole.

A Few Memories

Edward Teller

My earliest recollection of Niels Bohr goes back to my first visit to his institute in Copenhagen in 1929. At tea I was seated next to Bohr and, as a presumptuous novice, I began a conversation by speculating about the future. Said I, "Someday matrix mechanics will be the common way of thinking. This will indeed be a benefit because then the contradictions will disappear from the orderly discussions of science."

As I spoke, I noticed that Bohr's eyelids slowly closed. Since he appeared to have fallen asleep, I tried to finish my little speech as rapidly as possible. There then followed an interval of what seemed to me an eternity but which was probably only twenty seconds.

Bohr's lips moved slowly, and in a barely audible tone, with his eyes still closed, he whispered, "You might as well say we are not sitting here at all but merely dreaming it."

A year passed before I understood and came to agree with what Bohr had said. Yet today I would append a twofold postscript to the incident. First, what I suggested in 1929 came to pass. Among physicists, Bohr's ideas about classical measurements are rarely discussed. Perhaps, too many people take the mathematical formalism of quantum mechanics for granted, and too few of the new generation remember the deep and careful thought that went into describing the relationship between classical theory, with its strict causality, and the new insights. Bohr made a remarkable contribution to the balance that exists in the complementary concepts.

The second postscript relates to a conference which took place about thirty years later in the Pocono Mountains of New York State. The topic was the theory of elementary particles, and Bohr made a statement which may yet prove true. He said, "We will never understand anything until we have found some contradictions."

Edward Teller, born in Budapest, received his professional education in Germany and did theoretical work in many areas of atomic and nuclear physics. He emigrated to the United States in 1935 and has been associated mostly with the University of California. He is well known for his work in the application of nuclear energy to both peaceful and military purposes.

Today, in the arguments concerning the existence or nonexistence of quarks as independently extant particles, we are approaching contradictions. However, I doubt that we have as yet satisfied Bohr's standards for a contradiction.

Another of my memories is of a conference that was held in 1934 in Copenhagen, a meeting of philosophers which Bohr attended. Most of those gathered were positivists; and when Bohr presented his remarks, everyone agreed with him completely. The day after the conference, Bohr returned to the institute almost in despair. We were concerned, but the cause did not come out until Bohr explained. He said, "Whoever talks about Planck's constant and does not feel at least a little giddy obviously doesn't appreciate what he is talking about."

One final story is a purely personal one that took place at the institute in the early 1930s. Some of us, including Bohr, were having a discussion about the spectrum and states of molecular oxygen. Bohr had some opinions, the details of which I have now forgotten, but which were in obvious conflict with the facts that were known. In this special detailed case, I knew the situation and tried to explain it. Unfortunately, I could not do so to Bohr's satisfaction.

He began his objection: "Teller, of course, knows a hundred times more about this than I." With a lack of politeness occasionally seen among twenty-year-olds, I interrupted (with some difficulty): "That is an exaggeration."

Bohr instantly stopped and stared at me. After a pause, he declared, "Teller says I am exaggerating. Teller does not want me to exaggerate. If I cannot exaggerate, I cannot talk. All right. You are right, Teller. You know only ninety-nine times more than I do." He then proceeded with his original argument having dispensed with any possibility of further interruption.

I have never forgotten, nor have I often neglected to mention, Bohr's wisdom: if you cannot exaggerate, you cannot talk.

Let me close these remarks about the state of the theory by telling you of an exchange between Pauli and Bohr, after Pauli had given a colloquium on some ideas related to particle physics. Pauli to Bohr: "You will probably think that what I said is crazy." Bohr to Pauli: "Yes, but unfortunately it is not crazy enough."
Abraham Pais, "Particles, Fields, Quantum Mechanics," 1963

A Reminiscence from 1932

C. F. von Weizsäcker

In the New Year of 1932 Werner Heisenberg visited our family for about ten days in Oslo, where, since the early summer of 1931, my father had been the German ambassador.[1] I was studying in Leipzig, but had already visited my parents in Oslo in August 1931 and again at Christmas. To us, coming to Norway from Germany at the time of the economic crisis and the unemployment and the political agony of the Weimar Republic, Norway seemed an Isle of the Blessed, even though the inhabitants, the descendants of the Vikings and their bards, and the compatriots of Björnson, Ibsen, and Hamsun, did not always recognize that they lived in a paradise.

On 7 January 1932, Heisenberg and I returned to Leipzig. In Copenhagen he interrupted his journey for half a day to visit Niels Bohr. He telephoned in advance, asking if he might bring a student with him, and Bohr readily and cordially assented. That was my first meeting with him. Its consequences for me proved to be almost as important as my first meeting with Heisenberg five years earlier, also in Copenhagen. We sat in Bohr's study at the institute. Bohr wished to talk to Heisenberg about the philosophical problems of quantum theory. Their conversation lasted for three hours while I sat in silence. Later I wrote in my diary, "I have seen a physicist for the first time — he suffers as he thinks."

Bohr was forty-five years old; Heisenberg was thirty; I was nineteen. For me, Bohr was an old man. He was of medium height, perhaps two centimeters shorter than myself, with a slight stoop. There remained little of the former athlete in his appearance. His head, often bent thoughtfully and as if rather shyly, seemed to be split into two halves. There was his narrow high forehead under sparse gray hair that he would tug at while thinking, and from the furrowed brow one could see the enormous intensity of his thought. The lower part of the face was rather full, with thick lips and

Freiherr C. F. von Weizsäcker was educated in Berlin, Göttingen, and Leipzig. Among his chief academic appointments have been positions as Director of the Max Planck Institute at Göttingen and as Professor of Philosophy at Hamburg. He did important work in nuclear theory. He is now associated with the Max Planck Institut für Sozialwissenschaften at Starnberg, Germany.

183

In appearance, Niels Bohr reminded you of a peasant, with hairy hands and a big heavy head with bushy eyebrows. I still remember his eyes, which could hold you with all the power of the mind behind them; and then suddenly a smile would break over his face, turning it all into a joke. He was a yachtsman, he went skiing in the mountains of Norway, and I have seen him chop down trees, wielding a long-handled woodcutter's axe with the strength and precision of a professional.

O. R. Frisch,
What Little I Remember, 1979

rather pendulous cheeks; here he looked like the ordinary friendly Dane. The frequent shy smile that illuminated his face would pull the two halves together. His eyes? I must tell a story about them. A year later I was returning, in Bohr's company, from a skiing visit to Heisenberg. Bohr and I said goodbye in the Munich station although we were taking the same train to Berlin. He was traveling in a sleeping car in the front of the train, while I was sitting in third class at the back of the train, where I had learned to sleep quite well. At the Berlin station I had to go the whole length of the platform, and, arriving at the gates, I saw my mother in intimate conversation with Bohr; this amazed me, since they had never met before. Afterward she explained to me: "I was simply looking for you, but then a man approached whose eyes I just had to look at, and I remembered pictures of Bohr and spoke to him and it was indeed he." His eyes, deepset beneath bushy eyebrows, seemed to look straight at things and at the same time through them to an unfathomable distance. Yet with other people his eyes appeared shy and kind in a way which I had never seen elsewhere. I might also say a word about hands. Heisenberg was a slim blond young man and had the sinewy artistic hands of a pianist. Bohr had the rather fleshy, broad, strong, and sure hands of a woodcarver.

The words "I have seen a physicist for the first time" were a provocative comment on Heisenberg. With him I existed in a state of tension such as can only arise when one is very close to another person. In Berlin in April 1927, in a taxi, he told me of the uncertainty principle saying, "I think I have refuted the law of causality"; in that moment I decided to study physics to understand this. Now he was my tutor and my older friend. In all technical things in theoretical physics he was hopelessly superior to me; whenever he began to calculate, I stopped and waited for the result. But I found that he did not concern himself with the philosophical problems in physics for the sake of which I had been tempted by him to study the subject: neither the epistemological-ontological problems, nor the ethical ones concerning the intellectual dichotomy of life — to use Heisenberg's own words, the dichotomy between "the things which mean something and the things about which one can reach agreement." (For instance, music means something and on mathematics one can agree intellectually, yet they are very close to each other.) I am quite sure that he liked me because I made demands that he should face these problems, but his attitude to them was pedagogic and defensive. He said, "Physics is an honest trade; only after you have learned it have you the right to philosophize about it."

In this situation, meeting Bohr was a liberation for me. Here was a physicist, acknowledged to be great, who, unlike all other physicists I had come to know, did not dodge the painful consequences of his understanding. Others were very proud when they succeeded in proving something; but either they did not notice or did not know the meaning of what they had proved, or they invented an epistemology whose psychological purpose was not to notice it, or they split their life into two halves and in the other

184

half they made music or something of that sort. Bohr knew it. His concept of complementarity was invented in order to be able to speak about these things. This was so, for instance, when he talked of the exclusive nature of the relationship between the analysis of a concept and its immediate use. He compared thinking with a *Riemann surface* which can be made intelligible to a nonmathematician by the picture of a spiral staircase: one goes around the singular point (the axis of the staircase) and comes back to the same point on the surface but at a different level. This was an abstract theory that he could not easily formulate in words; it was at the same time an existential suffering. His stumbling way of talking, which has been so often described in a friendly but ironic way, would become less and less intelligible the more important the subject became, and this came from that suffering. He constantly appealed to the understanding of his partner in conversation, with very little hope of success but always with tireless optimism. On a later occasion he started a philosophical conversation with me by going to the blackboard and writing a single word: "Thinking." Then he turned to me and said, "I only wanted to say that I have written down something here which is quite different from writing down any other word."

What Bohr spoke about to Heisenberg on that day in 1932 I no longer know. I can only sense the smell of it. In the first six months of 1932, as is understandable in a nineteen-year-old, everything for me was emotionally confused, not just this one connection. But in the ups and downs of the wave, a question constantly appeared and reappeared like a lighted buoy in the ocean: What had Bohr meant? What must I understand to be able to tell what he meant and why he was right? I tortured myself on endless solitary walks. With my youthful ambitions to be omniscient, I believed, after I had met Bohr, that now or never I must make the breakthrough to my personal philosophy.

Around the twelfth of March that year I joined Heisenberg in his ski hut some half an hour's climb from Bayrischzell in southern Germany. That year only the two of us were there, but the following year Niels Bohr and his son Christian were there, together with Felix Bloch. It was for me the first chance to get acquainted with Bohr's philosophy of everyday life. The oft-told story of washing glasses I remember in this way: Bohr was looking proudly at some drinking glasses he had just washed, and said: "If one were to say to a philosopher that, using dirty water and a dirty cloth, one could clean dirty glasses, he would not believe it."

Our ski descent on that excursion I remember well. Heisenberg took eight minutes, Bloch took twelve, I took eighteen, and Bohr forty-five. In Danish woodlands one cannot learn alpine skiing, but Bohr never gave in.

During the Whitsun vacation of 1932 Heisenberg and I went into the Thüringer Wald to the village of Brotterode, high above Gotha. We enjoyed long walks at night and, in the morning, the views over the mountain peaks; also, we were working hard. In January 1932 Chadwick had discovered the neutron. This caused a small revolution in Heisenberg's expecta-

He utters his opinions like one perpetually groping and never like one who believes himself to be in possession of definite truth.
Albert Einstein, quoted in Harry Woolf, *Some Strangeness in the Proportion*, 1979

185

tion about the further development of physical theory. Bohr's fundamental realization in 1912 was that the explanation of the stability of atoms required not only a new model of the atom but also new basic laws of physics. In 1925, at the age of twenty-three, Heisenberg had given a definitive shape to these laws under the name of quantum mechanics. But both Bohr and Heisenberg expected rapid advances beyond this step, and believed then that the problem of how electrons could exist in the nucleus would require a new fundamental theory. Chadwick's discovery of the neutron led Heisenberg to the idea that the nucleus consisted only of protons and neutrons, which meant that the nucleus could be described by the existing quantum mechanics. Thus, the stability of the nucleus was to be explained just the other way round from the way in which, twenty years before, Bohr had explained the stability of the outer atom: not by new fundamental laws, but simply by a new model. This was something of a disappointment in the premature hope for a new great revolution.

My own research was on a much less interesting subject. I was torturing myself with a rather dull, and not very easy, piece of work for my doctoral thesis. One day I filled a scrap of paper with a sequence of sentences that I felt to be a direct inspiration. The first sentence that opened the gates was: "Consciousness is an unconscious act." Not until thirty years later did I learn where this inspiration had originated. In 1962 a student of mine, writing a dissertation about Bohr, told me: "It is to be found in William James, and in the winter of 1931–32 Bohr was constantly reading James."

I next saw Bohr again in September 1932. Under the silky blue September sky of Copenhagen, he regularly organized a scientific conference at his institute; from 1932 to 1938 I took part in these. They were scientifically the most productive meetings that I have ever attended, and the most human of all. Every year Bohr invited four to six of his closest friends and pupils to the conference, and asked each to bring along one or two colleagues. On the evening before the start of the conference, Bohr would invite the small circle of his friends to his flat and ask: "Well — what shall we talk about in the next few days?" So there were no prepared manuscripts, and everything had the highest degree of topicality; and because the conference in those good early years had twice as many hours as participants there was time to discuss every worthwhile problem.

At the 1932 conference Bohr gave a fundamental report on the current difficulties of atomic theory. I do not remember the content, but I do remember the frustration of the participants about the well known unintelligibility of his talk. With an expression of suffering, his head held to one side, he stumbled over incomplete sentences. Even the language that he used varied: it fluctuated between German, English, and Danish, and when the topic became quite important he mumbled with his hands in front of his face. We bad boys said he knew only three mathematical symbols: $\gg$, $\ll$, and $\approx$ ("much greater than," "much less than," and "just about the same"). That day for a time he had the chalk in his right hand and a sponge in his left; he would write down equations with the right hand and wipe them off almost at once with the left. Suddenly there sounded the forceful voice of his old friend Paul Ehrenfest: "Bohr! Give me the sponge!" With a rather pained smile Bohr handed the sponge to Ehrenfest, who kept it on his lap during the rest of the lecture.

But then when someone else — say, Heisenberg, Dirac, or Pauli — gave a report, Bohr would interrupt him with questions which were wrapped in the gingerbread of his helplessly amiable way of talking: "That is very interesting!" . . . "I don't intend to criticize; it is only to learn that I ask" . . . "We are in much greater agreement than you think." (With quite silly people he would just sit resignedly and say, "Oh, sehr, sehr . . . ") And then in the course of a couple of hours everything would become relentlessly clear.

I should like to insert here a recollection from a somewhat later time. At the 1933 Copenhagen conference the British physicist E. J. Williams reported on a method for making an approximate calculation of scattering cross-sections at high energies. There were long discussions of improvements on the Williams method, and several people took part, among them Lev Landau from Russia and myself. After all the others had left, Bohr asked me to write down the results of the discussion in a paper. He also wrote about it to Landau, who was quite inexhaustibly aggressive in oral discussion. Landau answered, after six weeks, with a letter which was one sentence long. Bohr showed me the letter and said, "Landau doesn't seem to write as much as he speaks." I wrote up the paper in the course of some weeks and gave it to Bohr's secretary. Bohr himself could not be seen very much, since at that time he was very absorbed and busy in advisory work for the Danish government in tireless attempts to help German refugees.

After a fortnight an interview with Bohr was arranged. He was late and looked infinitely tired. He pulled the paper from the pile and said "Oh, very good, very good; that is a very nice piece of work, now everything is clear . . . I hope that you will publish it soon!" I thought to myself, "The poor man! He probably has had almost no time to read the paper." He continued: "Just to clarify something, what is the meaning of the formula on page 17?" I explained it to him. Then he said "Yes, I understand that, but then the footnote on page 14 must mean . . . so and so." "Yes — that is what I meant." "But then . . . " And so it went on. He had read everything. After an hour he was getting fresher all the time, and I came to a point where I had difficulty with an explanation. After two hours he was flourishingly fresh and complete master of the situation, full of naive enthusiasm, while I felt that I was getting tired and was being driven into a corner. In the third hour, however, he said, triumphantly but without any trace of malice, "Now I understand it, now I understand the point . . . The point is that everything is exactly the opposite of what you said — that's the point!" With due reservations about the use of the word "everything," I agreed that it was so. When one has had such experiences

From one of the early colloquia the scene of a discussion between Bohr and Landau is imprinted on my mind: Bohr bending over Landau in earnest argument while Landau gesticulated at him, lying flat on his back on the lecture bench (neither seemed to be aware of the unconventional procedure).

O. R. Frisch, "The Interest Is Focusing on the Atomic Nucleus," 1967

Title page of the "Copenhagen Faust." The motto ascribed to Bohr ("Not to criticize . . .") was his usual prelude to an assault on an idea or a theory that he considered unsound.

with one's teachers several times, one has learned something that cannot be learned in any other way.

The year 1932 was the hundredth anniversary of the death of Goethe. The conference that year ended with a little play, essentially by Max Delbrück, which was a parody of *Faust,* applied to current theoretical physics.

There was a Walpurgisnacht, a classical one and a quantum-theoretical version; and because we were celebrating the discovery of the neutron, there was at the end the "Eternal Neutral." The prologue began in heaven. Three archangels appeared in the masks of the astrophysicists Arthur Eddington, James Jeans, and E. A. Milne.

The three "archangels" (from left: Eddington, Milne, Jeans).

"Eddington" began:

> As well we know, the Sun is fated
> In polytropic spheres to shine;
> Its journey, long predestinated,
> Confirms *my* theories down the line . . .

After presenting their arguments the three astrophysicists addressed "God the Lord," saying in unison:

> The vision of you brings elation
> (Though none of us can understand).
> As on their day of publication
> Your brilliant works are strange and grand.

Near them, on a lecture-room bench, was a high stool, and on it was a draped figure. At this point the sheet was lifted and revealed the Lord God,

Mephistopheles in the person of Wolfgang Pauli.

wearing a top hat. It was the person of Felix Bloch, but the mask was unmistakably the face of Niels Bohr. Mephistopheles was equally unmistakably meant to represent Wolfgang Pauli and was acted by Léon Rosenfeld, who jumped on the table, sat down at the feet of the Lord, and began:

Since you, O Lord, yourself have now seen fit
To visit us and learn how each behaves,
And since it seems you favor me a bit,
Well, now you see me here [*turns to the audience*] — among the slaves.

I am not going to repeat the whole text, but that is how it started.[2] Our laughter about Bohr was the escape route which enabled us to say that, although often we could not understand him, we admired him almost without reservation and loved him without limit.

The Como Lecture

In what is generally known as "the Como lecture," Bohr presented an extensive survey of the current state of quantum theory. The lecture was actually delivered twice: first on 16 September 1927 at the conference held in Como, Italy, in commemoration of the great physicist Alessandro Volta; and again shortly afterward at the Fifth Solvay Conference, held in Brussels from 24 to 29 October. The latter occasion marked the beginning of the second phase of the Bohr-Einstein dialogue on the fundamentals of the quantum-mechanical description of nature (see "The Bohr-Einstein Dialogue").

In the body of this long lecture, published in *Nature* in 1928,[1] Bohr dealt with the following topics: the quantum postulate and causality, quantum of action and kinematics, measurements in quantum theory, the correspondence principle and matrix theory, wave mechanics and the quantum postulate, the reality of stationary states, and the problem of elementary particles. The "general point of view" to which Bohr refers was his idea of complementarity, presented for the first time in this lecture. We reproduce here the opening of the lecture, as Bohr delivered it in Como.[2]

A.P.F.

Although it is with great pleasure that I follow the kind invitation of the presidency of the congress to give an account of the present state of the quantum theory in order to open a general discussion on this subject, which takes so central a position in modern physical science, it is with a certain hesitation that I enter on this task. Not only is the venerable originator of the theory [Max Planck] present himself, but among the audience there will

Title page of the proceedings of the
1927 Como Conference.

ONORANZE AD ALESSANDRO VOLTA

NEL PRIMO CENTENARIO DELLA MORTE

ATTI

DEL

CONGRESSO INTERNAZIONALE

DEI FISICI

11–20 SETTEMBRE 1927 – V

COMO – PAVIA – ROMA

PUBBLICATI A CURA DEL COMITATO

VOLUME SECONDO

BOLOGNA
NICOLA ZANICHELLI
MCMXXVIII

be several who, due to their participation in the remarkable recent development, will surely be more conversant with the details of the highly developed formalism than I am. Still I shall try by making use only of simple consideration and without going into any details of a technical mathematical character to describe to you a certain general point of view which I believe is suited to give an impression of the general trend of the development of the theory from its very beginning and which I hope will be helpful in order to harmonize the apparently conflicting views taken by different scientists. No subject, indeed, may be better suited than the quantum theory to mark the development of physics in the century passed since the death of the great genius whom we are here assembled to commemorate. At the same time, just in a field like this where we are wandering on new paths and have to rely upon our own judgment in order to escape from the pitfalls surrounding us on all sides, we have perhaps more occasion than ever at every step to be remindful of the work of the old masters who have prepared the ground and furnished us with our tools.

Supplement to NATURE

No. 3050   APRIL 14, 1928

New Problems in Quantum Theory.

FIFTEEN years have elapsed since Niels Bohr first published a series of papers which were the beginning of a new epoch in the development of the quantum theory. Adopting the atomic model proposed by Rutherford, in which electrons circle round a massive nucleus under the action of a Coulomb force of electric attraction, Bohr gained immediate success in interpreting the spectrum of hydrogen and of ionised helium. For his purpose he was compelled to assume the existence of 'stationary states,' and the emission of monochromatic radiation in the transition between two such states of an atomic system.

In one sense the new method raised as many difficulties as it removed, and to some of the more conservative physicists the account of Bohr's atom read like a fairy tale. Further progress in the interpretation of line spectra was made through the generalisations of Wilson and Sommerfeld, but in spite of the inclusion of a widening circle of facts and the fulfilment of predictions, it came to be realised that a more radical procedure was necessary before a consistent and complete theory could be evolved. In the forward movement few have been more active than Bohr himself. The employment of a spinning electron by Goudsmit and Uhlenbeck removed many discrepancies, and it seems as if some form of magnetic electron is likely to be accepted as a fundamental constituent of an atomic system. The magneton of S. B. McLaren with its quantum of angular momentum may be regarded as the prototype of all such magnetic electrons.

Within the last few years the matrix mechanics of Heisenberg, Born, and Jordan, the quantum algebra of Dirac, and the undulatory mechanics of Schrödinger, have led to remarkable theoretical developments. The new wave mechanics gave rise to the hope that an account of atomic phenomena might be obtained which would not differ essentially from that afforded by the classical theories of electricity and magnetism. Unfortunately, Bohr's statement in the following communication of the principles underlying the description of atomic phenomena gives little, if any, encouragement in this direction.

In classical mechanics it is assumed that the position of a particle (such as an electron) can be determined at a specified instant of time by means of its co-ordinates. As the time varies it is supposed to be possible to trace the path of the particle through space, or to determine its 'world line' in the four-dimensional world. Further, it is assumed that the concept of causality may be applied in considering the effect of the action of external forces. Thus in classical physics we have a causal space-time co-ordination, based on the assumption that the methods or tools of measurement do not affect the phenomena which are observed.

In the new quantum theory the outlook is changed, for any attempt to observe the position or motion of an electron involves illumination by light, and this implies interaction between the electron and the light employed in making the measurement. The position and the path of an electron become vague. Thus there is introduced in the new quantum mechanics an indefiniteness which contrasts with the clear-cut concepts of classical mechanics. Bohr asserts that in any phenomenon which we may attempt to observe there is an essential discontinuity, or rather individuality, which may be symbolised by Planck's constant h. The causal space-time co-ordination of atomic phenomena must on this view be abandoned, and we are left with a somewhat vague statistical description.

The strange conflict which has been waged between the wave theory of light and the light quantum hypothesis has resulted in a remarkable dilemma. But now we have a parallel dilemma, for a material particle manifests some of the attributes of wave motion. Can these apparently contradictory views be reconciled? According to Bohr, the pictures ought to be regarded not as contradictory but as complementary. Radiation in free space is not open to observation, and is a mere abstraction. An isolated material particle likewise can never be observed and is also an abstraction. It is only through their interaction with other systems that the properties of these abstractions can be defined and observed.

It must be confessed that the new quantum mechanics is far from satisfying the requirements of the layman who seeks to clothe his conceptions in figurative language. Indeed, its originators probably hold that such symbolic representation is inherently impossible. It is earnestly to be hoped that this is not their last word on the subject, and that they may yet be successful in expressing the quantum postulate in picturesque form.

Introduction by H. S. Allen to the text of Bohr's Como lecture. The brief essay gives an interesting and informative summary of the state of theoretical physics at that time.

IV. IN THE WORLD
OF NUCLEAR PHYSICS

Niels Bohr and Nuclear Physics

Roger H. Stuewer

Niels Bohr was never far from nuclear physics; his friendship with Ernest Rutherford guaranteed that. He was a twenty-six-year-old postdoctoral student at Cambridge when he first saw Rutherford in the fall of 1911, and was immediately captivated by Rutherford's "great human personality." A short time later he talked with Rutherford in Manchester at the home of Lorrain Smith, a colleague and friend of Bohr's recently deceased father. The die was cast. Bohr transferred from Cambridge to Manchester in March 1912. He returned to Copenhagen in July to marry Margrethe Nørlund, and in August he introduced his bride to the Rutherfords on his wedding trip.

Thus began a friendship that deepened increasingly over the next quarter-century, until Rutherford's death on 19 October 1937. Rutherford became almost a second father to Bohr, and when Bohr learned of Rutherford's death he left the Galvani meeting in Bologna and traveled immediately to England, where he was asked to stand with the family at Rutherford's grave in Westminster Abbey. Two weeks later, Peter Kapitza told Bohr, in an extraordinarily moving letter from Moscow, that he always had been a little jealous of Bohr, because he sensed that Rutherford had liked Bohr most of all of his pupils. Bohr's next to last publication was his Rutherford Memorial Lecture.[1]

From the beginning, Bohr kept abreast of developments in nuclear physics. During his stay in Manchester he came to realize, particularly through discussions with George de Hevesy, that Rutherford's nuclear atom, proposed less than a year earlier, implied that a sharp distinction could be drawn between phenomena associated with an atom's electronic distribution and those associated with its nucleus. He concluded that the nuclear charge would change by fixed amounts in radioactive alpha decay and beta

Roger Stuewer began his career as a nuclear physicist. He is now Professor of the History of Science and Technology at the University of Minnesota, Minneapolis.

197

decay, and that the nuclear mass, but not the charge, could be changed by adding or subtracting electron-proton pairs. Rutherford dissuaded Bohr from publishing these ideas because he felt they still rested upon too insecure an experimental foundation, but their subsequent vindication offered clear proof of Bohr's penetrating mind. The same was true for Bohr's analysis, stimulated by a paper written shortly before by C. G. Darwin in Manchester, of the slowing down of charged particles when passing through matter. This seminal work was published,[2] and it constituted a milestone on the road to Bohr's discovery of his atomic model early in 1913.

The avenues of research opened up by that discovery commanded Bohr's attention for the next two decades, but they were by no means his exclusive concern. Bohr discussed many questions in nuclear physics with Rutherford during a second extended stay in Manchester in the early years of the First World War, from the autumn of 1914 to the summer of 1916. Furthermore, after Rutherford moved to Cambridge in 1919, Bohr visited him almost every year, while Rutherford went twice to Copenhagen, first at the opening of Bohr's institute in 1920, and again in 1932. Information was also exchanged through visits of Bohr's and Rutherford's colleagues; and if that were not enough, Bohr and Rutherford wrote to each other frequently: some 170 letters were exchanged in the quarter-century between 1912 and 1937.

That, of course, was only a small fraction of Bohr's total correspondence. Wolfgang Pauli, Werner Heisenberg, and virtually every other atomic and nuclear physicist of the period exchanged letters, in some cases hundreds of them, with Bohr. Moreover, beginning in 1929 Bohr organized small conferences in Copenhagen. Thus, even though Bohr did not publish on nuclear physics throughout most of the 1920s, he possessed a thorough knowledge of developments in the field, and his prodigious memory and style of work — constant discussion, analysis, drafting and redrafting of manuscripts — enabled him to respond with deep understanding to fundamental challenges and problems as they arose.

Bohr was drawn unexpectedly into nuclear physics in 1928. In June of that year, twenty-four-year-old George Gamow left Leningrad and arrived at Max Born's institute in Göttingen, where he came across a paper of Rutherford's in the physics library. In that paper, published in the September 1927 issue of the *Philosophical Magazine,* Rutherford had proposed an essentially classical theory of alpha decay, and before Gamow had put the paper down he saw that Rutherford's theory was incorrect: alpha decay had to be understood as a quantum-mechanical *tunneling* process. That same conclusion was reached independently and virtually simultaneously by R. W. Gurney and E. U. Condon at Princeton. It constituted the first proof that the ordinary laws of quantum mechanics, which had recently been formulated, apply to nuclear as well as to atomic phenomena.

Gamow brought this news to Bohr personally at the end of August 1928. Impressed by Gamow's work, and by Gamow personally, Bohr arranged fellowship support to enable Gamow to spend the entire academic year 1928–29 in Copenhagen. Bohr also made arrangements for him to visit Rutherford in Cambridge for about a month in January and February of 1929. Gamow thrived in both places. Even before visiting Cambridge, he had extended his theory to the inverse case in which charged particles penetrate a nucleus from the outside — an extension that helped stimulate J. D. Cockcroft and E. T. S. Walton to build a proton accelerator at Cambridge and disintegrate the lithium nucleus in 1932. Gamow also conceived the liquid-drop model of the nucleus during his first months in Copenhagen. Bohr unquestionably understood this idea immediately owing to his familiarity with surface-tension phenomena from his prize-winning research in this field as a student. Gamow discussed his *Tröpfchenmodell* with Paul Ehrenfest in Leiden en route to Cambridge, and with Rutherford after his arrival there. Rutherford invited Gamow to present the model, as well as his theory of alpha decay, at a meeting of the Royal Society in London on February 7. Subsequently, in 1930, Gamow applied his liquid-drop model to a calculation of nuclear *mass defects* and an analysis of nuclear stability, which attracted the attention of, among others, C. F. von Weizsäcker, who in 1935 utilized and extended Gamow's work in proposing his semiempirical nuclear-mass formula.[3]

By the end of 1928 another fundamental problem had arisen in nuclear physics: How should the bewildering behavior of the electrons in nuclei be understood? In 1926, following George Uhlenbeck and Samuel Goudsmit's discovery of electron spin, Ralph de Laer Kronig pointed out that if electrons were present in nuclei along with protons, as generally assumed, they should possess the same magnetic moment as those in atoms. Hence, they should produce hyperfine spectral-line splittings as large as ordinary Zeeman splittings, in flat contradiction to observation. Two years later, Kronig also noted certain measurements suggesting that the nitrogen nucleus, with a net charge of 7 units, possessed a total spin of 1 (in units of $h/2\pi$), whereas its composition out of 14 protons and 7 electrons implied that it should possess half-integer, not integer spin.

This contradictory behavior became even more flagrant as other puzzles emerged. For example, the calculated energy of an electron confined to nuclear dimensions, as estimated from Heisenberg's uncertainty principle, greatly exceeded known nuclear-binding energies. Actually, it appears that this difficulty was not widely appreciated prior to about 1930. Another one, however, was well known; this was the so-called Klein paradox, discovered by Oskar Klein in Copenhagen at the end of 1928. This paradox may be formulated as follows: How can an electron be confined within a nuclear potential well if, as follows from the Dirac equation of the electron, any electron of sufficiently high energy impinging upon a high and steeply rising potential barrier has a high probability of escaping simply by being transformed from a particle of positive mass into one of negative mass? The clarity and precision of this paradox led Bohr and others to debate its

In wave mechanics there are no impenetrable barriers, and, as the British physicist R. H. Fowler put it after my lecture on that subject at the Royal Society of London . . . "Anyone at present in this room has a finite chance of leaving it without opening the door — or, of course, without being thrown out of the window."

George Gamow,
My World Line, 1970

meaning intensely, both privately, in conversations and in correspondence, and publicly, in the literature.

Yet another closely related and puzzling fact was established by C. D. Ellis and W. A. Wooster at the Cavendish Laboratory in 1927: beta particles (electrons) are not emitted from radioactive nuclei with discrete energies; rather, they are emitted directly with a continuous distribution of energies up to some maximum value. Bohr became increasingly convinced that this observation meshed with the other puzzles associated with nuclear electrons, and he proposed a bold solution: he argued that to account for them, an entirely "new physics" would have to be developed whose laws lay outside those of the "existing quantum physics." By mid-1929 he became more specific. He became convinced that the ordinary laws of conservation of energy and momentum would have to be relinquished in this new physics. This was, in fact, an old idea of Bohr's, one which even predated the Bohr-Kramers-Slater paper of 1924, where it found its most influential expression. Bohr knew, however, that in 1929 he could not simply resurrect that idea, which had been disproved by direct experiments. Rather, he emphasized that he was not calling for an abandonment of the conservation laws in "ordinary quantum theory." Instead, the limit of validity of the conservation laws would coincide with the limit of applicability of ordinary quantum theory.

While Heisenberg embraced Bohr's ideas and Gamow increasingly questioned them, Pauli, P. A. M. Dirac, and Rutherford opposed them from the start. Pauli conveyed his opposition to Bohr by letter as early as 17 July 1929; Rutherford did the same on November 19; and Dirac soon followed. Rutherford, for example, told Bohr bluntly: "I have heard rumors that you are on the war path and wanting to upset Conservation of Energy, both microscopically and macroscopically. I will wait and see before expressing an opinion, but I always feel 'there are more things in Heaven and Earth than are dreamed of in our Philosophy.'"

Bohr was undeterred; he continued to explore the ramifications of nonconservation. From 27 April to 4 May 1930, he delivered the Scott Lectures in Cambridge, and on May 8 the Faraday Lecture in London. An expanded version of the latter, a *tour de force,* appeared in print two years later.[4] Its final section was devoted to nuclear physics. In it Bohr maintained that at present "we may say that we have no argument, either empirical or theoretical, for upholding the energy principle in the case of β-ray disintegrations, and are even led to complications and difficulties in trying to do so."

That was the challenge to which Pauli responded. On 4 December 1930, in a letter addressed principally to Hans Geiger and Lise Meitner, he very tentatively suggested that the emission of an electron in beta decay might be accompanied by the emission of what he called a "neutron" of spin ½ and finite mass (on the order of the mass of the electron) "in such a way that the sum of the energies of neutron and electron is constant." He discussed this idea again in Pasadena and Ann Arbor in mid-1931, and in Rome in Oc-

Pauli wrote a letter to Bohr . . . which contained a number of penetrating questions. Bohr did not feel that he could answer that letter easily. He was working on other things, so he asked Mrs. Bohr to write Pauli a nice letter with some family gossip and to explain that he himself would write on Monday. Three or four weeks later there came a kind letter from Pauli to Mrs. Bohr, thanking her for the letter and saying that her husband had been very wise in saying that he would write on Monday without specifying on which Monday. But, Pauli added, "He should not in any way feel tied to write on a Monday; a letter written on any other day would be equally welcome." I have some reason to believe that that letter of Pauli's which Bohr was going to answer on Monday was a letter containing the first suggestion of the neutrino as a means to solve the problem of apparent nonconservation of energy.

H. B. G. Casimir,
"Some Recollections," 1977

tober 1931 at another conference organized by Enrico Fermi. This conference was the first international conference devoted specifically to nuclear physics. Bohr also attended, and during the course of it he and Pauli directly debated the validity of the conservation laws in beta decay, without coming to agreement.

A few months later, the entire field of nuclear physics was transformed. Between the end of 1931 and the end of 1932, deuterium, the neutron, and the positron were discovered, and the Cockcroft-Walton accelerator and cyclotron were invented. It became possible to carry out entirely new experiments, and to invent entirely new theories of nuclear structure. James Chadwick's discovery of the neutron in February 1932 became particularly significant in both respects. By the time the second international conference on nuclear physics was held — the Seventh Solvay Conference, which met in October 1933 in Brussels — the face of nuclear physics had been altered fundamentally.

The face of Europe, too, had changed. Adolf Hitler had become chancellor of Germany on 30 January 1933, and a few months later, on April 7, the Nazi Civil Service Law went into effect, sending tens of thousands of Jewish citizens, including over a thousand scholars, into permanent exile. Bohr had opened his institute in Copenhagen as an international center of research more than a decade earlier, and now, in 1933, former visitors, and many others whom he had never seen before, appealed to him for help. Felix Bloch and Victor Weisskopf were among those who received offers of new positions in the United States while staying temporarily in Copenhagen. But many more experienced Bohr's deep humanitarianism as well.

In October 1934 Ernest Rutherford opened the third international conference on nuclear physics, held in London and Cambridge. By that time a number of basic questions, which had been unresolved only a year earlier at the Solvay Conference, had been settled. One concerned the nature of the neutron. Chadwick, its discoverer, being strongly influenced by a model that Rutherford had proposed in 1920, had been inclined to regard the neutron not as a new elementary particle but as an electron-proton compound. Werner Heisenberg, in his fundamental theoretical papers of 1932, had straddled the fence on this question. The first step toward its resolution was taken by Frédéric and Irène Joliot-Curie, who presented calculations and cloud-chamber photographs at the 1933 Solvay Conference from which they concluded that the mass of the neutron is slightly greater than the mass of the proton, not less, as Chadwick believed. Later, in mid-1934, Chadwick himself, following up an idea of Maurice Goldhaber's, and with Goldhaber's assistance, came to this same conclusion on the basis of their pioneering deuteron-photodisintegration experiments. It followed that the neutron indeed is an elementary particle, but one which could decay, so it seemed, into a proton and an electron.

That conclusion itself had to be revised in light of the fundamental theory of beta decay developed by Enrico Fermi. At the 1933 Solvay Confer-

201

In the World of Nuclear Physics

Participants in the Seventh Solvay Conference, Brussels, October 1933.

Seated, from left: Erwin Schrödinger, Irène Joliot-Curie, Niels Bohr, Abram Joffé, Marie Curie, Paul Langevin, O. W. Richardson, Ernest Rutherford, Théophile De Donder, Maurice de Broglie, Louis de Broglie, Lise Meitner, James Chadwick.

Standing: Émile Henriot, Francis Perrin, Frédéric Joliot-Curie, Werner Heisenberg, H. A. Kramers, Ernest Stahel, Enrico Fermi, E. T. S. Walton, P. A. M. Dirac, Peter Debye, N. F. Mott, Blas Cabrera, George Gamow, Walther Bothe, P. M. S. Blackett (at back), M. S. Rosenblum, Jacques Errera, Édmond Bauer, Wolfgang Pauli, J. E. Verschaffelt, Max Cosyns (at back), Édouard Herzen, J. D. Cockcroft, C. D. Ellis, Rudolf Peierls, Auguste Piccard, Ernest Lawrence, Léon Rosenfeld.

ence, Pauli again proposed the existence of his new neutral particle, which he now called, following a suggestion of Fermi's, a "neutrino." The emission of this particle along with an electron in beta decay, Pauli argued at some length, would permit the preservation of the conservation laws, as opposed to the view of Bohr (who was also in the audience). Stimulated by Pauli's remarks, Fermi placed Pauli's hypothetical neutrino, now considered to be a particle of spin ½ but of negligible or zero mass, at the heart of his theory of beta decay. He assumed that inside the nucleus a neutron can decay into a proton, an electron, and a neutrino, thereby creating the observed beta-decay electron and at the same time preserving the conservation laws. Fermi submitted his theory for publication in January

Enrico Fermi walking with Bohr on the Appian Way, near Rome, in 1931.

1934.[5] Bohr, initially skeptical, overcame his doubts over the next two years and abandoned his nonconservation views.

Just as Fermi was submitting his theory of beta decay for publication, the Joliot-Curies astonished physicists everywhere by reporting, on 15 January 1934, their discovery of artificial radioactivity: alpha particles incident upon aluminum 27, they found, produce not a stable new nucleus but a radioactive isotope of phosphorus. No response to their discovery was more fruitful than Fermi's. If alpha particles could produce such remarkable new effects, might not neutrons do the same? Enlisting the help of his team in Rome, Fermi bombarded element after element of increasing atomic number, all without success — until fluorine was inserted. At that point his primitive Geiger-Müller counter responded, indicating that an isotope of nitrogen had been produced through an (n, α) reaction and then underwent beta decay. That was in March 1934. Fermi considered this discovery so significant that he chose to report on it, rather than on his new theory of beta decay, at the London-Cambridge conference in October.

Bohr did not attend this conference. He and his family were in deep mourning following the loss of his oldest son, Christian, on a sailing trip at sea. By the time he recovered sufficiently to resume work, Fermi had made another astonishing discovery. On the morning of 22 October 1934, Fermi suddenly and inexplicably decided to substitute a paraffin wax filter for one of lead between his neutron source and target element — and observed enormously increased activity. By the afternoon Fermi understood: the incident neutrons had been slowed down greatly through collisions with protons inside the paraffin filter, and these slow neutrons, for some still unknown reason, were much more efficient in producing artificial radioactivity than fast ones. Among the elements exhibiting this behavior was uranium, the heaviest naturally occurring element. By adding a neutron to a uranium nucleus, a heavier isotope of uranium would be produced which then could undergo beta decay, yielding a new nucleus one unit higher in atomic number. Thus, it was both experimentally and logically supportable for Fermi to conclude that, in this case, he had produced a transuranic element.

Fermi's experiments, begun in 1934 and continued in 1935, exerted a direct influence on Bohr. They led Bohr in large measure to make his most significant contribution to nuclear physics in this period: his theory of the *compound nucleus*. On the basis of Bohr's published and unpublished papers and J. A. Wheeler's and O. R. Frisch's retrospective accounts, Rudolf Peierls has arrived at a plausible reconstruction of the development of Bohr's thought on this subject. It appears that three seminars in Copenhagen punctuated the development. The first consisted of a lecture by H. A. Bethe in September 1934 in which Bethe presented a single-particle theory of nuclear reactions — a theory that left Bohr intuitively dissatisfied. Eight months later, in April 1935, a second seminar on Fermi's latest experiments was given by Christian Møller, who had just returned to Copenhagen from

Rome. This seminar was witnessed by Wheeler, who vividly described how Bohr, after about a half hour, suddenly took the floor and offered a new interpretation of Fermi's results. This seminar was followed by a third one at the end of 1935 which involved a similarly dramatic interruption by Bohr as witnessed by Frisch. Thus, the pieces of the puzzle evidently fell into place in Bohr's mind between September 1934 and the end of 1935, although the precise way in which they did remains unknown. What is known, however, is the end result: Bohr saw that Fermi's large slow-neutron capture probabilities meant that the incident neutron was not interacting with a single nuclear particle in the target element. Rather, it was interacting with a large number of them, undergoing repeated collisions, and thus forming a long-lived compound nucleus consisting of the incident neutron plus target nucleus. This compound nucleus then, in a second distinct stage, would lose its energy of excitation by decaying in any one of several freely competing ways, such as by emitting gamma radiation, or by the expulsion of a neutron through the chance concentration of energy on one close to the nuclear surface.

Bohr saw that he held the key to a general theory of nuclear reactions. The formation of such a semistable compound nucleus accounted for the striking tendency of charged particles to react with nuclei as soon as they established contact with them. It accounted for Fermi's high slow-neutron capture probabilities, and the fact that the capture probability increases as the neutron velocity decreases, as Fermi and his coworkers had also observed. It explained how, in decaying, the compound nucleus could live long enough for radiative (gamma-ray) transitions to dominate over neutron emission, as observed. Finally, it accounted for the observation by Fermi and others that when slow neutrons strike intermediate to heavy target nuclei, many sharp "resonances," closely spaced in energy, are produced. Bohr saw that during the formation of the compound nucleus these neutrons excite high-energy states — on the order of 10 million electron volts (MeV) above the ground state — which then decay through radiative transitions, producing the observed resonance lines. Furthermore, because of the complementary relationship between energy-width and lifetime embodied in Heisenberg's uncertainty principle, the extreme sharpness of these lines afforded direct confirmation of the long life of the compound nucleus: the incident neutron was not traversing the target nucleus rapidly; here, too, it was interacting strongly with many nuclear particles. Then, after its formation, the compound nucleus could decay either through the emission of gamma radiation, or through the emission of a neutron or a proton, or indeed through any competing process consistent with conservation of energy.

Bohr understood that these processes, in general, had to be governed by quantum laws, but that an essentially classical collision model serves as an adequate approximation for their description. In fact, Bohr had small wooden models made for him in the institute workshop to demonstrate his

205

Bohr's compound-nucleus model, shown here for the case of an incident neutron (no Coulomb barrier).

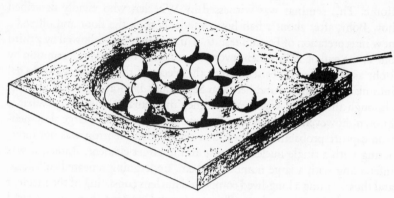

theory (see figure). These consisted of shallow circular dishes (the target nuclei) filled with tiny steel balls (their constituent particles), either with or without an outer embankment (Coulomb barrier) to demonstrate the case either of an incident charged particle or of a neutron. Bohr was so convinced of the correctness of his theory (and of his model) that he speculated far beyond the limit of current observations. He claimed that it would hold

Niels Bohr (right) with Fritz Kalckar in 1934, at the Copenhagen Conference.

not only for incident neutrons or protons of 100 MeV in energy, where several nuclear particles would be discharged; it would hold even for incident energies of 1000 MeV, where one could expect "an explosion of the whole nucleus."

These ideas crystallized in Bohr's mind as he debated them, especially with Fritz Kalckar, at the end of 1935. His theory was published in *Nature* on 29 February 1936, with a German translation soon following in *Die Naturwissenschaften*.[6] Meanwhile, he had sent copies of his manuscript to quite a number of people, and he had lectured on his theory in Copenhagen, London, and Cambridge. Rutherford, just after Bohr left Cambridge, wrote to Chadwick and to Max Born on February 17 and 22, describing Bohr's conception of the target nucleus as a "mush" of particles.

Just at this same time, Gregory Breit and Eugene P. Wigner, collaborating in Princeton, submitted for publication their fundamental paper on slow-neutron resonances: it was received by *The Physical Review* on 15 February 1936, and published in the April 1 issue.[7] Like Bohr, Breit and Wigner were impressed by the failure of the single-particle theories of Bethe and others to account for, especially, Fermi's slow-neutron observations. They proposed, instead, that the incident neutron passes into "quasi-stationary (virtual) energy levels of the system nucleus+neutron" which "then jump into a lower level through the emission of γ-radiation or perhaps in some other fashion." This was the essential idea that Bohr also had proposed in more general terms. Unlike Bohr, however, Breit and Wigner developed their theory quantitatively, deriving their famous resonance formulas for the neutron-capture and neutron-scattering cross-sections, and showing that the sum or total cross-section varies inversely with the velocity of the incident neutrons, as Fermi and his coworkers had observed. In general, physicists soon understood that Breit and Wigner's theory supported Bohr's and showed how Bohr's could yield quantitative results. It was to a large extent for this reason that Bohr's theory of the compound nucleus dominated the theory of nuclear reactions for almost two decades.

Another reason for its dominance was Bohr's continuing discussion and development of his theory. Assisted by Kalckar, Bohr began to plan and compose a comprehensive three-part paper on nuclear transmutations. The first part was to consist of general theoretical remarks, the second of a detailed theory of nuclear collisions, and the third of an analysis of the available experimental data. By January 1937 part one was already set in type. Bohr, however, declined to release it for publication, primarily because he and Kalckar were then on the verge of leaving Copenhagen for the United States to attend the third conference on theoretical physics in Washington, D.C., organized by George Gamow, Edward Teller, and Merle Tuve, and scheduled to begin on 15 February 1937. Bohr, who was accompanied on this trip by his wife and son Hans, knew that it would present many opportunities to refine his ideas further through discussions. This proved to be the case. Bohr lectured on his compound-nucleus theory at various universities on the East Coast, always with the help of his little wooden model.

After the Washington conference, he traveled with Hans Bethe to Duke University in Durham, North Carolina, where both gave invited papers on the compound nucleus at a meeting of the American Physical Society on 19 February 1937. The following day the meeting moved to the University of North Carolina in Chapel Hill, where John Wheeler, who had spent the academic year 1934–35 with Bohr in Copenhagen, then held a position. These travels, in fact, were just the beginning of a six-month around-the-world trip that also took the Bohrs to the University of Oklahoma; to the West Coast of the United States; to Japan, where Bohr lectured on his theory at the Imperial University in Tokyo; and to Russia, where he lectured on it in Moscow.

By the time Bohr returned to Copenhagen in July 1937, he saw that it would be largely superfluous for him and Kalckar to complete and publish the second and third parts of their joint treatise, because in April and July the final two parts of Hans Bethe's enormously influential article in *Reviews of Modern Physics*—"Nuclear Dynamics, Theoretical" and "Nuclear Dynamics, Experimental"—had appeared, the latter written jointly with M. Stanley Livingston.[8] Bohr knew that this work was in progress and, under these circumstances, he and Kalckar decided simply to bring the first part of their paper up to date by joining an addendum to it, and to submit it for publication alone. This they did in October 1937—the very month that Bohr, deeply shaken, was called from Bologna to England to attend Rutherford's funeral.

In their paper,[9] Bohr and Kalckar first reiterated the basic ideas underlying Bohr's theory of the compound nucleus, promising that "many properties of nuclear matter" could be compared to "the properties of ordinary liquid and solid substances." This promise was fulfilled when they analyzed why the spacing between energy levels in the excited compound nucleus decreases rapidly with increasing excitation energy, becoming at some point "practically continuous." These energy levels arise from "some quantized collective type of motion of all its constituent particles," which means that they have "very much the same character as that of the quantum states of a solid body, well known from the theory of specific heats at low temperatures." To develop this analogy quantitatively, Bohr and Kalckar compared the excitations of the compound nucleus to "the oscillations in volume and shape of a sphere under the influence of an elasticity ϵ or surface tension ω." They found, however, large discrepancies with experiment, even for heavy compound nuclei, leading them to conclude that these "simple considerations" could "at most serve as a first orientation as regards the possible origin of nuclear excitations."

A detailed theory would have to take into account "the specific character of the interactions between the individual nuclear particles." However, the close coupling of these particles made it difficult to draw reliable conclusions on the exchange character or spin dependence of such specifically nuclear forces. "In particular," Bohr and Kalckar wrote, "any attempt of accounting for the spin values by attributing orbital momenta to the individual nuclear particles seems quite unjustifiable." Rather, it was necessary to "assume that any orbital momentum is shared by all the constituent particles of the nucleus in a way which resembles that of the rotation of a solid body." Bohr and Kalckar went on to show that transitions between such rotational energy levels might account for the fine-structure spectra of heavy nuclei.

Bohr and Kalckar also discussed the process of absorption and reemission of neutrons by a target nucleus. The former process resembled "the adhesion of a vapor molecule to the surface of a liquid or solid body," the latter "the evaporation of such substances at low temperatures." This analogy had been emphasized recently by Jacob Frenkel and Lev Landau. It also had been elaborated by Victor Weisskopf in a fundamental paper (submitted to *The Physical Review* from Copenhagen in March 1937) in which he introduced the concept of a nuclear temperature and explored its consequences in detail.[10] Finally, Bohr and Kalckar discussed slow-neutron reactions (emphasizing the "decisive" work of Breit and Wigner), the theory of radioactive alpha decay, and the inverse penetration problem, all of which they showed to be consistent with Bohr's theory of the compound nucleus.

Several aspects of Bohr's thought at this time (October 1937), as displayed in his and Kalckar's paper, are worthy of comment. First, while Bohr was touching upon the liquid-drop model of the nucleus, it seems that he was more inclined at this time to view the nucleus as an elastic solid. Perhaps he was influenced here to some degree by Rutherford, who on more than one occasion had argued that heavy nuclei exhibit a crystal-like structure. Whatever the case may have been, it seems clear that Bohr had not yet fully embraced the liquid-drop model.

That may have been the reason, in fact, why Bohr did not cite or mention Gamow as the originator of the liquid-drop model in his and Kalckar's paper. As we have seen, Gamow conceived this model in Copenhagen at the end of 1928, and he subsequently applied it to calculations of mass-defect curves and nuclear stability. The only point, it seems, at which Bohr noted Gamow's contribution was in the discussion following Heisenberg's paper at the October 1933 Solvay Congress, where Heisenberg treated le modèle de la "goutte" de Gamow at some length. In 1935 von Weizsäcker adopted Gamow's model in arriving at his semiempirical mass formula, and two years later he again discussed Gamow's model at length in his book *Die Atomkerne*—a work cited by Bohr and Kalckar. The fact, however, that Bohr and Kalckar did not cite or mention Gamow directly had at least one unfortunate consequence, historically speaking: when Bethe discussed the liquid-drop model in the second part of his *Reviews of Modern Physics* article, published in April 1937, he based his discussion on Bohr and Kalckar's paper in its prepublication form, and since then virtually everyone has

Bohr in a discussion with Wolfgang
Pauli, L. W. Nordheim, Léon
Rosenfeld, and others.

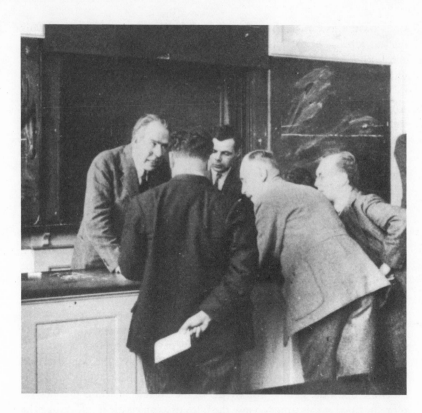

assumed, incorrectly, that Bohr conceived the liquid-drop model along
with his theory of the compound nucleus. Bethe, incidentally, in citing
Bohr and Kalckar's paper, slyly assigned it a publication date of 1939 — a
reasonable guess, given Bohr's legendary procrastination in publication.

Finally, as we have seen, Bohr and Kalckar argued that the collective
behavior of the compound nucleus is incompatible with any attempt to
assign orbital momenta to its individual constituent particles. In other
words, it was fruitless to search for a *shell structure* of the nucleus analogous
to that of the atom, and hence it was no surprise (Bohr might have said) that
Walter Elsasser's and Kurt Guggenheimer's attempts in 1933–34 had not
succeeded. Bohr, moreover, was adamant on this point: Peierls has pointed
out that when Bohr visited Japan he dissuaded the Japanese physicist Taka-
hiko Yamanouchi (who had just tried to apply group theory to nuclear
structure) from carrying out a serious shell-model investigation. It seems,
therefore, that the very success of Bohr's theory of the compound nucleus,
coupled with his great authority and persuasiveness, was a key factor in
discouraging shell-model studies after 1936 or 1937.

On 6 January 1938, Kalckar, not yet twenty-eight years old, suddenly
died. Bohr was deeply affected by the loss of his young colleague. Nevertheless, he continued working. He applied his theory of the compound

nucleus to an understanding of, for example, nuclear photoeffects. He also continued a program begun in 1932, lecturing on his principle of complementarity as extended to fields outside physics, first biology and then, in August 1938, anthropology. He argued that "different human cultures are complementary to each other"—a humanitarian view that stood in sharp contrast to the brutal racial policies of the Nazis.

Bohr also kept the lamp of reason burning in Copenhagen by continuing to hold small conferences whenever possible. In addition, he oversaw and raised funds for a strong experimental program in nuclear physics at his institute. He first came face to face with the wave of the future in May 1933, when he visited Charles C. Lauritsen's laboratory at the California Institute of Technology before attending Chicago's "Century of Progress" meeting the following month. Over the next few years, he persuaded the Carlsberg and Rockefeller foundations to provide monetary support, and the Thrige Foundation to provide a big magnet, for the construction of a cyclotron in Copenhagen, which went into operation in 1938. Soon thereafter, he secured additional financial support from the Carlsberg and Rockefeller foundations for the construction of a high-pressure Van de Graaff accelerator.

Meanwhile, another chain of events had begun. On the thirteenth or fourteenth of July 1938, Lise Meitner, imperiled in Berlin after the *Anschluss* of Austria on March 7, was spirited into the Netherlands by Dirk Coster, and from there went to Stockholm. Five months later, in mid-December, Otto Hahn and Fritz Strassmann, pursuing the research program they had begun with Meitner, cautiously concluded that when neutrons bombard uranium a transuranic element is not formed (as Fermi had believed), but that one of the products is the intermediate element barium. Hahn informed Meitner about this deeply puzzling situation by letter on 19 December 1938, and three days later he and Strassmann reported their discovery to *Die Naturwissenschaften*.[11] Thus began the events that would engulf Bohr, and change the world.

Meitner, puzzled by Hahn's letter, answered it on December 21, questioning him further about his and Strassmann's results. After receiving Hahn's assurance that they were correct, she left Stockholm for Kungälv, just north of Gothenburg, to spend the Christmas holidays with friends and with her physicist nephew Otto R. Frisch, who was then working at Bohr's institute in Copenhagen. Discussing Hahn and Strassmann's results one day, and inspired by an awareness of the liquid-drop model of the nucleus, Frisch and Meitner calculated that the repulsive force of the high surface charge of the uranium nucleus almost completely cancels the attractive force of its surface tension, so that a bombarding neutron might induce vibrations and hence instability. Furthermore, they found that if the uranium nucleus were split into two roughly equal parts, a transformation of rest mass into kinetic energy would occur which, according to Einstein's relationship, would amount to about 200 MeV, an enormous figure. With

A couple of days later I travelled back to Copenhagen in considerable excitement. I was keen to submit our speculations — it wasn't really more at the time — to Bohr, who was just about to leave for the U.S.A. He had only a few minutes for me; but I had hardly begun to tell him when he smote his forehead with his hand and exclaimed: "Oh what idiots we all have been! Oh but this is wonderful! This is just as it must be! Have you and Lise Meitner written a paper about it?" Not yet, I said, but we would at once; and Bohr promised not to talk about it before the paper was out. Then he went off to catch his boat.

O. R. Frisch,
What Little I Remember, 1979

this in mind, Meitner returned to Stockholm and Frisch to Copenhagen, where Frisch sought Bohr out and caught him for five minutes on Tuesday, 3 January 1939. During the next four days Bohr considered Frisch and Meitner's interpretation while Frisch and Meitner began composing a joint note to *Nature* over the telephone. On Saturday morning, January 7, Frisch gave Bohr two pages of his and Meitner's incomplete manuscript at the train station as Bohr was leaving for Gothenburg; that same day he was to embark for the United States on the Swedish-American Liner *Drottningholm.* By that time, Bohr was completely familiar with Frisch and Meitner's work.

Frisch's initial five-minute conversation with Bohr, however, had been sufficient for Bohr to grasp the essentials. In fact, as Frisch wrote to Meitner, Bohr was astonished that he himself had not considered the splitting of the uranium nucleus as a possibility earlier, since it followed so directly from the current conceptions of nuclear structure. However, as we have seen, Bohr had had in mind a quite different possibility as to how a nucleus might break up, as expressed in his compound-nucleus paper of 1936. Furthermore, in his and Kalckar's paper of 1937 he had been much less inclined to view the nucleus as a liquid drop than as an elastic solid. Thus, to the extent that Bohr was still committed to these views on 3 January 1939, he would not have seen the possibility that Meitner and Frisch saw.

Now, however, as Bohr boarded his ship for the United States, he knew that a new world had been opened up in nuclear physics. It is easy to imagine the excitement with which he explored every aspect of the new discovery with Léon Rosenfeld, his traveling companion from the University of Liège. The two arrived in New York on Monday, 16 January 1939, after a nine-day voyage. They were met by John Wheeler, who took Rosenfeld to Princeton, and by Enrico and Laura Fermi, who took Bohr to Columbia. Bohr had promised Frisch that, to protect their priority, he would not discuss Frisch and Meitner's interpretation with anyone. Unfortunately, while on board ship, he had forgotten to convey the need for confidentiality to Rosenfeld, who let the cat out of the bag in Princeton — much to Bohr's distress when he himself arrived in Princeton a few days later and learned what had happened.

Events then accelerated. On January 20, Bohr wrote to Frisch commenting on Frisch and Meitner's manuscript and also enclosing a draft of a note to *Nature* he himself had written describing how their interpretation meshed with his own conception of the formation and decay of the compound nucleus. Meanwhile, in Copenhagen, George Placzek had prodded Frisch into carrying out experiments, and between January 13 and 16 Frisch observed the predicted ionization pulses from the fission fragments — the historic term "fission" having been suggested to Frisch by an American biochemist then in Copenhagen. Frisch immediately drafted a second note for *Nature* under his name alone and submitted both notes for publica-

THE NEW YORK TIMES, SUNDAY, JANUARY 29, 1939.

Atom Explosion Frees 200,000,000 Volts; New Physics Phenomenon Credited to Hahn

By The Associated Press.

WASHINGTON, Jan. 28.—American scientists heard today of a new phenomenon in physics—explosion of atoms with a discharge of 200,-000,000 volts of energy.

Theoretical physicists attending a meeting sponsored by the Carnegie Institution of Washington and George Washington University said that Dr. Enrico Fermi of the University of Rome told yesterday that this had been accomplished by Dr. G. Hahn of Berlin.

The report so stirred the limited circle of scientists with facilities to carry on such experiments that work on attempts to duplicate Dr. Hahn's accomplishment has begun at the Carnegie institution's terrestrial magnetism laboratory and at Columbia University.

Scientists at the meeting said the discovery was comparable in significance to the original discovery of radioactivity thirty years ago.

They said that it was too soon to discuss possible applications of the new 200,000,000-volt force, which is thirty times more powerful than radium, but pointed to the fact that radium is now the most efficient weapon used for the treatment of cancer. Like radium, it may be twenty or twenty-five years before the phenomenon could be put to practical use and it might not be practical at all, they said.

Dr. Fermi related that Dr. Hahn bombarded a synthetic element known as "ekauranium" with neutrons, the slow-moving particles of the atom, and produced barium, the substance used in making X-ray pictures of the stomach and intestines.

The only way that this could occur, according to physicists, would be for the ekauranium atom to split apart to form barium and the rare element masyrium.

In causing such a split a force of 200,000,000 volts would be generated since atoms are held together by electrical forces many hundred times more powerful than the force of gravity which holds the stars, planets, sun, earth and moon in their orbits.

PRESIDENT'S GUESTS FOR BIRTHDAY DINNER

'Cuff Links Club' Members Included in Monday's Ceremony

WASHINGTON, Jan. 28 (AP).—President Roosevelt, following a long-standing custom, included members of his "Cuff Links Club" among the more than a score of guests invited to his birthday dinner Monday at the White House.

Other guests will be Mr. and Mrs. James Roosevelt, Mr. and Mrs. Elliott Roosevelt, Mr. and Mrs. Franklin D. Roosevelt, Jr., Hall Roosevelt, Secretary and Mrs. Morgenthau, Secretary Hopkins, Dr.

man of New York; Kirke Simpson, Charles McCarthy and James P. Sullivan.

The President will be 57 years old, but there will be only twenty-one candles on his cake. That, too, is an old Roosevelt custom.

Mrs. Roosevelt after the dinner will make the rounds of seven capital birthday balls, which, like hundreds of similar celebrations elsewhere, will raise money for the national campaign to stamp out infantile paralysis. More than $45,000 in dime contributions have reached the White House.

Endorses Auto Union Rebels

Support of the insurgent group in the United Automobile Workers of America, now attacking the leader-

REPUBLICAN GROUP BACKS LOTTERIES

Westchester Clubs' Platform Calls Also for Pari-Mutuels to Ease Tax Burden

LA GUARDIA SLAP IMPLIED

Plank Urges a Penalty for Official Who Fails to Back Party Candidate Openly

Special to THE NEW YORK TIMES.

WHITE PLAINS, N. Y., Jan. 28.—Legislation paving the way for legalization of lotteries in New York State is urged in the 1939 platform of the Young Men's Republican Clubs of Westchester County adopted today at the organization's sixth annual convention in the County Center here.

In a plank designed to "relieve property owners of the disproportionate tax burden they now bear," there is recommended legalization of the pari-mutuel system at recognized race tracks, as previously urged by the organization, and repeal of the constitutional prohibition against lotteries, giving the Legislature power to prohibit or establish and regulate such lotteries.

Plank on Relief Funds

The same plank urges that the State finance relief and reimburse municipalities for relief expenditures on a 100 per cent basis, instead of 40 per cent, as at present, and further urges strict economy in county and local government.

Delegates recommended legislation assuring Westchester a "more equitable" share of funds derived from gasoline taxes and automobile license fees, and suggested that the State pay the cost of improving the Cross County Parkway at Yonkers. The platform opposes the plan under which the county is to

publican thereafter whole-hear publican n the support thereafter. name, how in discussi

The platf opposition and altern posed use Westcheste roadbed fo urged tha taken to re way, by e authority i

State Sen of Erie Cou Republican bring disgr youth into Harry W tor was e county club Bailey of elected w Tuckahoe, Seidenstein ond vice-Warren of vice presid Larchmont Robert F. secretary, Jr., of Oss

C. Wayl sistant pro and Repub didate in in a speech that the P was using to create pi American leaders. for the adm due influen because the mits was ment agen

MURPHY

His Aid A Teleg

The An Union yes ney Gener vestigate a seizure an souri State

The discovery of fission as reported in *The New York Times*, 29 January 1939

The paper was composed by several long-distance telephone calls, Lise Meitner having returned to Stockholm in the meantime. I asked an American biologist [Arnold], who was working with Hevesy, what they call the process by which single cells divide in two; "fission," he said, so I used the term "nuclear fission" in that paper.

O. R. Frisch,
What Little I Remember, 1979

tion. Unfortunately, however, partly because he was overtired, he then delayed six days, until January 22, before writing to Bohr, informing Bohr of his experimental confirmation of fission, and enclosing copies of both notes to *Nature.* On January 24, Bohr, still having received no word from Frisch, wrote to him once again. Bohr realized that containment was now no longer possible. The issue of *Die Naturwissenschaften* containing Hahn and Strassmann's article had arrived in Princeton a few days earlier, and hence word of their discovery soon would spread far and wide. Moreover, the fifth conference on theoretical physics organized by Gamow, Teller, and Tuve, scheduled for January 26–28, was about to be convened in Washington, D.C. Bohr had no choice but to speak out. The report on the conference, dated 1 February 1939, states: "Certainly the most exciting and important discussion was that concerning the disintegration of uranium . . . Professors Bohr and Rosenfeld had arrived from Copenhagen the week previous with this news . . . Professors Bohr and Fermi discussed the excitation energy and probability of transition from a normal state of the uranium nucleus to the split state." The experimental team at the Carnegie Institution's Department of Terrestrial Magnetism demonstrated the fission process for Bohr and Fermi after the close of the conference on January 28.

On February 2, at long last, Bohr received Frisch's letter of January 22 with its two notes enclosed; he replied the following day. At the Washington conference, Bohr had made certain that Meitner and Frisch's priority in interpretation had been recognized, and he told Frisch that he had also succeeded in getting that information into a *Science Service* news release on January 30. Bohr enclosed a final version of his own note to *Nature.* As it turned out, Meitner and Frisch's note appeared in *Nature* on February 11, Frisch's on February 18, and Bohr's on February 25.[12]

Bohr immersed himself in the fission problem in Princeton, where he had a visiting appointment at the Institute for Advanced Study. Already by February 7 he had prepared a letter to the editor of *The Physical Review,* partly to reiterate the history of the discovery and interpretation of nuclear fission for the record, but mainly because on one of the preceding mornings he had reached the conclusion, through an ingenious argument, that not U^{238} but the rare isotope U^{235} is the nucleus primarily responsible for fission.

Bohr was stimulated by George Placzek, who had just joined Bohr and Rosenfeld in Princeton, into thinking very carefully and very deeply about the existing experimental evidence on fission. Observations had shown — so went the essence of Bohr's argument, which he presented to Rosenfeld, Placzek, and Wheeler — that fast neutrons produce fission in uranium; this could be ascribed to the abundant isotope U^{238}, following the formation of a compound nucleus of mass 239. Neutrons of lesser, intermediate energy did not produce fission, so it could be assumed that here the U^{238} isotope was no longer contributing, nor was the rare U^{235} isotope because of its low

Copenhagen, 22.1.1939.

Nature 2-2 1919

Dear Professor Bohr.

Please excuse my writing english, but although I speak danish quite easily I find it rather difficult to write it, especially about scientific things. (And then my machine has no danish types).

Enclosed I send you copies of two letters which I sent to NATURE a few days ago. The first one you have seen before you left; it has increased somewhat by adding details but I hope otherwise to have followed, more or less, the suggestions you gave me on our talk in Carlsberg. The second paper contains the report of an experiment, which I decided to undertake on Thursday Jan.12th; I was so lucky as to get a positive result the next day, which I confirmed, and got details of, during the next three days, and on monday night I sent off the letters. Yesterday I got the proofs and sent them back last night; so I hope both papers will come out very soon. (Of course, there is no "tavshedspligt" any longer!); Hahns paper came out the day you left.) It was great fun to get this experiment done so quickly, but now of course there are a lot of things for more detailed study. The next thing, I shall try to do the experiment with a very thin layer of uranium (or thorium) so as to have practically no energy loss of the particles emerging from it; this should permit the determination of energy groups and thereby of the mode of division (mass ratio of the two parts; whether there is a nearly unique "division line" or whether a broad range of mass ratios occurs). Further- more in this experiment should permit the determination of the cross- section of the uranium nucleus effective with respect to these "fission" processes (I wonder how you like this word; it was suggested by the bio- chemist Mr.Arnold, who told me it was the usual term for the division of bacteria). From my present experiments with a thick layer of uranium

and under the assumption of a range of 4 mms in air for the particles resulting from the fission, a cross-section of $1.5 \cdot 10^{-25} cm^2$ is obtained, which agrees exactly with the value found by Hahn, Meitner etc, for all the activities obtained in uranium (including the "transuranium" activi- ties, which are the strongest). The cross-section is, however, much less then the total cross-section (about $15 \cdot 10^{-25}$) of uranium. I was first surprised at this, but one should perhaps expect it from the "liquid drop" model, since a particle hitting a droplet will excite many different oscillations, of which only the ones corresponding to the lowest order spherical harmonics will favour the division of the droplet in two.

Of course one must see if elements beside uranium and thorium show this phenomenon. I have had lead in the chamber, with negative result, but this experiment (and experiments with Bi, Tl, Hg, Au etc) will be repeated more carefully and, perhaps, with the fast neutrons from Li + D from our High-tension tube, which will be ready to work in a few days, from what Bjerge tells me. We also thought of looking for fission induced by the hard gamma-rays from Li + H. Furthermore we intend to carry out some experiments of the kind indicated at the end of my letter to NATURE, probably in collaboration with Prof.Meitner, by sending her the irradiated samples (air mail) for physical and chemical investigation.

So it seems that this new phenomenon opens possibilities for quite a bit of work. But of course the other things are not being neglected. The cyclotron is working again after a period of general repair; I think Dr.Jacobsen has written to you about this. The magnet for the neutron magnetic moment is being cast these days. Dr.Simons, of Helsingfors, has started work; he is going to measure the scattering cross-section of protons (in water) relative to resonance neutrons of silver and iodine, which have energies high in comparison to the chemical binding of the protons.

With my best regards, also to Erik, yours sincerely O.R.Frisch

Frisch's letter of 22 January 1939 to Bohr, describing the fission hypothesis.

abundance. Slow neutrons, however, again produced fission—and here it could be ascribed to the rare U^{235} isotope, because the great increase in fission probability (owing to its inverse dependence upon neutron velocity) more than compensated for the decrease arising from the low isotopic abundance. Moreover, since the compound nucleus formed in this case would be of mass 236, an even number, the binding energy of the incident neutron, and hence the excitation energy of the compound nucleus, could be expected to be appreciably higher than that for the case of a compound nucleus of mass 239, an odd number. It could be concluded that the isotope mainly responsible for fission, and entirely responsible for fission by slow neutrons, was the rare isotope U^{235}. More than one year later, in mid-April 1940, John R. Dunning and his colleagues at Columbia University, bombarding a small sample of U^{235} separated out mass-spectroscopically by Alfred O. Nier at the University of Minnesota, proved that Bohr's conclusion was correct.

Bohr concluded his paper of 7 February 1939 by noting that he and Wheeler were currently collaborating on a "closer discussion" of the mechanism of nuclear fission. That closer discussion occupied the entire balance of time remaining to Bohr in Princeton. The two prepared a short paper which Wheeler delivered at the Washington meeting of the American Physical Society between April 27 and 29. Two months later, on June 27, after Bohr had sailed for Europe, Wheeler submitted their final twenty-five-page paper for publication in *The Physical Review*.[13] This work remains one of the finest collaborative efforts in physics. The two authors, sharing a deep mutual respect, combined their physical intuitions, knowledge of physics, and mathematical skills to produce a classic of the literature.

Bohr and Wheeler's starting point was Meitner and Frisch's comparison of the target nucleus to a liquid drop for which the mutual repulsion of its surface charges nearly cancels the attractive nuclear forces at its surface, analogous to those associated with surface tension. The relatively small amount of energy added by an incident neutron then produces an excited compound nucleus, which during its decay becomes deformed. If this deformation is large enough, it will divide into two smaller droplets, releasing a large amount of energy. Bohr and Wheeler's great achievement was to analyze this process quantitatively. They began by appropriately modifying the semiempirical mass formula first proposed by Gamow and later extended by von Weizsäcker. This enabled them to calculate the energy released when various heavy nuclei divide in various ways, in general accompanied by the emission of neutrons. They also calculated the energy released when a given fission fragment subsequently undergoes beta decay, finding that in a few cases the new nucleus thus produced is sufficiently excited to permit the emission of a neutron—a so-called "delayed neutron," whose origin had not previously been understood.

To predict whether a given compound nucleus actually would or would not undergo fission, Bohr and Wheeler introduced an entirely new concept, that of the "fission barrier." They had to determine how small departures from sphericity affect the repulsive and attractive forces at the surface of a given compound nucleus. Would the forces change in opposition to each other in such a way that for some critical energy of excitation the compound nucleus would elongate beyond a point of no return and split into two smaller droplets? Was there an energy barrier which, if transcended, would inevitably result in fission? Wheeler recalled that the very concept of such a well defined barrier seemed strongly counterintuitive to some of his colleagues, for how could such discreteness be compatible with the continuousness of a liquid? Nevertheless, Bohr and Wheeler found, by modifying and carrying to higher approximation certain old but beautiful calculations of Lord Rayleigh, that the fission barrier did indeed emerge as a well defined quantity, and one which held the key to understanding the conditions under which heavy nuclei were stable or unstable against fission.

A harmonious picture of the mechanism of nuclear fission took shape. The theory explained why nuclei of intermediate atomic weight are stable against division into two roughly equal parts, and why uranium is about at the limit of stability, so that heavier nuclei are not found in nature. It explained why, for excitations below the fission barrier, a given compound nucleus was more likely to decay by some competing process such as the emission of gamma radiation, well described by the Breit-Wigner formula. It established a quantitative foundation for Bohr's earlier conclusion that slow neutrons can cause the rare isotope U^{235}, an even-odd nucleus, to undergo fission, whereas fast neutrons are required for U^{238} or Th^{232}, even-even nuclei. It led to a simple channel formula for the fission width, a concept that became of central importance in reaction theory. Furthermore, the very analogy to the division of a drop of liquid, which is generally accompanied by the formation of a few tiny drops between the two larger droplets, offered a basis for understanding the origin of the two or three neutrons emitted in the fission process, as had been reported by Fermi and his coworkers at Columbia University and by Joliot-Curie and his colleagues in Paris. Finally, the theory provided insight into the possibility that protons, deuterons, and gamma rays, in addition to neutrons, might produce fission in certain cases. Thus, while certain features of the fission process were not yet understood in detail — for example, why the two main fragments are generally unequal in size — Bohr and Wheeler had mapped out the entire terrain, establishing, as they concluded, "a satisfactory picture of the mechanism of nuclear fission."

John T. Tate, editor of *The Physical Review*, received Bohr and Wheeler's paper on 28 June 1939, the day after Wheeler had submitted it. Wheeler also sent a copy to Bohr in Copenhagen, who replied on July 20, expressing his admiration for all of the work that Wheeler had done on it and, inevitably, suggesting some changes — but remarkably few. One pertained to an important insight into how cross-sections should be calculated — a problem that Bohr, Peierls, and Placzek meanwhile had explored in Copenhagen.

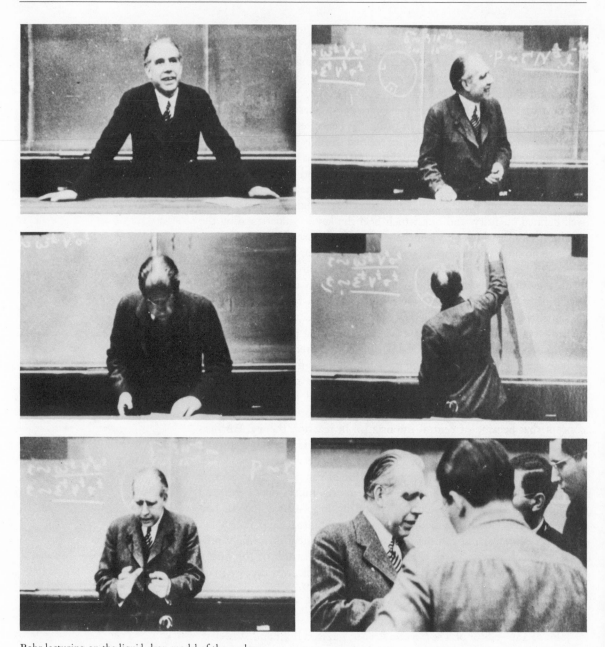

Bohr lecturing on the liquid-drop model of the nucleus.

Bohr enclosed a copy of their paper, which he had submitted to *The Physical Review* on July 4, to enable Wheeler to take it into account.

By that time, many people were beginning to think about practical means of releasing nuclear energy. The idea of a chain reaction had occurred to Leo Szilard as early as 1934, and now in early 1939 he, Wigner, and Fermi pressed it vigorously. A little later, in Birmingham, England, Peierls began to calculate the amount of uranium that would be required for criticality. Bohr, too, pursued similar questions after his return to Copenhagen: in a manuscript dated 5 August 1939 he examined the necessity of introducing a light element as a neutron moderator to achieve a chain reaction in natural uranium, while recognizing that the situation would be entirely different if pure or highly concentrated U^{235} were employed. Less than one month later, on September 1, the German armed forces invaded Poland. In another seven months, on 9 April 1940, they occupied Denmark.

Despite these shattering events, Bohr continued to work on fission-related problems. On 20 October 1939, he and Wheeler submitted a letter to the editor of *The Physical Review,* composed through correspondence, showing that fast neutrons could be expected to produce fission in protactinium more easily than in thorium but less easily than in U^{235}, in line with experiments carried out at Columbia. That fall and winter Bohr also gave several lectures on nuclear fission. One, which he delivered in December before the Society for Dissemination of Natural Science in Copenhagen, was particularly memorable.to those who heard it, among them Christian Møller and Stefan Rozental, because Bohr outlined in it the entire problem of how to liberate nuclear energy on a practical scale, including the ominous possibility of its use in a weapon of enormous destructive power.

Bohr also was thinking deeply about the problem of the statistical distribution of the fission fragments — why, in general, the two main fragments differ considerably in mass. On 16 December 1939, he sent a manuscript on this subject to Wheeler, who was to go over it and then submit it for joint publication. Under the wartime conditions then prevailing in Europe, communication by sea mail was uncertain, with delivery times of a month or more quite common. In fact, the ensuing revisions and correspondence between Bohr and Wheeler on this manuscript extended into mid-1940, by which time the growing secrecy surrounding fission-related research delayed its publication indefinitely.

Not all of Bohr's work, however, suffered that fate. In the summer of 1940 Bohr submitted three papers to *The Physical Review* which were subsequently published. In the first and third (received July 9 and September 3), he analyzed the process by which fission fragments are scattered and stopped in their passage through matter, and he showed that his analysis was consistent with cloud-chamber photographs of fission-fragment tracks recently taken in Copenhagen. In the second paper (received August 12), he showed that in some cases the compound nucleus formed by fast-neutron bombardment — for example, U^{239} — may be sufficiently excited to undergo fission even after emitting a neutron. Bohr's continuing deep interest in nuclear fission, of course, paralleled that of physicists everywhere. Less than a year after its discovery, Louis A. Turner of Princeton University could submit an extensive survey to *Reviews of Modern Physics* in which he noted that almost 100 papers had already appeared on the subject.

After the occupation of Denmark, Bohr continued to work in isolation in Copenhagen. In May 1941 he submitted a paper both to *The Physical Review* and to *Nature* in which he discussed the possibility of producing fission in thorium and uranium by deuteron bombardment. At the same time, he oversaw the construction of a 2-million-volt high-pressure Van de Graaff accelerator and other new equipment at the institute. However, the harshness of the German occupation forces increased more and more until, at the end of September 1943, Bohr and his family were forced to flee Copenhagen. Soon, Bohr and his physicist son Aage became deeply involved in the British and American efforts under way to secure the use of nuclear energy for military purposes. Yet foremost in Bohr's mind were the political implications of the development of nuclear weapons, and throughout the balance of the war and thereafter they were Bohr's principal concern. He took every opportunity he could to press his vision of a peaceful world based upon cooperation and trust between individuals and nations.

During the early postwar period, Bohr also found time and energy for research. In 1948 he published a classic 144-page monograph, which had been long in preparation, on the penetration of charged particles through matter, a subject that had attracted him from his earliest postdoctoral days. Later, from the last quarter of 1949 through the first month of 1950, Wheeler was again in Copenhagen, where he gained important insights into the fundamental question of how the many-particle behavior of nuclei, as described by Bohr's compound-nucleus model, might be reconciled with their single-particle behavior, as represented in the recently enunciated shell model. After Wheeler returned to Princeton, Bohr and Wheeler continued to work — and Wheeler (and David Hill and Wheeler) to publish — on aspects of this question, which was central to the collective model subsequently developed by Aage Bohr, Ben Mottelson, and James Rainwater.

At the center of Bohr's contributions to nuclear physics was his quest for harmony and unity. Some themes, such as the behavior of liquid drops and the passage of charged particles through matter, were sustained in his thought from beginning to end. But his vision was much more comprehensive, for at the center of the center, whether surfacing in wave and particle, compound nucleus, biological life, or human cultures, was his principle of complementarity. This powerful concept crossed all boundaries, bringing meaning to his physics and to his life.[14]

Physics in Copenhagen in 1934 and 1935

John A. Wheeler

The first sight of Bohr's institute surprises anyone aware of how important it has been to the development of physics over the years.[1] How could so much come out of a building so small, smaller than many a house? The impression of surprise was increased for this young arrival in September 1934 only by seeing how few there were inside. Only once or twice a year, at times of conferences, did the attendance climb to as high as twenty-five. Naturally there was the familiar, active experimental group in nuclear physics working in a laboratory addendum to the building. It included T. Bjerge, K. J. Brostrøm, George de Hevesy, J. C. Jacobsen, and E. H. Rasmussen. Old collaborators stopped in at various times during the year, including Werner Heisenberg, Oskar Klein, Harrie Massey, Ivar Waller, and C. F. von Weizsäcker. There were also special visitors from time to time in consequence, direct or indirect, of Hitler's accession to power on 30 January 1933 and of the growing shadow over Germany. They came for a few days or a few weeks or a few months on their way from positions in Germany to positions elsewhere, often still to be found. Among them were Hans Bethe, Max Delbrück, James Franck (a prince of a man), Hilde Levi, George Placzek, and Edward Teller. Learn Danish in one month? That was one of the achievements of Otto Robert Frisch, a visitor for the year who had quickly made himself another highly productive member of the experimental group. Theorist colleagues there for most of the year were few but hospitable: Bohr himself, Fritz Kalckar, Christian Møller, Milton S. Plesset, Stefan Rozental, E. J. Williams, and myself, supplemented for extended periods by Bohr's wonderfully scholarly and long-term collaborator Léon Rosenfeld. The library was quiet; the cleaning woman each day cleaned the neat gray trim and aired the room all out, no matter how cold this operation

John A. Wheeler first became famous when he, together with Bohr, wrote the classic 1939 paper on the theory of nuclear fission. He taught for many years at Princeton University and is now at the University of Texas at Austin. His research interests have been largely in fundamental cosmological questions involving relativity and quantum mechanics.

made the user. It was the standard place for work and for occasional two- or three-person discussions.

E. J. Williams was mentor and guide in my first months at the institute. He was of less than average height, compact, vigorous. Coming from Aberystwyth, he had used only his native Welsh until near the end of his days in secondary school. At Cambridge he had done experimental work until Rutherford forced him into theory by instituting a rule that no one should be in the laboratory beyond the end of the afternoon. Williams explained to me about the ways of life at the institute and of the Niels Bohr whom I had first seen and heard speak at the "Century of Progress" Chicago World's Fair in 1933. He told me of the agony Bohr had experienced on his arrival at Cambridge as a young postdoctoral worker, shy, and at that time ill at ease in English. Not knowing what to say when he had his first meeting with J. J. Thomson, director of the Cavendish Laboratory, and having read thoroughly and appreciated a book of Thomson's (his *Electricity and Matter*), in his embarrassment he could only blurt out a catalogue of the principal errors in the book — foretaste of Bohr's disastrous encounter with Winston Churchill on 16 May 1944. It was no wonder, Williams explained, that close relations did not develop between the two men. The twenty-seven-year-old Bohr found it more stimulating to go to Manchester and work with the magnetic forty-year-old Rutherford. At Copenhagen, Williams told me, it had been Bohr's habit each year to start to give a course of lectures. Each time, however, an exciting issue had developed in the course of the first session or two, and Bohr had put off further lectures until the question should have been entirely cleared up. A few months before I arrived Williams had visited the Soviet Union in company with Bohr, and he regaled me with an account of his own exciting motorcycle trip in the outskirts of Leningrad. Eleven years later, at age forty-two, his heart failed, and the man of so much warmth and drive was gone.

The summer before my arrival in Copenhagen Niels Bohr had lost his oldest son, Christian, overboard from their sailboat in a storm. He and the family were too deeply affected for him to go to the London conference, or to show up regularly at the institute the first part of the fall. One day, however, as I arrived on my bicycle at the usual hour at the institute, I noticed a workman tearing down the vines that had grown too thickly over the gray stucco front, and on closer view saw that it was Bohr himself, following his usual modest but direct approach to a problem. I had talked with him briefly on a couple of earlier occasions; but now full institute activity was to start. How did he pick the topic? How did he make headway with it? And how did he communicate his conclusions?

"He sees farther ahead than any man alive": that was why — with the backing of Gregory Breit — I had asked to spend with Bohr the second of my two fellowship years. Bohr distilled the central issue out of dialogue with those who were themselves distillers of issues, former collaborators and special visitors. The single-hearted attention that Bohr gave to such a colleague showed nowhere better than in the way the two of them would walk up and down outside the institute. They might share a less private discussion at the lunch to which so many of us brought the open-faced sandwiches — smørrebrød — purchased down the street. However, soon the talk would focus again more sharply on the issue that had been, or was in the course of being, "smoked out." Bohr would take the visitor away to his office, often carrying along his "right-hand man" of the moment — Léon Rosenfeld or E. J. Williams in my time. Bohr would go round and round the table as he talked or joked, expostulated or reflected, his whole soul taken up in the action. He would stop to make an especially strong point — or to listen briefly. His words were forceful. His voice was soft. His glance was piercing as he looked up from time to time and stared into one's eyes. His mood would change from time to time as dictated by the discussion itself: for making a point, "How could one possibly believe . . ." or "There is not the slightest evidence that . . ." If in doubt, his head would tip to one side as he spoke to one position; to the other, as he spoke to the opposite position.

Explanation was never dry pedagogy, but a one-man tennis match in which Bohr hit the ball from one court, then ran to the other fast enough to hit it back — the more times, the more enjoyable the game: "Such-and-such an effect leads one to expect thus-and-so . . . Indeed one sees thus-and-such, but then so-and-so observed such-and-such . . . Then we were in immense difficulty. Just at this point so-and-so pointed out that the proper formulation of the principle is thus-and-such . . . This discovery brought the whole subject into order. But then so-and-so realized that this extended principle stands in absolute contradiction to the stability of such-and-such . . . We were all lost until we found that the new formulation was really absolute nonsense . . . What fools we've been! We have only to recognize such-and-such and we see that absolutely everything has to be exactly as it is." From time to time, to make a point or lighten the atmosphere, there would come a joke. One of Bohr's favorites was his definition of a "great truth": a truth whose opposite is also a great truth.

Of all ways to tell the visitor of some new finding at the institute, and convince him of it, and tell him how to convince others, it would be difficult to imagine a single one both more modest, and more effective, than this "explanation-by-tennis"; but for the colleague it was only a warmup for the real tennis match. In it he himself knocked the ball back and forth with Bohr. The spirit might have been the game for the game's sake, but the excitement came from expectation of the unexpected. The best witness to the level of the dialogue was the level of the participants, from Kramers to Heisenberg and from Bethe to Pauli. How else could a man hold up his end in such an encounter except to make his own point of greatest concern?

Complementarity had a place in almost every discussion. *Contraria sunt*

[Bohr] had a soft voice with a Danish accent, and we were not always sure whether he was speaking English or German; he spoke both with equal ease and kept switching.

O. R. Frisch,
What Little I Remember, 1979

223

Niels Bohr's coat of arms, designed in 1947 when he was awarded the Danish Order of the Elephant, normally given only to members of royal families and presidents of foreign states. It was hung near the king's coat of arms in the church of Frederiksborg Castle at Hillerod. Bohr chose the Chinese yin-yang symbol, which stands for the two opposing but inseparable elements of nature. The Latin motto reads: "Opposites are complementary."

complementa (contrary things are complementary) was the motto that Bohr later chose for his coat of arms. Years after Bohr's death Heisenberg told me how, right after the publication of his own paper on the uncertainty principle, he was out sailing with Bohr and was explaining to their sailing partner, Niels Bjerrum, the contents of the article. After hearing him out, Bjerrum turned to Bohr and said, "But Niels, this is what you have been telling me ever since you were a boy." Bohr's lifetime advocacy of complementarity owed much to the philosophy of his great teacher, Harald Høffding, just as Einstein's continuing rejection of complementarity, and 1917–1929 rejection of the big bang, were influenced by his youthful admiration for the thought of Benedict Spinoza, advocate of determinacy and of a universe that goes on from everlasting to everlasting.

Sometimes a week or two would go by without a meeting. When it came, three or four of us would gather with Bohr in his office or another room to discuss some then worrisome point. To bring about a seminar it took a visitor, perhaps an experimental physicist and former collaborator from Poland, or a new and important paper reported by someone at the institute, perhaps Jacobsen or Rosenfeld. The attendance amounted to a dozen or two. The language was usually German, occasionally English. The joy was to have something that "wouldn't fit." The central idea of the institute was clear: "No progress without a paradox." Most seminars were successful in the sense that Bohr broke in halfway through or sooner to solve the puzzle or explain the central point at issue. He would get to his feet and, reflecting as he kept talking and pacing up and down before the blackboard, encourage himself by saying every now and then, "Now it comes, now it comes."

Suddenly it really would come, and he would give the explanation to us all as another tennis match. It was more reminiscent of soccer, however, in which he had been a national hero, to see the way he plunged into the middle of things, found the central point, seized on it, and delivered it with great force to all assembled. Only rarely did the worst happen: nothing came up that surprised anyone, and Bohr had to utter those dreaded words, "It was an interesting seminar."

Usually the new problem became a focal point of discussion in the following days. Those days could almost have been numbered odd and even. One day was a day of building. If so-and-so is true, such-and-such follows. That will give us the chance to understand thus-and-so. That means it will be absolutely central to measure this-and-this cross section. Then we will be able to predict such-and-such with great assurance. No criticism. That was reserved for the next day. If at its end anything survived, that battle-tested core became the starting point of yet another day of building — and so on, up to a conclusion that could be played out as a complete tennis match.

I never saw Niels Bohr make progress with an idea except in dialogue or dictation or sudden revelation out of the depths of the subconscious. Always the end desired was a harmonious account of a wide range of experience. For this purpose he kept a continuous slow fire under about fifteen topics. They ranged from the angular momentum of light to dispersion relations for reaction cross-sections in the continuum, and from stopping power to superconductivity. Preliminary drafts of papers on each he stored in a little cabinet at Carlsberg in his home office, just off the Pompeian court, where he did his dictation on any issue of importance.

The opposite of a scatter-shot investigator, Bohr used to concentrate on the biggest physics questions in sight. In the early 1930s one such question was: Does quantum electrodynamics predict correctly the loss of energy by electrons of 100 MeV and more to radiation, and the production of electrons by photons of 100 MeV and more? No. That in brief was the theme of a lecture I had heard Robert Oppenheimer give to a packed audience at a meeting of the American Physical Society in Chicago on 29 December 1933. At the September 1934 London and Cambridge International Conference of Physics, I had heard Hans Bethe likewise argue that "quantum theory apparently goes wrong for energies of about 10^8 volts" Month after month in the fall and winter of 1934–35 Bohr hammered away at this question with E. J. Williams and anyone else of us around who could contribute, establishing simple and battle-tested arguments that quantum electrodynamics cannot and does not fail at high energies. No finding did more to establish the climate of opinion that led — through the beautiful experiments of Carl D. Anderson and Seth Neddermeyer — to the discovery of the mu meson.

If the argument for upholding electrodynamics took months to incubate and hatch, the compound nucleus model of nuclear reactions (the other

If one forgets the occasional employment of a pair of skis, the bicycle was [Bohr's] favorite mode of locomotion. Its relatively slow progress, based on a balance of dynamic variables that one cannot explain adequately in a few words, is, I think, a good introduction to the character of Niels Bohr.

Edward Teller, "Niels Bohr and the Idea of Complementarity," 1969

225

great Bohr achievement of the academic year 1934 – 35) took minutes. I will never forget that springtime excitement. Christian Møller, just back from a visit to Rome, was starting to report in our seminar some of the beautiful results that Enrico Fermi, Eduardo Amaldi, Bruno Pontecorvo, Emilio Segré, and Franco Rasetti had discovered about the absorption of slow neutrons in selected atomic nuclei. He had only barely outlined the Rome findings when Bohr rushed forward to explain how they worked a complete change in all our ideas of nuclear transformations. Our picture should be, Bohr explained and emphasized, not a long mean free path, with the energy concentrated on one nucleon, but a short mean free path, with the energy spread among many nucleons. Then and there he outlined the "compound nucleus" model of nuclear reactions. It formed the center of attention of his work — and the work of many others — in the weeks and years that followed.

Human issues claimed the upper hand with Bohr from time to time. How to make, despite the economic depression, a new position in Denmark for a promising young Danish scientist? How to manage a temporary haven for a refugee scientist from Hitler's Germany and find him or her a position in a larger country? What could be done, in concert with Rutherford, to persuade Stalin to let Peter Kapitza return to Cambridge?

No great human concern left Bohr indifferent. From the rise in expectations of the humblest inhabitant of a Third World country to the plight of blacks in some parts of the United States, and from the Achilles' heel of gangster governments to the mutual enrichment of diverse cultures, every broad problem that combined importance for the world with "doability" was the subject of deep thought and intense conversation with everyone who cared, from king to cleaning woman and from ambassador to fellow scientist. Nothing has done more to convince me that there once existed friends of mankind with the human wisdom of Confucius and Buddha, Jesus and Pericles, Erasmus and Lincoln than walks and talks under the beech trees of Klampenborg Forest with Niels Bohr.

Some Recollections of Bohr

Rudolf Peierls

I probably spent less time with Bohr than most of the others who have contributed to this book, but many short visits to Copenhagen, and encounters with him elsewhere, have left me with very vivid impressions of his personality and his mannerisms.

The first occasion was in 1930, when I left Zurich to visit Copenhagen during Easter vacation. For part of this time the Bohrs were at their country house in Tisvilde, and at least one afternoon there was spent kicking a soccer ball around, though not in an organized game. Of course Bohr, a soccer enthusiast, dominated the scene; Werner Heisenberg, George Gamow, and Bohr's sons participated with great energy, while Wolfgang Pauli, Lev Landau, and I stood around, trying to keep out of the way.

It was very characteristic of Bohr to give his full and serious attention to the matter in hand, whether it was an important problem in physics or philosophy or an ephemeral problem raised by a game, a film, or a piece of detective fiction.

Similarly he never seemed to be influenced by the rank or status of the person with whom he was talking. On a visit to our house in 1938 he talked with our son, then aged three, without condescension, and with genuine interest in the conversation.

Of course, that first visit in 1930 was not all soccer; there was much physics discussed. I remember that Paul Dirac had then put forward his hole theory to deal with the states of negative energy and Bohr would not accept this as a consistent solution. I tried to convince him that it was acceptable, and, in retrospect, it seems presumptuous of a raw Ph.D. of twenty-two to argue with the great man on a point of principle. But he listened patiently and considered my words, though he was not convinced.

Bohr and I had a very different conversation about a year later. Landau

Sir Rudolf Peierls was born in Berlin and received his university education in Germany. Since 1933 he has made his home in England, holding chairs at the University of Birmingham and then at Oxford, where he is now Professor Emeritus. His research has embraced many different fields, but particularly nuclear theory and statistical mechanics.

Cartoon by George Gamow showing Bohr talking to a bound and gagged Landau. Léon Rosenfeld has commented on this episode as follows: "When I arrived at the institute on the last day of February 1931, for my annual stay, the first person I saw was Gamow. As I asked him about the news, he replied in his own picturesque way by showing me a neat pen drawing he had just made. It represented Landau, tightly bound to a chair and gagged, while Bohr, standing before him with upraised forefinger, was saying 'Bitte, bitte, Landau, muss ich nur ein Wort sagen!' ("Please, please, Landau, may I just say a word?") I learned that Landau and Peierls had just come a few days before with some new paper of theirs which they wanted to show Bohr, 'but' (Gamow added airily) 'he does not seem to agree—and this is the kind of discussion which has been going on all the time.' Peierls had left the day before, 'in a state of complete exhaustion,' Gamow said. Landau stayed for a few weeks longer, and I had the opportunity of ascertaining that Gamow's representation of the situation was only exaggerated to the extent usually conceded to artistic fantasy."

About a year later the same episode was incorporated in the "Copenhagen Faust," with the following dialogue:

THE LORD   Keep quiet, *Dau!* . . . Now, in effect
              The only theory that's correct
              Or to whose lure I can succumb
              Is

LANDAU   Um! Um-um! Um-um! Um-um!

THE LORD   Don't interrupt this colloquy!
              I'll do the talking. Dau, you see,
              The only proper rule of thumb
              is

LANDAU   Um! Um-um! Um-um! Um-um!

and I had written a paper about the uncertainty relations in relativistic quantum mechanics, and when we came to Copenhagen we found Bohr strongly opposed to our ideas. We decided to publish our paper anyway,[1] and then had the problem of whether to acknowledge our discussions with Bohr in the paper. We felt that this should not be done without permission, since it might imply that Bohr approved of the paper. So I went to ask him. He somehow misunderstood our purpose. He became very angry and said he was only trying to do his best to be helpful, but if we queried whether this was sufficient help to be acknowledged, he would insist that we not mention his name. We were unable to correct this impression and felt terrible. We did, of course, put in the acknowledgment, and the incident was soon forgotten. However, his opposition to our paper persisted, and was elaborated in his work with Léon Rosenfeld.

Bohr often expressed views on general matters. On the role of scientists in the discussion of political and other public matters, he used to say that scientists were not more objective or less prejudiced than other intellectuals, but that there was a difference: if one is a physicist, one has necessarily gone through the experience of making a statement and being proved wrong. To a social scientist or a philosopher, this might never happen.

Bohr was opposed to nationalism and excessive patriotism. He used to say that if there had to be such sentiments, he preferred them in the version common in English-speaking countries, expressed in the phrase "My country, right or wrong!" A patriotic German or Frenchman might never even imagine that his country could be wrong.

Many of Bohr's mannerisms are famous. Some arose from his intolerance of statements he regarded as incorrect, and this intolerance, coupled with his reluctance to hurt people's feelings, led on at least one occasion to the memorable remark: "I am not saying this in order to criticize, but your argument is sheer nonsense."

He could express his thoughts in many languages, but he always used words in a very personal way. One could often spot people who had spent some time at his institute by their having picked up some of his characteristic phrases — for example, "It is not the meaning . . . " (in German, "Die Meinung ist nicht . . . "), in the sense "It is not my intention to . . . "

Bohr's spoken or written words were not always easy to follow. As he liked to say, truth was complementary to clarity — and in his papers he leaned far toward the side of truth! The early drafts of his papers were often easier to read than the final product, because in revising the text he would add all the clauses and reservations that seemed to be necessary for the whole truth. Papers usually started with many handwritten drafts, followed by a number of typed ones, and a large number of proofs. He had an irrepressible tendency to keep improving things until the last possible moment. On one occasion, when he was inspecting the site where an extension to the institute was being constructed, the old foreman said: "Professor Bohr, look at the wall over there. If you want to move it again, you'd better be quick, because in a few hours the concrete will have set!"

The long and abortive birth pangs of a well-known paper by Bohr, George Placzek, and me on nuclear reactions were only partly due to Bohr's preoccupation with wording. Although the three of us agreed about the results, there was a disagreement about method: Bohr wanted to discuss the

It used to be recounted that Bohr had a horseshoe nailed over his front door for good luck, and that someone said to him: "But surely you don't believe such superstitions?" To which Bohr is said to have replied: "Oh no, but I am told that it works even if you don't believe it." (However, it seems that the story may have been apocryphal and an invention of George Gamow's.)

After H. B. G. Casimir

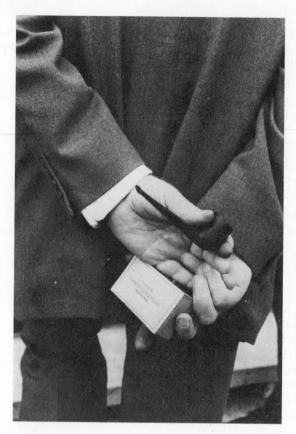

Niels Bohr, pipe smoker.

situation in qualitative terms, which Placzek and I did not regard as compelling, whereas we wanted to use exact expansions, which Bohr regarded as too formal. The outbreak of World War II interrupted these protracted discussions, and the paper was never published.[2]

Other characteristic incidents relate to the many long discussions at the institute, in which Bohr was so absorbed by the subject that he tended to be absent-minded about other matters. He usually had a cigar (later a pipe) in his hand, which would go out as he talked. He would pat all his pockets and then turn apologetically to his listeners: "Do you perhaps have a match?" Someone would produce a box, which he would use and then pocket. Five minutes later, the cigar was cold again, and after another patting of pockets the request would be repeated: "Do you have a match?" We learned to come to these sessions with a good supply of matches.

Lighting a cigar was not easy for Bohr, because he liked to light it holding it in his hand while talking. I long treasured as a souvenir a piece of chalk with one end blackened by soot. Evidently he had been deep in discussion, holding both the chalk and the cigar, and had applied the match to the wrong object.

I would not want it thought that these little disrespectful trivia sum up my recollections of Niels Bohr. I am mindful of the saying "No man is a hero to his valet," and perhaps I am writing here from the viewpoint of the valet. Yet for me Bohr remains a hero. His greatness as a scientist and thinker, his influence on all those who worked with him, and his generous kindness are well known. My little stories, which are perhaps less well known, help me see him as a human being and hence increase my affection for him.

Niels Bohr and His Institute

Hans A. Bethe

Hans Bethe was born in Strassburg (Alsace-Lorraine) and received his early training at Munich. In 1933 he came to the University of Manchester, but shortly afterward moved to the United States, where he was on the faculty of Cornell University from 1935 until his retirement. He received the Nobel Prize for physics in 1967 for his many contributions to theoretical nuclear physics, especially his discoveries concerning the mechanism of energy production in stars.

In contrast to most of my contemporaries among theoretical physicists, I did not make frequent and long visits to the Niels Bohr Institute during my early career. It was not until six years after obtaining my Ph.D. that I visited Copenhagen, for about four weeks in the summer of 1934.

I had relatively little direct contact with Bohr, but I heard him deliver lectures and participate in discussions at the colloquia. I remember particularly an incident during a colloquium on joint work by Edward Teller and Fritz Kalckar. The lecturer was Kalckar, who spoke in German, and the talk was far from clear. After a while Bohr said, "Let Teller continue; he can speak in English." It was a typical remark, indicating politely that probably Teller could give a clearer presentation of the physics.

At that time, Bohr was deeply involved in his work with Léon Rosenfeld on the observables in an electromagnetic field. It was too difficult for me to understand these intricacies. I was very much preoccupied by my own concerns, especially the offer of an assistant professorship from Cornell University, which promised to give me a permanent job. I accepted immediately. In physics, I was most interested in nuclear physics, especially the simplest nuclei. Bohr was not yet very interested in the nucleus, so he and I had little personal interaction on scientific matters. But of course I was very stimulated by all the people at the institute.

My main contact among older people was James Franck, to whom Bohr had given a temporary position prior to his permanent appointment at the University of Chicago. Among my own generation, I associated much with Edward Teller, and with Victor Weisskopf. I gave a talk at the colloquium on nuclear collisions, using the approach which would now be called *direct reactions*. Bohr did not seem to be much impressed, and subsequent histori-

232

cal research has shown that this occasion was probably not what stimulated him to conceive the compound nucleus.

Immediately after the discovery of fission — that is, early in 1939 — Bohr spent several months at the Institute for Advanced Study in Princeton. The next contact I remember having with Bohr after that visit was a little drawing he sent to the United States after he had escaped from Denmark to Sweden in 1943. The drawing, attributed to Heisenberg, had been made when Heisenberg had visited Bohr in Copenhagen. Bohr sent it to Los Alamos, where several of us, including Oppenheimer, Teller, and myself, puzzled over its meaning. As far as we could see, the drawing represented a nuclear power reactor with control rods. But we had the preconceived notion that it was supposed to represent an atom bomb. So we wondered: Are the Germans crazy? Do they want to drop a nuclear reactor on London?

Some time later (I believe early in 1944) Bohr himself came to Los Alamos, accompanied by his son Aage. I was charged with explaining to them the status of the project. It was a great pleasure to do this, and Bohr asked me many questions. At the end he concluded that we at Los Alamos knew what we were doing, that we were sure to succeed, and that we did not need his help.

However, once in 1945, Robert Bacher did ask for Bohr's help. This was on the question of the initiator for the first plutonium bomb — the device that was to inject neutrons at the right moment, and not before. There were two separate designs, one favored by Fermi, and the other by nearly all those who had worked on the design. It was difficult to decide against the opinion of Fermi, so Bacher asked Bohr's still higher authority. Bohr decided in favor of the design that most of us preferred.

Bohr spent the bulk of his time in the United States on his political aims. His talks with Roosevelt, and especially his disastrous talk later with Churchill, have often been described. At Los Alamos, Bohr talked repeatedly with Oppenheimer on his ideas for international control of atomic energy and weapons. This surely laid the groundwork for Oppenheimer's later work, on the commission headed by David Lilienthal, regarding the establishment of an international control organization. As is well known, this program was endorsed by Dean Acheson, and presented to the United Nations by Bernard Baruch, on behalf of the U.S. government. It was not accepted.

After the Second World War, I paid several visits to Copenhagen, some in connection with scientific conferences, others for purely scientific reasons. Bohr was most interested in improving the support for science in the Nordic countries. From that time stems the foundation of NORDITA. He was also able to get greater support for physics from the Danish government. Initially, he was less interested in CERN. But once the support for physics in the Nordic countries was assured, he threw his weight wholeheartedly into the debate in favor of CERN.

During each visit to Copenhagen, I was invited to Bohr's home in

Carlsberg, usually with only a few others. Before dinner, Bohr would take me for a stroll through the beautiful grounds and we would discuss the state of nuclear weapons and other armaments. Clearly the Lilienthal plan was no longer feasible. We discussed some of the concrete problems involved in fostering understanding and disarmament; as I recall, these included nuclear test ban efforts, in which I was very much involved.

But Bohr's main interest was his idea that there should be an "open world"—a notion on which he wrote a manifesto. He believed that by removing secrecy and fostering visits across national boundaries, the world could achieve better understanding. In fact, around 1955, a step in this direction was taken at the first international conference on nuclear power at Geneva. Since then, the Bohr Institute and NORDITA have been meeting places for scientists from East and West. But the world remains far from open.

I was looking forward to talking about these matters with Bohr again in 1964, when I would spend several months in Copenhagen. I made the visit, but unfortunately Bohr had died the year before.

After the war, Bohr's scientific interests turned to quantum electrodynamics. Especially important was the Pocono conference in 1949. Bohr, like all the rest of us, listened hour after hour to Julian Schwinger's talk about his formulation of quantum electrodynamics. Bohr was obviously very impressed; Schwinger's work started from standard formulations of relativistic quantum theory and electrodynamics, and then developed an entire scheme of *renormalization*. Richard Feynman, however, did not fare well. He spoke much more briefly and introduced entirely new concepts, such as electrons going backward in time. All this was apparently confusing to Bohr (and to most other people who had not grown up with the theory, as I had). So Bohr gave Feynman a hard time, and I had to console Feynman afterward when we got back to Cornell. Not long after this, Freeman Dyson proved that the theories of Feynman and of Schwinger were fundamentally identical.

Some ten years later, Feynman and Bohr met again at a Solvay conference in Brussels. Feynman gave a beautiful talk on all the experimental confirmations of renormalization theory in quantum electrodynamics, especially the precision measurement and calculation of the *Lamb shift*. Bohr became very unhappy: Are there then no paradoxes left in quantum electrodynamics? That would be terrible, because then we would have no way to search for a better theory.

Niels Bohr would probably have liked the unification of electrodynamics and weak interaction theory by Stephen Weinberg, Abdus Salam, and Sheldon Glashow. Their solution, I believe, was very much the one he would have liked to see. He would also have liked the richness of the attempts to develop a grand unified theory, and his deep intuition might well have contributed an essential idea to this theory.

SCIENCE

Vol. 86   Friday, August 20, 1937   No. 2225

SCIENCE: A Weekly Journal devoted to the Advancement of Science, edited by J. McKeen Cattell and published every Friday by

THE SCIENCE PRESS

New York City: Grand Central Terminal

Lancaster, Pa.   Garrison, N. Y.

Annual Subscription, $6.00   Single Copies, 15 Cts.

SCIENCE is the official organ of the American Association for the Advancement of Science. Information regarding membership in the Association may be secured from the office of the permanent secretary, in the Smithsonian Institution Building, Washington, D. C.

TRANSMUTATIONS OF ATOMIC NUCLEI[1]

By Professor NIELS BOHR

INSTITUTE OF THEORETICAL PHYSICS, UNIVERSITY OF COPENHAGEN

It has been pointed out on an earlier occasion[2] that in order to understand the typical features of nuclear transmutations initiated by impacts of material particles it is necessary to assume that the first stage of any such collision process consists in the formation of an intermediate semi-stable system composed of the original nucleus and the incident particle. The excess energy must in this state be assumed to be temporarily stored in some complicated motions of all the particles in the compound system, and its possible subsequent breaking up with the release of some elementary or complex nuclear particle may from this point of view be regarded as a separate event not directly connected with the first stage of the collision

[1] Abstract of lectures given in the spring of 1937 at various universities in the United States. The illustrations are reproductions of three slides shown in these lectures.

[2] N. Bohr, *Nature*, 137: 344, 1936.

process. The final result of the collision may therefore be said to depend on a competition between all the various disintegration and radiation processes from the compound system consistent with the conservation laws.

A simple mechanical model which illustrates these features of nuclear collisions is reproduced in Fig. 1, which shows a shallow basin with a number of billiard balls in it. If the bowl were empty, then a ball which was sent in would go down one slope and pass out on the opposite side with its original energy. When, however, there are other balls in the bowl, then the incident one will not be able to pass through freely but will divide its energy first with one of the balls, these two will share their energy with others, and so on until the original kinetic energy is divided among all the balls. If the bowl and the balls could be regarded as perfectly smooth and elastic, the collisions would continue until a sufficiently large part of the kinetic energy happened again to be concentrated upon a ball close to the edge. This ball would then escape from the basin, and if the energy of the incident ball were not very large, the remainder of the balls would be left with insufficient total energy for any of them to climb the slope. If, however, there were even a very small friction between the balls and the basin or if the balls were not perfectly elastic, it might very well happen that none of the balls would have a chance to escape before so much of the kinetic energy were lost as heat through friction that the total energy would be insufficient for the escape of any of them.

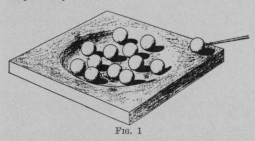

Fig. 1

Such a comparison illustrates very aptly what happens when a fast neutron hits a heavy nucleus. On account of the large number of particles which in this case constitute the compound system and their strong interaction with one another, we must in fact expect from this simple mechanical analogy that the lifetime of the intermediate nucleus is very long compared with the time taken by a fast neutron to cross a nucleus. This explains, first of all, that although the probability for a heavy nucleus to emit electromagnetic radiation in such a time is extremely small, nevertheless there is on account of the long life of the compound nucleus a not quite negligible probability that the system instead of releasing a neutron will emit its excess energy in the form of electromagnetic radiation. Another experimental fact, which is easily understood from such a picture, is the surprisingly large probability of inelastic collisions, resulting in the emission of a neutron with a much smaller energy than the incident one. Indeed from the above considerations it is clear that a disintegration process of the compound system, which claims a smaller amount of energy concentrated on one single particle, will be much more likely to occur than a disintegration, in which all the excess energy has to be concentrated on the escaping particle.

At first sight such simple mechanical considerations might be thought to contradict the fact, so well established from the study of the radioactive γ-ray spectra, that nuclei like atoms possess a discrete distribution of energy levels. For in the above discussion it was essential that the compound system would be formed for practically any kinetic energy for the incident neutron. We must realize, however, that in the impacts of high-speed neutrons we have to do with an excitation of the compound system far greater than the excitation of ordinary γ-ray levels. While the latter at most amounts to a few million volts, the excitation in the former case will considerably exceed the energy necessary for the complete removal of a neutron from the normal state of the nucleus, which from mass defect measurements can be estimated to be about eight million electron volts.

Fig. 2 then illustrates in a schematic way the general character of the distribution of energy levels for a heavy nucleus. The lower levels, which have a mean energy difference of some hundred thousand volts, correspond to the γ-ray levels found in radioactive nuclei. For increasing excitation the levels will rapidly come closer to one another and will for an excitation of about 15 million volts, corresponding to a collision between a nucleus and a high-speed neutron, probably be quite continuously distributed. The character of the upper part of the level scheme is illustrated by the two lenses of high magnification placed over the level diagram, one in the above-mentioned

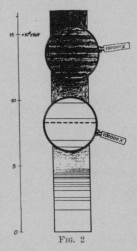

Fig. 2

region of continuous energy distribution and the other in the energy region corresponding to the excitation which the addition of a very slow neutron to the original nucleus would give for the compound system thus formed. The dotted line in the middle of the field of the lower magnifying glass represents the excitation energy of the compound nucleus when the kinetic energy of the incident neutron is exactly zero, and the distance from this line down to the ground state is therefore just the binding energy of the neutron in the compound system.

Information about the level distribution in the energy region near this line can be obtained from experiments on the capture of very slow neutrons with energies of a fraction of a volt. Thus if the kinetic energy of the incident neutron just corresponds to the energy of one of the stationary states of the compound system, quantum mechanical resonance effects will occur, which may give effective cross sections for capture of the neutrons several thousand times larger than ordinary nuclear cross sections. Such selective effects have actually been found for a number of elements, and it has further been found that the breadth of the resonance region in all these cases is only a small fraction of a volt.[3] From the relative incidence of selective capture among the heavier elements and from the sharpness of the resonances, it can be estimated that the mean distance of levels in this energy region is of the order of magnitude of about 10–100 electron volts. In the field of the lower magnifying glass in Fig. 2 there are indicated a number of such levels, and the circumstance that one of these levels is very close to the dotted line corresponds to the possibility of selective capture for very slow neutrons in this particular case.

The distribution of energy levels indicated in Fig. 2 is of a very different character from that with which we are familiar in ordinary atomic problems where on account of the small coupling between the individual electrons bound in the field round the nucleus the excitation of the atom can in general be attributed to an elevated quantum state of a single particle. The nuclear level distribution is, however, just of the type to be expected for an elastic body, where the energy is stored in vibrations of the system as a whole. For, on account of the enormous increase in the possibilities of combination of the proper frequencies of such motions with increasing values of the total energy of the system, the distance between neighboring levels will decrease very rapidly for high excitations. Indeed, considerations of the above character are well known from the discussion of the specific heat of solid bodies at low temperatures.

Thermodynamical analogies can also be applied in a fruitful way for the discussion of the disintegration of the compound system with release of material particles. Especially the case of emission of neutrons, where no forces extend beyond proper nuclear dimensions, exhibits a very suggestive analogy to the evaporation of a liquid or solid body at low temperature.

[3] The phenomenon of selective capture of slow neutrons, which shows an interesting formal analogy with optical resonance, has especially been studied in a paper of G. Breit and E. Wigner (*Phys. Rev.*, 49: 642, 1936). Estimates from experimental evidence of the breadth of the levels were first given by O. R. Frisch and G. Placzek (*Nature*, 137: 357, 1936) and have been discussed in details in a recent paper by H. Bethe and G. Placzek (*Phys. Rev.*, 51: 450, 1937).

In fact, it has been possible from the approximate knowledge of the level system of nuclei at low excitations to get an estimate of the "temperature" of the compound nucleus, which leads to evaporation probabilities for neutrons consistent with the lifetimes for the compound system in fast neutron collisions derived from the analysis of experiments.[4]

Fig. 3 illustrates the course of a collision between a fast neutron and a heavy nucleus. To follow the simple trend of the arguments, an imaginary thermometer has been introduced into the nucleus. As the figure shows, the scale on the thermometer is in billions of degrees centigrade, but in order to get a more familiar measure for the temperature energy, one has also added another scale to the thermometer showing the temperature in millions of electron volts. The figures give the different stages of the collision process. To begin with, the original nucleus is in its normal state and the temperature is zero. After the nucleus has been struck by a neutron with about ten million volts kinetic energy, a compound nucleus is formed with 18 million volts energy, and the temperature is raised from zero to

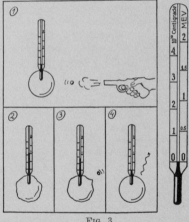

FIG. 3

roughly one million volts. The irregular contour of the nucleus symbolizes the oscillations in shape corresponding to the different vibrations excited at the temperature in question. The next figure shows how a neutron escapes from the system and the excitation, and accordingly the temperature, is somewhat lowered. In the last stage of the process the remaining part of the energy is emitted in the form of electromagnetic radiation and the temperature drops down to zero.

The course of the collision described above is the

[4] The idea of applying for the probability of neutron escape from compound nuclei the usual evaporation formula was first proposed by J. Frenkel (*Sow. Phys.*, 9: 533, 1936). A more detailed investigation on the basis of general statistical mechanics is given in a paper by V. Weisskopf (*Phys. Review*, in print).

most probable one if the energy of the incident neutron is large, but for lower energies of the neutron the probabilities of escape and of radiation will become of the same order of magnitude, giving rise to a considerable probability for capture. If we finally go down to the region of very slow neutrons it is known experimentally that the probability for radiation is even very much larger than the probability of escape. It will, however, be clear that in this case the analogy between neutron escape and evaporation will be quite inadequate, because the mechanism of escape, like the formation of the compound system, involves here specific quantum mechanical features which can not be analyzed in such a simple way.

A quantitative comparison between ordinary evaporation and neutron escape can in fact be carried through only in case of excitation energies of the compound system, very large compared with the energy necessary for the removal of a single neutron, for only in this case will the excitation of the residual nucleus left after the escape of a neutron be nearly equal to that of the compound system, as is assumed in the usual evaporation phenomena where the change in the heat content of the bodies concerned during the escape of a single gas molecule is negligibly small. The above considerations can therefore be applied in this simple form only when the change in the temperature in going from the second to the third stage in Fig. 3 is comparatively small.

Although the conditions for the application of the evaporation analogy are in general not strictly fulfilled in the experiments on fast neutron impacts so far carried out, there are still a great number of more qualitative consequences derivable from the analogy, which are very useful in the discussion of such collision processes. For instance, the above-mentioned large probability of energy loss in collisions between fast neutrons and nuclei just corresponds to the fact, that the molecules released in ordinary evaporation do not take the whole energy of the hot body, but that they in general come off with the much smaller energy per degree of freedom corresponding to the temperature of the evaporating body. It should further be expected from the thermodynamic analogy that the released particles would have an energy distribution around this mean value which corresponds to the Maxwellian distribution. If the energy of the incident neutron is several times larger than the binding energy per particle, it can moreover be predicted that not one single particle but several particles, each with an energy small compared to that of the incident particle, will leave the compound system in successive separate disintegration processes. Nuclear reactions of this type have actually been experimentally found to take place in a number of cases.

The above considerations can also be applied to the release of charged particles like protons and α-particles from the compound system, but it must be kept in mind that in this case the latent heat of evaporation is not simply the binding energy of the charged particle, but that we have to add to this the electrostatic energy due to the mutual repulsion of the escaping particle and the residual nucleus. This repulsion will moreover have the effect of speeding up the particles after their escape from the nucleus, and the mean kinetic energy of the charged particles will thus be larger than that of the neutrons by an energy amount corresponding to this repulsion. We should, therefore, expect that the most probable energy of the emitted particles would be approximately equal to the sum of the temperature energy and the electrostatic repulsion, and that the probability for the emission of charged particles with still larger energies would, as in the case of neutrons, decrease exponentially according to a Maxwellian distribution. This preference for nuclear processes, where the escaping charged particle takes only a part of the available energy, leaving the residual nucleus in an excited state, is in fact one of the most striking features of a great number of reactions in which protons or α-particles are emitted from the compound system.

So far we have mainly been concerned with nuclear processes initiated by impacts of neutrons. Similar considerations concerning the formation of an intermediate state will, however, apply for collisions between charged particles and nuclei; but in this case it must be taken into account that the repulsive electric forces acting between the positively charged nuclei may often for small kinetic energies of the incident particle prevent or make less probable the contact necessary for the establishment of the compound nucleus. The combined action of this electrostatic repulsion of nuclear particles at great distances and their strong attraction at small distances can in fact be simply described by saying that the nucleus is surrounded by a so-called "potential barrier" which the incident charged particles have to pass in order to come in contact with the nucleus. As is well known from the explanation of the laws governing the spontaneous α-ray disintegration of radioactive nuclei, a charged particle may in quantum mechanics have a probability of penetrating through such a potential barrier, even in cases where the particle on classical mechanics, on account of its insufficient energy, would be stopped at the surface of the barrier. This quantum mechanical effect gives also a familiar explanation of the experimental fact that slow protons, when striking not too heavy nuclei, have been found to have a considerable probability of producing nuclear disintegrations, even for energies where classically the particles would be prevented by the electric repulsion from coming in contact with the bombarded nucleus.

Another interesting feature in collisions between charged particles and lighter nuclei is the remarkable resonance effects found for disintegrations caused by impacts of protons and α-particles. As in the case of selective effects of slow neutrons, such resonances must be ascribed to the coincidence of the sum of the energies of the incident particle and the original nucleus with a stationary state of the compound system corresponding to some quantized collective type of motion of all its constituent particles.[5] Especially in case of α-particle impacts, much information concerning the distribution of highly excited levels in lighter nuclei has been derived from such resonance effects. In contrast to the dense distribution of levels found in heavier nuclei, the spacing of the levels in this case is as large as several hundred thousand volts for an excitation considerably higher than ten million volts. This result can, however, be readily understood if one realizes that the lowest excited levels are farther away from each other for light nuclei than for heavier and that therefore the number of possible combinations of these levels in a given energy region is much smaller in the first case than in the second.

Not only the distances between the resonance levels, but also their half value breadths, are in general much larger in lighter nuclei than in heavier, indicating that the lifetime of the compound system is very much shorter in the former case than in the latter. This comes first of all from the circumstance that the resonances in heavy nuclei are found only for very slow particles, where the probability for escape is extremely small, so that the lifetime of the compound system is only limited by the probability of emission of electromagnetic radiation, whereas in lighter nuclei the lifetime is in general entirely determined by the possibility of releasing comparatively fast particles. Quite apart from this, we should, however, expect that the lifetime of a heavy nucleus—even if the nucleus were highly enough excited to emit fast particles— would be much longer than of a light nucleus on account of the lower temperature to be ascribed to a heavy nucleus than to a lighter one for a given excitation energy.

In fact, it would appear that quite simple considerations such as those here outlined enable us to account in a general way for the peculiar features of nuclear reactions initiated by collisions. Likewise it seems possible to explain the characteristic differences between the radiation properties of nuclei and those of atoms by means of similar considerations based also essentially on the extreme facility of energy exchange between the closely packed nuclear particles as compared to the approximately independent binding of each electron in the atom. The closer discussion of such problems will, however, claim more detailed considerations, which lie outside the scope of the present brief report.[6]

[5] Besides the total energy of the compound system also its spin and other symmetry properties may, as often pointed out, be of importance for the analysis of resonance phenomena. How such considerations can be brought into connection with the general picture of nuclear reactions here presented is discussed in a paper by F. Kalckar, I. R. Oppenheimer and R. Serber to appear shortly in *Physical Review*.

[6] A more comprehensive account of the development of the ideas here presented will be published shortly in the *Proceedings* of the Copenhagen Academy by Mr. F. Kalckar and the writer.

SEPTEMBER 1, 1939　　　　PHYSICAL REVIEW　　　　VOLUME 56

The Mechanism of Nuclear Fission

NIELS BOHR

University of Copenhagen, Copenhagen, Denmark, and The Institute for Advanced Study, Princeton, New Jersey

AND

JOHN ARCHIBALD WHEELER

Princeton University, Princeton, New Jersey

(Received June 28, 1939)

On the basis of the liquid drop model of atomic nuclei, an account is given of the mechanism of nuclear fission. In particular, conclusions are drawn regarding the variation from nucleus to nucleus of the critical energy required for fission, and regarding the dependence of fission cross section for a given nucleus on energy of the exciting agency. A detailed discussion of the observations is presented on the basis of the theoretical considerations. Theory and experiment fit together in a reasonable way to give a satisfactory picture of nuclear fission.

INTRODUCTION

THE discovery by Fermi and his collaborators that neutrons can be captured by heavy nuclei to form new radioactive isotopes led especially in the case of uranium to the interesting finding of nuclei of higher mass and charge number than hitherto known. The pursuit of these investigations, particularly through the work of Meitner, Hahn, and Strassmann as well as Curie and Savitch, brought to light a number of unsuspected and startling results and finally led Hahn and Strassmann[1] to the discovery that from uranium elements of much smaller atomic weight and charge are also formed.

The new type of nuclear reaction thus discovered was given the name "fission" by Meitner and Frisch,[2] who on the basis of the liquid drop model of nuclei emphasized the analogy of the process concerned with the division of a fluid sphere into two smaller droplets as the result of a deformation caused by an external disturbance. In this connection they also drew attention to the fact that just for the heaviest nuclei the mutual repulsion of the electrical charges will to a large extent annul the effect of the short range nuclear forces, analogous to that of surface tension, in opposing a change of shape of the nucleus. To produce a critical deformation will therefore require only a comparatively small energy, and by the subsequent division of the nucleus a very large amount of energy will be set free.

Just the enormous energy release in the fission process has, as is well known, made it possible to observe these processes directly, partly by the great ionizing power of the nuclear fragments, first observed by Frisch[3] and shortly afterwards independently by a number of others, partly by the penetrating power of these fragments which allows in the most efficient way the separation from the uranium of the new nuclei formed by the fission.[4] These products are above all characterized by their specific beta-ray activities which allow their chemical and spectrographic identification. In addition, however, it has been found that the fission process is accompanied by an emission of neutrons, some of which seem to be directly associated with the fission, others associated with the subsequent beta-ray transformations of the nuclear fragments.

In accordance with the general picture of nuclear reactions developed in the course of the last few years, we must assume that any nuclear transformation initiated by collisions or irradiation takes place in two steps, of which the first is the formation of a highly excited compound nucleus with a comparatively long lifetime, while

[1] O. Hahn and F. Strassmann, Naturwiss. **27**, 11 (1939); see, also, P. Abelson, Phys. Rev. **55**, 418 (1939).

[2] L. Meitner and O. R. Frisch, Nature **143**, 239 (1939).

[3] O. R. Frisch, Nature **143**, 276 (1939); G. K. Green and Luis W. Alvarez, Phys. Rev. **55**, 417 (1939); R. D. Fowler and R. W. Dodson, Phys. Rev. **55**, 418 (1939); R. B. Roberts, R. C. Meyer and L. R. Hafstad, Phys. Rev. **55**, 417 (1939); W. Jentschke and F. Prankl, Naturwiss. **27**, 134 (1939); H. L. Anderson, E. T. Booth, J. R. Dunning, E. Fermi, G. N. Glasoe and F. G. Slack, Phys. Rev. **55**, 511 (1939).

[4] F. Joliot, Comptes rendus **208**, 341 (1939); L. Meitner and O. R. Frisch, Nature **143**, 471 (1939); H. L. Anderson, E. T. Booth, J. R. Dunning, E. Fermi, G. N. Glasoe and F. G. Slack, Phys. Rev. **55**, 511 (1939).

The opening pages of the Bohr-Wheeler paper.

the second consists in the disintegration of this compound nucleus or its transition to a less excited state by the emission of radiation. For a heavy nucleus the disintegrative processes of the compound system which compete with the emission of radiation are the escape of a neutron and, according to the new discovery, the fission of the nucleus. While the first process demands the concentration on one particle at the nuclear surface of a large part of the excitation energy of the compound system which was initially distributed much as is thermal energy in a body of many degrees of freedom, the second process requires the transformation of a part of this energy into potential energy of a deformation of the nucleus sufficient to lead to division.[5]

Such a competition between the fission process and the neutron escape and capture processes seems in fact to be exhibited in a striking manner by the way in which the cross section for fission of thorium and uranium varies with the energy of the impinging neutrons. The remarkable difference observed by Meitner, Hahn, and Strassmann between the effects in these two elements seems also readily explained on such lines by the presence in uranium of several stable isotopes, a considerable part of the fission phenomena being reasonably attributable to the rare isotope U^{235} which, for a given neutron energy, will lead to a compound nucleus of higher excitation energy and smaller stability than that formed from the abundant uranium isotope.[6]

In the present article there is developed a more detailed treatment of the mechanism of the fission process and accompanying effects, based on the comparison between the nucleus and a liquid drop. The critical deformation energy is brought into connection with the potential energy of the drop in a state of unstable equilibrium, and is estimated in its dependence on nuclear charge and mass. Exactly how the excitation energy originally given to the nucleus is gradually exchanged among the various degrees of freedom and leads eventually to a critical deformation proves to be a question which needs not be discussed in order to determine the fission probability. In fact, simple statistical con-

siderations lead to an approximate expression for the fission reaction rate which depends only on the critical energy of deformation and the properties of nuclear energy level distributions. The general theory presented appears to fit together well with the observations and to give a satisfactory description of the fission phenomenon.

For a first orientation as well as for the later considerations, we estimate quantitatively in Section I by means of the available evidence the energy which can be released by the division of a heavy nucleus in various ways, and in particular examine not only the energy released in the fission process itself, but also the energy required for subsequent neutron escape from the fragments and the energy available for beta-ray emission from these fragments.

In Section II the problem of the nuclear deformation is studied more closely from the point of view of the comparison between the nucleus and a liquid droplet in order to make an estimate of the energy required for different nuclei to realize the critical deformation necessary for fission.

In Section III the statistical mechanics of the fission process is considered in more detail, and an approximate estimate made of the fission probability. This is compared with the probability of radiation and of neutron escape. A discussion is then given on the basis of the theory for the variation with energy of the fission cross section.

In Section IV the preceding considerations are applied to an analysis of the observations of the cross sections for the fission of uranium and thorium by neutrons of various velocities. In particular it is shown how the comparison with the theory developed in Section III leads to values for the critical energies of fission for thorium and the various isotopes of uranium which are in good accord with the considerations of Section II.

In Section V the problem of the statistical distribution in size of the nuclear fragments arising from fission is considered, and also the questions of the excitation of these fragments and the origin of the secondary neutrons.

Finally, we consider in Section VI the fission effects to be expected for other elements than thorium and uranium at sufficiently high neutron velocities as well as the effect to be anticipated in

[5] N. Bohr, Nature **143**, 330 (1939).
[6] N. Bohr, Phys. Rev. **55**, 418 (1939).

thorium and uranium under deuteron and proton impact and radiative excitation.

I. Energy Released by Nuclear Division

The total energy released by the division of a nucleus into smaller parts is given by

$$\Delta E = (M_0 - \Sigma M_i)c^2, \qquad (1)$$

where M_0 and M_i are the masses of the original and product nuclei at rest and unexcited. We have available no observations on the masses of nuclei with the abnormal charge to mass ratio formed for example by the division of such a heavy nucleus as uranium into two nearly equal parts. The difference between the mass of such a fragment and the corresponding stable nucleus of the same mass number may, however, if we look apart for the moment from fluctuations in energy due to odd-even alternations and the finer details of nuclear binding, be reasonably assumed, according to an argument of Gamow, to be representable in the form

$$M(Z, A) - M(Z_A, A) = \tfrac{1}{2}B_A(Z - Z_A)^2, \qquad (2)$$

where Z is the charge number of the fragment and Z_A is a quantity which in general will not be an integer. For the mass numbers $A = 100$ to 140 this quantity Z_A is given by the dotted line in Fig. 8, and in a similar way it may be determined for lighter and heavier mass numbers.

B_A is a quantity which cannot as yet be determined directly from experiment but may be estimated in the following manner. Thus we may assume that the energies of nuclei with a given mass A will vary with the charge Z approximately according to the formula

$$M(Z, A) = C_A + \tfrac{1}{2}B_A{}'(Z - \tfrac{1}{2}A)^2 + (Z - \tfrac{1}{2}A)(M_p - M_n) + 3Z^2e^2/5r_0A^{\frac{1}{3}}. \qquad (3)$$

Here the second term gives the comparative masses of the various isobars neglecting the influence of the difference $M_p - M_n$ of the proton and neutron mass included in the third term and of the pure electrostatic energy given by the fourth term. In the latter term the usual assumption is made that the effective radius of the nucleus is equal to $r_0A^{\frac{1}{3}}$, with r_0 estimated as 1.48×10^{-13} from the theory of alpha-ray disintegration. Identifying the relative mass values given by expressions (2) and (3), we find

$$B_A{}' = (M_p - M_n + 6Z_Ae^2/5r_0A^{\frac{1}{3}})/(\tfrac{1}{2}A - Z_A) \qquad (4)$$

and

$$B_A = B_A{}' + 6e^2/5r_0A^{\frac{1}{3}} = (M_p - M_n + 3A^{\frac{2}{3}}e^2/5r_0)/(\tfrac{1}{2}A - Z_A). \qquad (5)$$

The values of B_A obtained for various nuclei from this last relation are listed in Table I.

On the basis just discussed, we shall be able to estimate the mass of the nucleus (Z, A) with the help of the packing fraction of the known nuclei. Thus we may write

$$M(Z, A) = A(1 + f_A) \\ \quad + 0 \begin{cases} A \text{ odd} \\ + \tfrac{1}{2}B_A(Z - Z_A)^2 - \tfrac{1}{2}\delta_A \begin{cases} A \text{ even, } Z \text{ even} \\ + \tfrac{1}{2}\delta_A \end{cases} A \text{ even, } Z \text{ odd} \end{cases}, \qquad (6)$$

where f_A is to be taken as the average value of the packing fraction over a small region of atomic weights and the last term allows for the typical differences in binding energy among nuclei according to the odd and even character of their neutron and proton numbers. In using Dempster's measurements of packing fractions we must recognize that the average value of the second term in (6) is included in such measurements.[7] This correction, however, is, as may be read from Fig. 8, practically compensated by the influence of the third term, owing to the fact that the great majority of nuclei studied in the mass spectrograph are of even-even character.

From (6) we find the energy release involved in electron emission or absorption by a nucleus unstable with respect to a beta-ray

TABLE I. *Values of the quantities which appear in Eqs. (6) and (7), estimated for various values of the nuclear mass number A. Both B_A and δ_A are in Mev.*

A	Z_A	B_A	δ_A	A	Z_A	B_A	δ_A
50	23.0	3.5	2.8	150	62.5	1.2	1.5
60	27.5	3.3	2.8	160	65.4	1.1	1.3
70	31.2	2.5	2.7	170	69.1	1.1	1.2
80	35.0	2.2	2.7	180	72.9	1.0	1.2
90	39.4	2.0	2.7	190	76.4	1.0	1.1
100	44.0	2.0	2.6	200	80.0	0.95	1.1
110	47.7	1.7	2.4	210	83.5	0.92	1.1
120	50.8	1.5	2.1	220	87.0	0.88	1.1
130	53.9	1.3	1.9	230	90.6	0.86	1.0
140	58.0	1.2	1.8	240	93.9	0.83	1.0

[7] A. J. Dempster, Phys. Rev. **53**, 869 (1938).

transformation:

$$E_\beta = B_A\{|Z_A - Z| - \tfrac{1}{2}\} - \delta_A \begin{cases} +0 & A \text{ odd} \\ -\delta_A & A \text{ even, } Z \text{ even} \\ +\delta_A & A \text{ even, } Z \text{ odd} \end{cases}. \quad (7)$$

This result gives us the possibility of estimating δ_A by an examination of the stability of isobars of even nuclei. In fact, if an even-even nucleus is stable or unstable, then δ_A is, respectively, greater or less than $B_A\{|Z_A - A| - \tfrac{1}{2}\}$. For nuclei of medium atomic weight this condition brackets δ_A very closely; for the region of very high mass numbers, on the other hand, we can estimate δ_A directly from the difference in energy release of the successive beta-ray transformations

$$\text{UX}_\text{I} \rightarrow (\text{UX}_\text{II}, \text{UZ}) \rightarrow \text{U}_\text{II},$$
$$\text{MsTh}_\text{I} \rightarrow \text{MsTh}_\text{II} \rightarrow \text{RaTh}, \text{RaD} \rightarrow \text{RaE} \rightarrow \text{RaF}.$$

The estimated values of δ_A are collected in Table I.

Applying the available measurements on nuclear masses supplemented by the above considerations, we obtain typical estimates as shown in Table II for the energy release on division of a nucleus into two approximately equal parts.[8]

Below mass number $A \sim 100$ nuclei are energetically stable with respect to division; above this limit energetic instability sets in with respect

TABLE II. *Estimates for the energy release on division of typical nuclei into two fragments are given in the third column. In the fourth is the estimated value of the total additional energy release associated with the subsequent beta-ray transformations. Energies are in Mev.*

ORIGINAL	TWO PRODUCTS	DIVISION	SUBSEQUENT
$_{28}\text{Ni}^{61}$	$_{14}\text{Si}^{30,\ 31}$	-11	2
$_{50}\text{Sn}^{117}$	$_{25}\text{Mn}^{58,\ 59}$	10	12
$_{68}\text{Er}^{167}$	$_{34}\text{Se}^{83,\ 84}$	94	13
$_{82}\text{Pb}^{206}$	$_{41}\text{Nb}^{103,\ 103}$	120	32
$_{92}\text{U}^{239}$	$_{46}\text{Pd}^{119,\ 120}$	200	31

to division into two nearly equal fragments, essentially because the decrease in electrostatic

[8] Even if there is no question of actual fission processes by which nuclei break up into more than two comparable parts, it may be of interest to point out that such divisions in many cases would be accompanied by the release of energy. Thus nuclei of mass number greater than $A = 110$ are unstable with respect to division into three nearly equal parts. For uranium the corresponding total energy liberation will be $\sim$210 Mev, and thus is even somewhat greater than the release on division into two parts. The energy evolution on division of U^{239} into four comparable parts will, however, be about 150 Mev, and already division into as many as 15 comparable parts will be endothermic.

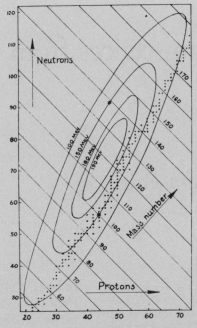

FIG. 1. The difference in energy between the nucleus $_{92}\text{U}^{239}$ in its normal state and the possible fragment nuclei $_{44}\text{Ru}^{100}$ and $_{48}\text{Cd}^{139}$ (indicated by the crosses in the figure) is estimated to be 150 Mev as shown by the corresponding contour line. In a similar way the estimated energy release for division of U^{239} into other possible fragments can be read from the figure. The region in the chart associated with the greatest energy release is seen to be at a distance from the region of the stable nuclei (dots in the figure) corresponding to the emission of from three to five betarays.

energy associated with the separation overcompensates the desaturation of short range forces consequent on the greater exposed nuclear surface. The energy evolved on division of the nucleus U^{239} into two fragments of any given charge and mass numbers is shown in Fig. 1. It is seen that there is a large range of atomic masses for which the energy liberated reaches nearly the maximum attainable value 200 Mev; but that for a given size of one fragment there is only a small range of charge numbers which correspond to an energy release at all near the maximum value. Thus the fragments formed by division of uranium in the *energetically* most favorable way lie in a narrow band in Fig. 1, separated from the region of the stable nuclei by an amount which corresponds to the change in nuclear charge associated with the emission of three to six betaparticles.

Reminiscences from the Postwar Years

Abraham Pais

Abraham Pais was born in Amsterdam and obtained his Ph.D. at the University of Utrecht in 1941. After World War II he joined the faculty of the Institute for Advanced Study in Princeton, where he became closely acquainted with Einstein; he is now Professor of Physics at Rockefeller University, New York City. He has wide interests in theoretical physics and in the history of quantum theory. His biography of Einstein, *Subtle Is the Lord,* has won wide acclaim.

In January 1946 I came for the first time to Copenhagen from my native Holland, as a Rask-Oersted Fellow.[1] I was the first of the postwar generation to come from abroad to Bohr's institute for a long period of study. The morning after my arrival I went to Mrs. Schultz, the secretary, who told me to wait in the journal room adjoining the library, where she would call me as soon as Professor Bohr was free to see me. I had sat there reading for a while when someone knocked at the door. I said come in. The door opened. It was Bohr. And my first thought was, what a gloomy face.

Then he began to speak.

I have later often been puzzled about this first impression. It vanished the very moment Bohr started to talk to me that morning, never to return. True, one might correctly describe Bohr's physiognomy as unusually heavy or rugged. Yet Bohr's face is remembered by all who knew him for its intense animation and its warm, sunny smile.

I did not see much of Bohr during the next month or so. After a brief trip to Norway, he was very busy with plans for the extension of his institute. I was soon invited for Sunday dinner at Carlsberg, however, and that evening I had my first opportunity to talk physics with Bohr in his study. I told him of things I had worked out during my years in hiding in Holland. These concerned the *self-energy* problem in quantum field theory. Briefly, in such a theory one considers a basic particle, like the electron, as an object without extension, a point. This assumption leads to the apparent difficulty that the electron thereby requires an infinite energy due to the electromagnetic field which it generates. At the time, I was concerned with the question of whether such infinities could be compensated for by a hypothetical coupling of the electron to another field of short range. While I was telling Bohr about this, he smoked his pipe, looked mainly at the floor, and rarely

244

looked up at the blackboard on which I was enthusiastically writing formulas. After I had finished, Bohr did not say much, and I left a bit disheartened with the impression that he could not have cared less about the whole subject. I did not know Bohr well enough at the time to realize that this was not entirely true. At a later stage I would have known right away that Bohr's curiosity was aroused, as he had remarked neither that this was very, very interesting nor that we agreed much more than I thought.

After this discussion we went back to the living room to rejoin the company. Then, as on later occasions, I felt fortunate to be for a while in the invigorating atmosphere of warmth and harmony which Mrs. Bohr and her husband knew how to create wherever they were in the world, but above all in their home.

It would be wrong to suppose that evenings at the Bohrs' were entirely filled with discussion of weighty matters. Sooner or later, for the purpose of illustrating some point, or just for the pleasure of it, Bohr would tell one or more stories. I believe that at any given time Bohr had about half a dozen favorite jokes. He would tell them; we would get to know them. Yet he would never cease to hold his audience. For me, to hear again the beginning of such a familiar tale would lead me to anticipate not so much the denouement as Bohr's own happy laughter upon the conclusion of the story.

During the following weeks it became clear that Bohr had become quite interested in the problems of quantum theory which I had mentioned to him. Every now and then he would call me to his office to have me explain one or the other aspect of them. He was particularly intrigued by those arguments which showed that many elementary-particle problems (such as the self-energy question mentioned earlier) are fundamentally quantum problems which cannot be dealt with by the methods of classical physics. This view did not have as wide an acceptance at that time as was to be the case two years later, when the modern version of field theory known as the *renormalization* program began its development.

Then, one day in May, Bohr asked me whether I would be interested in working together with him day by day for the coming months. I was thrilled, and accepted. The next morning I went to Carlsberg. The first thing Bohr said to me was that it would be profitable to work with him only if I understood that he was a dilettante. I reacted to this unexpected statement with a polite smile of disbelief. But evidently Bohr was serious. He explained how he had to approach every new question from a starting point of total ignorance. It is perhaps better to say that Bohr's strength lay in his formidable intuition and insight, not at all in erudition. I thought of his remarks of that morning some years later, when I sat next to him during a colloquium in Princeton. The subject was nuclear *isomers*. As the speaker went on, Bohr got more and more restless and kept whispering to me that it was all wrong. Finally he could not contain himself and wanted to make an objection. But after having half raised himself he sat down again, looked at me with unhappy bewilderment, and asked: "What is an isomer?"

The first subject of our joint work was the preparation of Bohr's opening address to the International Conference on Fundamental Particles to be held in July in Cambridge, England. It was the first meeting of its kind in the postwar era. Bohr planned to make a number of comments on the problems of quantum field theory alluded to earlier. I must admit that in the early stages of collaboration I did not follow Bohr's line of thinking a good deal of the time and was, in fact, often quite bewildered. I failed to see the relevance of such remarks as that Schrödinger was completely shocked in 1927 when he was told of the probability interpretation of quantum mechanics, or a reference to some objection by Einstein in 1928, which apparently had no bearing whatever on the subject at hand. But it did not take very long before the fog started to lift. I began to grasp not only the thread of Bohr's arguments but also their purpose. Just as in many sports a player goes through warmup exercises before entering the arena, so Bohr would relive the struggles that had taken place before the content of quantum mechanics was understood and accepted. I can say that in Bohr's mind this struggle started all over again every single day. This, I am convinced, was Bohr's inexhaustible source of identity. Einstein appeared forever as his leading spiritual sparring partner — even after Einstein's death, Bohr would argue with him as if he were still alive.

I can now explain the principal and lasting inspiration which I derived from discussions with Bohr. In Holland I had received a solid training as a physicist. It is historically inevitable that students of my generation received quantum mechanics served up ready made. While I may say that I had a decent working knowledge of the theory, I had not — and indeed hardly could have — fathomed how very profoundly the change from the classical to the quantum-mechanical way of thinking affected both the architects and the close witnesses of the revolution in physics which began in 1925. Through steady exposure to Bohr's "daily struggle" and his ever repeated emphasis on "the epistemological lesson which quantum mechanics has taught us" (to use a favorite phrase of his), I deepened my understanding not only of the history of physics but of physics itself. In fact, the many hours which Bohr spent talking to me about complementarity have had a liberating effect on every aspect of my thinking.

To us who knew him then, Bohr had become the principal consolidator of one of the greatest developments in the history of science. (It would be inappropriate to speak of Bohr the philosopher, at this stage of his life, as his attitude toward professional philosophy was always skeptical, to say the least.) It is true that, to the end, Bohr was one of the most open-minded physicists I have known, forever eager to learn of new developments from younger people and remaining faithful to his own admonition always to be prepared for a surprise. (In these respects he was entirely different from Einstein.) But, inevitably, his role in these quite new developments shifted from actor to spectator. Bohr created atomic physics and put his stamp on nuclear physics. With particle physics, the post-Bohr era began. The Cambridge paper of 1946 represents Bohr's farthest penetration into the more modern problems.

Bohr was an indefatigable worker, as I noticed most particularly during the six weeks I spent with him and his family at their country home in Tisvilde. When he was in need of a break in the discussions, he would go outside and apply himself to pulling weeds with what can only be called ferocity. At this point I can contribute a little item to the lore about Bohr the pipe smoker. It is well known that to him the operations of filling a pipe and lighting it were commutative, but the following situation is even more extreme. One day Bohr was weeding again, his pipe between his teeth. At one point, unnoticed by Bohr, the bowl fell off the stem. His son Aage and I were lounging in the grass, expectantly awaiting further developments. It is hard to forget Bohr's look of stupefaction when he found himself holding a thoughtfully lit match against a pipe without a bowl.

When I met Bohr again at the celebration of the bicentennial of Princeton University, he asked me to spend some time with him on the preparation of a talk for that occasion. I did so, and I know how well prepared he was with carefully structured arguments. However, I was amazed to find that Bohr actually gave the talk without a worked out manuscript before him. I should say that this amazement was due to the fact that till then I had never heard him speak publicly.

In attempting to describe the experience of listening to Bohr in public, I am reminded of a story about the violinist Eugene Issaye, who at one time had a member of a royal family as his pupil. Another musician of great renown (to whom I owe this tale) once asked Issaye how this pupil was doing. Whereupon Issaye opened his hands heavenward and sighed: "Ah, Her Royal Highness, she plays divinely bad."

However different the background in the two cases, these are the words which best characterize the situation: Bohr was divinely bad as a public speaker. This was not due to his precept never to speak more clearly than one thinks. Had he done so, the outcome would have been quite different, as Bohr was a man of the greatest lucidity of thought. Neither is it due to the fact that his voice did not carry far, which made it impossible for those at the back of a large audience to hear him. The main reason is that he was in deep thought as he spoke. I remember how that day he finished part of the argument, then said: "And . . . and . . . ," then was silent for at most a second, then said: "But . . ." and continued. Between the "and" and the "but" the next point had gone through his mind. However, he simply forgot to say it out loud and went on somewhere further down his road. To me, the story was continuous as I knew precisely how to fill the gaps Bohr had left open. But others were probably not so fortunate. And so it came to pass more than once that I saw an audience leave a talk by Bohr in a state of mild bewilderment, even though he had toiled hard in preparing himself in all detail. Still, when he would come up to me afterward with the characteristic remark, "Jeg håber det var nogenlunde" ("I hope it was

When Niels Bohr began his series of public lectures at MIT in 1957, people came from all over the Boston area and with great anticipation to hear the wisest of living men. But the microphone was erratic, Bohr's aspirated and sibilant diction mostly incomprehensible, and his thoughts too intricately evolved even for those who could hear. It must have been some way through the fifth lecture that those who still attended saw Bohr pause, take a drink of water, draw himself up, wag his great eyebrows, and say loud and clear, "And now we must ask: what is life?" The effect was electrifying. We sat on the edge of our seats. Bohr then said, "Let us rather ask . . . " but his diction became progressively muddier, his voice fainter, and the microphone whistled, and we all sank back again.

D. H. Frisch,
private communication

247

One of a series of Mickey Mouse cartoons by George Gamow. This one, captioned "Tyrefogtning" (bullfighting) shows Mickey Mouse (Bohr) saying: "Pauli, schweig! May I . . . may I . . . may I say . . . !" It refers to an incident of the same kind as that described by Abraham Pais, but more than twenty years earlier.

tolerable"), I could assure him that it was much more than that. In spite of all rhetorical shortcomings, this unrelenting struggle for truth was a powerful source of inspiration.

At the same time, it should be emphasized that Bohr's best way of communicating actually was the spoken word, but with just one or at most a few persons present. Bohr's needs for verbal expression were great, as the following incident illustrates. On a later occasion (1948), Bohr arrived in Princeton after a trip by sea from Denmark. For about a week, while on board ship, he had had no opportunity to discuss scientific matters; he was quite pent up. Pauli and I were walking in a corridor of the institute when Bohr first came in. When he saw us, he practically pushed us into an office, made us sit down, said "Pauli, schweig" ("Pauli, shut up"), and then talked for about two hours before either of us had a chance to interrupt him. Had Bohr's words been recorded, they would have constituted a fascinating document on the development of quantum theory.

I remember well my first direct experience of the impact of Einstein on Bohr. A few weeks after the Pauli episode, Bohr came to my office at the Institute for Advanced Study, of which I then was a temporary member. He was in a state of angry despair and kept saying, "I am sick of myself." I was concerned and asked what had happened. He told me he had just been downstairs to see Einstein. As always, they had become involved in an argument about the meaning of quantum mechanics, and, as remained true to the end, Bohr had been unable to convince Einstein of his views. There can be no doubt that Einstein's lack of assent was a very deep frustration to Bohr. It is our good fortune that this led Bohr to keep striving at clarification and better formulation. And not only that; it was Bohr's own good fortune, too.

In 1948 I saw a lot of Bohr, as he and his wife lived at 14 Dickinson Street, the same house in which I occupied the top floor. When I came home at night, the following charming little comedy would often be enacted. As I opened the door, Bohr would always be walking down the corridor, his back toward me, on his way to the kitchen. In that way he would let me notice him first. He would then turn around in apparent surprise and ask if I would care for a glass of sherry. And then we would settle down to talk about political problems. For at that period Bohr had become disillusioned with the official reactions to the atom. It was now his desire to make a direct attempt to get his views considered by those in positions of responsibility, and he was preparing a memorandum to this effect which was discussed over and over during those evenings. It formed the basis for Bohr's open letter to the United Nations in 1950.

Apart from this, Bohr spent most of his time putting the finishing touches on his article in the 1949 Einstein volume.[2] This paper is Bohr's masterpiece. Nowhere in the literature can a better access to his thinking be found, and it is and should continue to be required reading for all students of quantum mechanics. During that period I was witness to an amusing moment which involved both Bohr and Einstein.

One morning Bohr came into my office and began, "Du er så klog . . ." ("you are so wise"). I started to laugh (no formality or solemnity was called for in contact with Bohr) and said, "All right, I understand" — Bohr wanted me to come down to his office and talk, so we went there. It should be explained that Bohr at that time used Einstein's own office in Fuld Hall, while Einstein himself used the adjoining small assistant's office; he disliked the big one, and never used it. (A photograph in the 1949 Einstein anniversary volume of *Reviews of Modern Physics* shows Einstein sitting in the assistant's office.) After we had entered, Bohr asked me to sit down ("I always need an origin for the coordinate system") and soon started to pace furiously around the oblong table in the center of the room. He then asked me if I could put down a few sentences as they would emerge during his pacing. At such sessions, Bohr never had a full sentence ready. He would often dwell on one word, coax it, implore it, to find the continuation. This could go on for many minutes. At that moment the word was "Einstein." There Bohr was, almost running around the table and repeating: "Einstein . . . Einstein . . ." It would have been a curious sight for someone unfamiliar with Bohr. After a little while he walked to the window, gazed out, repeating every now and then: "Einstein . . . Einstein . . ."

At that moment the door opened very softly and Einstein tiptoed in. He signaled me with a finger on his lips to be very quiet, his urchin smile on his face. He was to explain a few minutes later the reason for his behavior. Einstein was not allowed by his doctor to buy tobacco. However, the doctor had not forbidden him to *steal* tobacco, and this was precisely what he set out to do now. Still on tiptoe he made a beeline for Bohr's tobacco pot, which stood on the table at which I was sitting. Meanwhile Bohr, unaware,

Father enjoyed the following little episode very much, as it characterizes both his own *modus operandi* and that of the famous physicist Dirac. While working on an article, father was busy walking round the table searching his mind for the correct expression. Dirac was sitting silent and unconcerned, until during a pause he interjected in his quiet and calm way: "When I was a boy, I was always taught never to start a sentence without knowing the end of it."

Hans Bohr,
"My Father," 1966

249

was standing at the window, muttering, "Einstein . . . Einstein." I was at a loss to know what to do, especially because I had at the moment not the faintest idea what Einstein was up to.

Then Bohr, with a firm "Einstein" turned around. There they were, face to face, as if Bohr had summoned him forth. It is an understatement to say that for a moment Bohr was speechless. I myself who had seen it coming, had felt distinctly uncanny for a moment, so I could well understand Bohr's own reaction. A moment later the spell was broken when Einstein explained his mission and soon we were all bursting with laughter.

The periods of closest contact I had with Bohr are those described above. In subsequent years I saw him often, either in Denmark or in the United States, but no longer for protracted periods of time.

In the fall of 1961 we were both present at the Twelfth Solvay Conference. It was the fiftieth anniversary of the first one, and Bohr gave an account, both charming and fascinating, of the developments during that period. He was present at the report I gave at that meeting, after which we walked along the corridor and spoke of the future of particle physics. It was the last time I talked with him.

Bohr's was a rich and full life. As I write these concluding words, I hear Bohr reciting two lines of one of his favorite poems:

> *Nur die Fülle führt zur Klarheit,*
> *Und im Abgrund wohnt die Wahrheit.*
>
> Only fullness leads to clarity,
> And the truth resides in the abyss.

V. BOHR AND POLITICS

Niels Bohr as a Political Figure

Ruth Moore

Niels Bohr's work as a scientist in the exploration of the atom changed our understanding of the world. It is not widely realized that when these discoveries began to affect the political course of the world, Bohr himself became politically active. In a very special way he exerted the kind of influence on policy usually wielded only by presidents and prime ministers; and if this part of Bohr's career is largely unknown, it is because much of it was necessarily secret. His help in rescuing scientists and other persecuted groups from Germany had to be hidden, and even the records destroyed.

It is doubtful that Bohr thought of himself as a politician, and certainly he was never elected or appointed to office. There was, however, no doubt about his political interest and awareness. His father, himself a scientist (a physiologist), and some of his father's colleagues who often met for dinner at the Bohr home, avidly talked politics as well as the science which brought them together. Niels and his brother, Harald, who as youngsters were permitted to sit at the side of the dining room and listen, drank in the discussion. It became a part of them.

World politics came to the fore when Niels Bohr went to England in 1914–1916 as an assistant to Ernest Rutherford. During World War I, Rutherford became increasingly preoccupied with anti-submarine research, and many of the young scientists went into the armed services. The dread losses of war were brought close to home when word came of the death of H. G. J. Moseley. Shortly before he enlisted, this young scientist had made brilliant X-ray experiments that confirmed the ideas of atomic number and of Bohr's own model of the atom. Bohr was learning early that science cannot be separated from the world as it is.

In 1916 Bohr's two-year leave of absence from his first post, as docent (lecturer) at the University of Copenhagen, was coming to an end. Ruther-

Ruth Moore, writer, has produced a number of books about various areas of science. Her biography of Niels Bohr (1966) presents a detailed and thorough study of his life.

ford strongly urged him to remain in England. The war still was on and the seas still were perilous, but Bohr felt that he had to return to Denmark, where he remained until the war ended on 11 November 1918. In a letter of thanksgiving to Rutherford, Bohr emphasized the changes that he believed would result from the terrible conflict. "People have got to look quite differently than before at the political side of life," he wrote. From that time on, Bohr did look "differently" at the "political side of life."

Bohr was made a professor at the early age of thirty-one. He dreamed, however, not of a simple university department, but of an institute of theoretical physics to reach beyond the university to all of Denmark and, indeed, to all the world. Friends of Bohr's joined in raising the "large sum" that would be needed for the first institute building — about $20,000 — and Bohr persuaded the city of Copenhagen to supply a site along the edge of one of the city's inner parks.[1]

The construction of the institute was only beginning when Bohr received an almost irresistible proposal from Rutherford. The young Dane was offered a "good post" at Manchester, helping, as Rutherford said, "to make physics boom." Bohr hesitated, but he always knew that Denmark would come first for him. "I feel that it is my duty here to do my best, though I feel very deeply the result will never be the same as if I could work with you," Bohr wrote to his revered friend. Bohr's decision essentially was political; he put his duty to his country, and to his countrymen who had supported the proposed institute, ahead of even the most brilliant of scientific opportunities.

Copenhagen continued to work in the closest collaboration with Manchester and Cambridge, and soon was a famed center in its own right. In 1922 Bohr was awarded the Nobel Prize, and scientists came from all parts of the world to take part in the excitement of the institute. It was, as Robert Oppenheimer said, "a heroic time." As the work went forward in the lecture rooms and laboratories, relatively little attention was given to the outside world.

The peaceful interlude was not to last long. In the United States the stock market crashed, and the ensuing economic paralysis and unemployment spread through the whole Western world. In Germany the "dark, hidden forces" of Adolf Hitler set out to destroy democratic government. Storm troopers marched through German streets. Bohr, with his political alertness, watched all this with rising concern. Although much of the world dismissed National Socialism as unimportant, Bohr could not be so unheedful. He felt from the beginning that it had to be taken seriously, and that action would be needed to help the German scientists.

Soon after Hitler assumed power, Bohr on his own initiative went to Germany, ostensibly to visit the German universities. Another, unexpressed aim was to find out how many German scientists would be affected by the Nazi racial laws and how many would need assistance. Bohr, himself half Jewish, recognized that lives might be endangered. In Hamburg, one of

A terrible shock to us all was the tragic message in 1915 of Moseley's untimely death in the Gallipoli campaign, deplored so deeply by the community of physicists all over the world, and which not least Rutherford, who had endeavored to get Moseley transferred from the front to less dangerous duties, took much to heart.

Niels Bohr,
Rutherford Memorial Lecture, 1958

those whom Bohr met was the young physicist Otto Frisch, the nephew of Bohr's friend and fellow physicist Lise Meitner. As Frisch related later, Bohr took him by the waistcoat button and asked if he would like to work in Copenhagen. Frisch wrote to his mother that he felt as though "the good Lord himself" had taken him in hand.[2]

When political action was called for, Bohr did not hesitate, and as he went from city to city in Germany he quietly organized the lines of escape. With the same kind of sensitivity he had shown with Frisch, he let James Franck, George de Hevesy, George Placzek, and many others know that a place would be waiting for them, if they should wish to leave.

Upon his return to Copenhagen, Bohr, his brother, and a number of close associates formed the Danish Committee for the Support of Refugee Intellectuals. They not only planned for the reception of refugees in Denmark, but wrote to friends in many other countries seeking permanent places for the escaping scientists.

Bohr decided to speak out against the Nazis' pseudoscientific credo of race and of Aryan cultural superiority. As a Dane, he could not make a direct attack; Denmark had a tradition of neutrality that had to be guarded. Bohr could make his point, however, by emphasizing the relativity of human judgments and the importance of considering opposing points of view. He had been invited to address the International Conference of Anthropological and Ethnological Sciences at a special meeting at Elsinore Castle.[3] Within the grim walls where Shakespeare had Hamlet make his soliloquy, Bohr likewise took up arms. It was, he declared, "purely a caprice of fate" that the culture of another people is theirs and "not ours." He argued that the different human cultures are complementary to one another, and have enriched civilization. At that point the German delegation walked out. Before an important segment of European intellectuals, Bohr had directly attacked National Socialism. It did not matter that he had not named his opponents. Friends warned Bohr that his speech would go into the Nazi files.

In September 1938, when the institute held its annual seminar, almost none of the German alumni dared attend. But Enrico Fermi, the leading physicist of Italy, was there. Bohr confidentially told him that he (Fermi) was under consideration for the Nobel Prize. Since Italian citizens had been ordered to convert all foreign currencies into lire, would Fermi prefer to have his name withdrawn until the prize money could be used without restriction? Fermi then told Bohr that he wanted to leave Italy. Mrs. Fermi was in jeopardy under the racial laws, and Fermi had long disliked Fascism. Fermi said that, if he should receive the prize, he would seek a position in the United States. Bohr at once invited him and his family to come to Copenhagen following the Nobel ceremonies and there await their transportation. Exactly this procedure was followed, and Fermi moved on to Columbia University and ultimately to a leading role in the development of the atomic bomb.

In the summer of 1933 Niels Bohr invited me to his usual summer conference in Copenhagen, which this time he proposed to use as a sort of labour exchange which might help those physicists who had to leave Germany in finding jobs abroad.

O. R. Frisch,
What Little I Remember, 1979

Bohr was being drawn even deeper into political opposition. In March 1938 the Hitler columns had overrun Austria. Up to this time, Lise Meitner, an Austrian by birth, had been able to continue as director of nuclear physics at the Kaiser Wilhelm Institute in Berlin. With the annexation of Austria, all the Nazi racial laws went into effect, and Meitner knew that she would have to leave Germany. Thirty years of distinguished work there had come to an end. She left, supposedly for a short holiday, and was admitted to Holland. Later, at Bohr's invitation, she went to Copenhagen and from there to the Nobel Institute in Stockholm to continue her work. Soon afterward came her historic meeting with her nephew Frisch, at which the concept of fission was developed, followed by Bohr's carrying of this news to the United States in January 1939, as described elsewhere in this book.

Toward the scheduled end of Bohr's visit to the United States, friends pleaded with him to remain there, but Bohr insisted on returning to his own country. The institute had to be kept open, refugees had to be helped, and Danish morale maintained. On the surface, in that summer of 1939, Copenhagen seemed quiet; but beneath the surface was endless worry. Czechoslovakia had been overrun, and on May 31 Hitler "presented" the Danes with a nonaggression pact. They signed — for the Danes, an eminently practical people, knew that it made no difference whether they did or not. On September 1 came the invasion of Poland. Britain and France at last stood firm. Hitler was notified by Great Britain that unless the attack was halted by 11 A.M. on September 3, a "state of war will exist between the two countries from that hour." France joined in, and war came to Europe.

On 8 April 1940 King Haakon of Norway warmly welcomed Niels Bohr to dinner at the royal palace. Bohr had come to Norway partly to lecture on several important recent discoveries in physics. But his trip to Norway had another purpose as well — a political one. There was danger that the war might spread to Denmark and Norway, and Bohr wanted to discuss the worsening situation. None suspected how close the catastrophe was. Immediately after the dinner, Bohr took the night ferry-train to Denmark. As it pulled into Elsinore in the morning, in the very shadow of Hamlet's castle, police were pounding on Bohr's compartment door, announcing the German invasion. Nazi planes flew overhead, German ships had moved into the Copenhagen harbor before dawn, and the Danish government, meeting in extraordinary session, yielded to the demand for nonresistance.

If Bohr had looked out of his ferry's portholes during the night, he might have seen other German ships moving toward Norway. Norway, too, was invaded on that fateful morning. The Norwegians fought back desperately, and when there was nothing else that could be done, retreated, still fighting, into their mountains where the struggle went on.

Bohr's train was allowed to go on to Copenhagen that morning. After checking on the safety of his family, he hurried to the Institute. There was nothing to be done about the laboratories, but he immediately began to destroy all the records on the refugee scientists who had come through Copenhagen. The records could not be permitted to fall into Nazi hands.

Bohr knew what else to do. He went to see the chancellor of the university to seek help in protecting the Polish-born members of the institute staff. He also went to call upon members of the government. He urged strong resistance if the occupiers attempted to put their racial laws into effect in Denmark. Bohr insisted that the Danes did not have to yield to racial persecution. On his own, Bohr again was taking the political initiative.

As soon as the news of the invasion became known, messages poured into Copenhagen begging Bohr to come to the United States. Again he, his wife Margrethe, and his brother Harald decided that they must stay in Denmark "as long as possible." The "possible" always had to be added, they knew. The institute would be kept in operation, although the last foreign student had left. The training of young Danes, including Bohr's son Aage, offered one of the best ways of maintaining science and preparing Denmark for another future, whatever it might be.

Bohr refused to cooperate with any undertaking tinged with Nazism. He boycotted all meetings directly or indirectly connected with the Germans. On the other hand, he joined publicly in the occasions at which the Danish people poured forth their pent-up feelings. The Bohrs usually were seated in the front row. Bohr had long been honored as a scientist; now he emerged as a national leader.

All the while the night of war grew blacker. The German forces rolled through Holland, Belgium, and France, unimpeded until they reached the beaches of Dunkirk. The Danes as well as the British were heartened by that miraculous rescue. It was, however, only a short respite. In August 1940 the Nazi air force began the attack intended to "bring Britain to her knees" and open the British Isles to invasion. The Danes watched many of the planes take off from Danish airstrips, and rejoiced when they did not return. The air attacks rocked Britain, but the Nazi losses in the huge daylight assault on London on September 15 were so heavy that the daylight raids were not renewed. The bombing continued for another fifty-seven nights, but the crisis had passed and soon the invasion was called off. When the Battle of Britain had been won, in what Churchill called "Britain's finest hour," the Danish spirit surged. The new determination found expression in two ways. The first was a definitive book on Danish culture.[4] If the Reich should endure for a thousand years, as Hitler boasted, the Danish spirit would last even longer. Bohr wrote the introduction; his aim was to evoke the very soul of Denmark.

The second new development was an increase in sabotage. Bombs began to explode in plants that supplied the German war machine. The Bohrs, of course, were not directly involved in this effort, but from their home, the House of Honor, which was separated from the main Copenhagen railroad tracks only by dense trees and shrubbery, they watched the wreckage being hauled away.

In June 1941 Hitler broke his misalliance with Stalin. German troops drove deep into Russia. All who knew about the possibilities of an atom

257

bomb worried increasingly that Hitler might obtain this ultimate weapon. A few American scientists who had earlier fled from Europe persuaded Einstein to write to President Roosevelt telling him of the potential bomb and its dangers. On December 7 when the Japanese launched their attack on Pearl Harbor, the president ordered full speed ahead on the production of an atomic bomb.

The British, for their part, had become convinced that a bomb was possible and quickly joined forces with the United States. They felt they needed Bohr, and a secret message urging him to come to Britain was sent to him. In his isolation in Denmark, Bohr knew nothing of the plans to develop the bomb, and felt still that he had to remain in Denmark "as long as possible." He regretfully declined by a message through the same underground route.

During 1942 and 1943 the war turned against the Germans. In response they cracked down so brutally on the Danish resistance that the government had to resign. In effect, Denmark was at war with the Nazis. Shortly after, word came to Bohr that his arrest had been ordered in Berlin and that he must leave at once. That same night, with the aid of the underground, he and Mrs. Bohr escaped to Malmö, Sweden. A small boat took them from Copenhagen out to a fishing boat, which safely made the crossing. Their family would come later by the same route.

Bohr hurried on to Stockholm; he had urgent business there. When he had been told that he must escape from Denmark, he had also been informed that the long-feared roundup of the Jews of Denmark would soon begin. It was Bohr's assignment to arrange with the Swedes for a safe reception of the refugees in Sweden. On the morning after his landing, Swedish intelligence officers took Bohr by a devious route to a meeting with the under secretary of state for foreign affairs. Bohr urged the Swedes to try to dissuade the Nazis from the arrests, but learned that the Germans had already been approached and had dismissed the suggestion of arrests as a mere rumor.

Two days later, on October 2, the "rumor" became reality. Arrests began, and the first victims were taken to a ship that would carry them to concentration camps. Bohr had to reach the king, Gustav V. Princess Ingeborg, a sister of the king of Denmark, obtained an appointment for him that same afternoon. Bohr appealed to the king to try to have the prison ships rerouted to Sweden and to "intern"—actually give asylum to—the prisoners. The king agreed to consult his government at once and advised Bohr to listen to the radio that night. The Nazis abruptly refused the rerouting, but the Swedish government formally announced that it would receive the Jewish people of Denmark. The rescue, then, would have to be made by other means. Operating every night for the next few weeks, small boats carried nearly 6,000 passengers out to Swedish vessels waiting at the limits of the territorial waters. Some 300 people were lost as the Nazis attacked the rescue effort, but the number taken to concentration camps was reduced to 472. The toll would have been vastly greater if a determined, humanitarian state and a scientist acting as a political leader had not gone into action.

As the refugees found safety, Bohr received a second invitation, this time from Churchill's scientific advisor, Lord Cherwell, to come to Britain. There was no longer any reason to hesitate; Bohr accepted for himself and the twenty-one-year-old Aage, who had recently graduated in physics and would serve as his assistant. There followed Bohr's involvement in the British and American atomic bomb projects, as described in the essay "Niels Bohr and Nuclear Weapons."

On 7 May 1945, while Bohr was still in the United States (in Washington), the German High Command surrendered unconditionally. Bohr hurried to the Danish Embassy and learned that the Nazi command in Denmark had given up a few days earlier and that a small Allied "holding" force had taken over. Denmark was free at last. But his jubilation was tempered; contrary to his own hopes and recommendations, it had been decided that the Russians should not be informed of the atomic bomb until it had first been used against Japan. "Bohr wanted to change the framework in which the problem would appear early enough so that the problem itself would be altered," said Oppenheimer, who had gone to the British mission office, where Bohr was working, to try to comfort him. "He was much too wise and he would not be comforted," Oppenheimer added.

Bohr was eager to return to Denmark, but he could not do so until the bomb had been tested and used. He was in England when the Hiroshima and Nagasaki bombs were dropped. Thus came to an end one of the few periods in which a handful of leaders might have drastically changed the course of history. Bohr, nevertheless, did not give up. He understood that the struggle for a safe world would have to be carried to the general population; perhaps more hope lay there. Bohr wrote a public letter to the London *Times*: "Civilization is presented with a challenge more serious, perhaps, than ever before, and the fate of humanity will depend on its ability to unite in averting common dangers.[5]

Back in Copenhagen, Bohr continued his political work. He made speeches in a number of cities and (during visits to the United States in 1946 and 1948) talked with the new leaders coming upon the scene, including General George Marshall, the new U.S. secretary of state, and Hendrik Kramers, Bohr's former student and associate who had become chairman of the United Nations Atomic Energy Commission.

Then, in September 1949, President Truman announced that an atomic explosion had occurred in the Soviet Union. Just four years after Hiroshima, the Russians had the bomb. Bohr was not surprised. But here was the arms race he had feared. The somber question was: What could be done? Bohr laid a new program before the United Nations — it called again for an open world. However, the Americans not long afterward exploded the first thermonuclear device, a fusion bomb a hundred times more destructive than the atom bomb. The Russians followed suit less than a year later, and the arms race was gathering speed.

I vividly remember the circumstances of my first meeting in Rutherford's office in the Cavendish with the young Robert Oppenheimer, with whom I was later to come into such close friendship. Indeed, before Oppenheimer entered the office, Rutherford, with his keen appreciation of talents, had described the rich gifts of the young man, which in the course of time were to create for him his eminent position in scientific life in the United States.

Niels Bohr,
Rutherford Memorial Lecture, 1958

On the peaceful side, there was the question of atomic energy for Denmark. The low, flat Danish peninsula had no fossil fuels of its own. To investigate the possibility of the new power for Denmark, the Danish Academy for Technical Sciences named a special committee. It was chaired, of course, by Niels Bohr. Bohr went to see the prime minister, who assigned minister for finance Vigo Kampmann to work with him.[6] As the two in Bohr's thorough way studied every phase of establishing a power plant in Denmark, they also worked on the first Atoms for Peace Conference in Geneva. At that conference, some of the benefits of atomic power were made available to all. It was a small step in the "right" direction, as Bohr saw it.

Meanwhile the Danish Atomic Energy plant was built under Bohr's chairmanship. It was called Risø, for the nearby town. On the day of its dedication, 6 June 1958, the big reactors, looking almost like an extension of the rocky headland on which they stood, were complete down to the grassy lawn and a kennel for the watchdog. A few days later, when the Danish parliament debated the heavy cost of Risø, the only questions dealt with the kennel and the cost of the flagpole from which the red and white Danish flag flew proudly. The people of Denmark were invited to visit the reactor plant, and thousands came.

In the world at large, the Cold War and the proliferation of nuclear armaments followed much as Bohr had feared. But even though Bohr's policies had not been adopted, his advocacy of a better course was appreciated and honored. No one can ever be certain that the Bohr way would have succeeded. But it can still be argued that he had identified a possible "path into the unknown" in world politics, just as he had done earlier in science. And from first to last, his life exemplified his humanitarian and political commitment. To use the words of Milton that John Wheeler quoted when Bohr received the first Atoms for Peace award in 1957, his was no "fugitive and cloistered virtue."

Energy from the Atom:
An Opportunity and a Challenge

The Second World War was brought to an end on 10 August 1945 by the surrender of Japan, following quickly after the atomic bomb attacks on Hiroshima and Nagasaki. The very next day there appeared in *The Times* of London the article below by Bohr.[1] Typically, it was a carefully thought-out statement that had clearly been prepared in readiness for the occasion, and it demonstrates his prescience in these matters — coming, as it did, nearly five years before his much longer "Open Letter to the United Nations," an abridged version of which is reprinted later in this volume.

A.P.F.

The possibility of releasing vast amounts of energy through atomic disintegration, which means a veritable revolution of human resources, cannot but raise in the mind of everyone the question whither the advance of physical science is leading civilization. While the increasing mastery of the forces of nature has contributed so prolifically to human welfare, and holds out even greater promises, it is evident that the formidable power of destruction which has come within reach of man may become a mortal menace unless human society can adjust itself to the exigencies of the situation. Civilization is presented with a challenge more serious, perhaps, than ever before, and the fate of humanity will depend on its ability to unite in averting common dangers and jointly to reap the benefit from the immense opportunities which the progress of science offers.

ATOMIC FORCE

COLLABORATORS IN RESEARCH

THE MORAL ISSUE

TO THE EDITOR OF THE TIMES

Sir,—As one who has been in touch with many aspects of the development of atomic bombing, I should like to supplement the official accounts in one respect, namely, the part played by the Scientific Advisory Committee of the War Cabinet (S.A.C.) in 1941.

On June 19 of that year the Minister of Aircraft Production, Colonel Moore-Brabazon (now Lord Brabazon), consulted me about the inquiry by Sir George Thomson's committee, which was attached to his department and was approaching a stage when the results of its work ought to be considered by the highest scientific authorities. We agreed that the right body was the S.A.C., which had been established by Mr. Neville Chamberlain in October, 1940. Sir John Anderson, Lord President, under whom the committee worked, agreed. We received our formal remit late in August, set up a special panel under my chairmanship, heard the evidence of many famous scientists, including some of those mentioned in Mr. Churchill's account, and others, and before the end of September we presented a report, covering all aspects of the question, clearing up some of the doubts and endorsing the feasibility of the project.

That authoritative and comprehensive report provided the basis for the invaluable work of the subsequent consultative council under Sir John Anderson, of which, as chairman of the S.A.C., I was a member until I left the Government in March, 1942. It also led to the establishment of the special division of the Department of Scientific and Industrial Research to direct the work, the whole of which was henceforward carried on under the powerful direction of Sir John Anderson.

I write this letter in justice to the distinguished scientists who took part in the S.A.C. inquiry, namely, Sir Henry Dale, Professor A. V. Hill, Sir Alfred Egerton, Sir Edward Appleton, and Sir Edward Mellanby, whose collective influence on the policy adopted was not mentioned in the official accounts.   Yours faithfully,
                                           HANKEY.
House of Lords, Aug. 10.

Sir,—Now that the dust that was Hiroshima and Nagasaki is subsiding there must be countless men and women, on both sides of the Atlantic, able to see more clearly and observe in its stark reality the unparalleled horror that is being perpetrated in their name.

The most ghastly aspect of modern warfare is the widespread slaughter of non-combatants, especially children. The allies have hitherto done their best to avoid it. They may have been unsuccessful, but at least they have tried. Here however there can be no pretence. By deliberate cataclysmic blows the hearts of two great cities, with all who lived within them, have been seared from the earth. "And by whom? By the same allies who barely a year ago were seething with righteous indignation at the indiscriminate brutality of the flying bomb. I quote a single extract from your columns in July, 1944: " The American people," wrote your New York Correspondent, " were profoundly shocked and horrified by Mr. Churchill's disclosures to-day of the actual number of victims from the flying bomb—for they count it nothing less—has claimed."

Its justification, so we are told, is that it will end the war sooner. It will certainly do that. So would poison gas or disseminated germs; so would any other diabolical invention calculated not merely to defeat the enemy's forces, but to annihilate his entire population —men, women and children—in batches of two hundred thousand at a time. For that is what the use of atomic force in war implies. The responsibility does not lie with the scientists. We honour them for their genius and thank them for their courage. No less are we grateful to those whose wisdom and foresight enabled use to forestall our enemies by the discovery of this awful power. But posterity will condemn those who have first used it for this purpose, thereby creating the most dreadful precedent in the history of mankind.

                              Yours, &c.,
                    JOHN A. F. WATSON.
Carlton Club, S.W.1, August 10.

ENERGY FROM THE ATOM

AN OPPORTUNITY AND A CHALLENGE

THE SCIENTIST'S VIEW

By Professor Niels Bohr

Professor Bohr is the Danish physicist who escaped through Sweden to England in 1943, and later went to America to help in the production of the atomic bomb. He won the Nobel Prize at the age of 37 for his work on atoms.

The possibility of releasing vast amounts of energy through atomic disintegration, which means a veritable revolution of human resources, cannot but raise in the mind of everyone the question whither the advance of physical science is leading civilization. While the increasing mastery of the forces of nature has contributed so prolifically to human welfare, and holds out even greater promises, it is evident that the formidable power of destruction which has come within reach of man may become a mortal menace unless human society can adjust itself to the exigencies of the situation. Civilization is presented with a challenge more serious, perhaps, than ever before, and the fate of humanity will depend on its ability to unite in averting common dangers and jointly to reap the benefit from the immense opportunities which the progress of science offers.

COOPERATION IN RESEARCH

In its origin science is inseparable from the collecting and ordering of experiences, gained in the struggle for existence, which enabled our ancestors to raise mankind to its present position among the other living beings which inhabit our earth. Even in highly organized communities where, within the distribution of labour, scientific study has become an occupation by itself, the progress of science and the advance of civilization have remained most intimately interwoven. Of course, practical needs are still an impetus to scientific research, but it need hardly be stressed how often technical developments of the greatest importance for civilization have originated from studies aimed only at augmenting our knowledge and deepening our understanding. Such endeavours know no national borders and where one scientist has left the trail another has taken it up, often in a distant part of the world. For long scientists have considered themselves as a brotherhood working in the service of common human ideals.

In no domain of science have these lessons received stronger emphasis than in the exploration of the atom, which just now is bearing consequences of such overwhelming practical implications. As is well known, the roots of the idea of atoms as the ultimate constituents of matter go back to ancient thinkers searching for a foundation to explain the regularity which, in spite of all variability, is ever more clearly revealed by the study of natural phenomena. After the Renaissance, when science entered so fertile a period, atomic theory gradually became of the greatest importance for the physical and chemical sciences, although, until half a century ago, it was generally accepted that, due to the coarseness of our senses, any direct proof of the existence of atoms would always remain beyond human scope. Aided, however, by the refined tools of modern technique, the development of the art of experimentation has removed such more difficult than ever to disentangle the contributions of individual workers.

The grim realities which are being revealed to the world in these days will no doubt, in the minds of many, revive terrifying prospects forecast in fiction. With all admiration for such imagination, it is, however, most essential to appreciate the contrast between these fantasies and the actual situation confronting us. Far from offering any easy means to bring destruction forth, as it were by witchcraft, scientific insight has on the contrary made it evident that use of nuclear disintegration for devastating explosions demands most elaborate preparations, involving a profound change in the atomic composition of the materials found on earth. The astounding achievement of producing an enormous display of power on the basis of experience gained by the study of minute effects, perceptible only by the most delicate instruments, has, in fact, besides a most intensive research effort, required an immense engineering enterprise, strikingly illuminating the potentialities of modern industrial development.

Indeed, not only have we left the time far behind where each man, for self-protection, could pick up the nearest stone, but we have even reached the stage where the degree of security offered to the citizens of a nation by collective defence measures is entirely insufficient. Against the new destructive powers no defence may be possible, and the issue centres on world-wide cooperation to prevent any use of the new sources of energy which does not serve mankind as a whole. The possibility of international regulation for this purpose should be ensured by the very magnitude and the peculiar character of the efforts which will be indispensable for the production of the new formidable weapon. It is obvious, however, that no control can be effective without free access to full scientific information and the granting of the opportunity of international supervision of all undertakings which, unless regulated, might become a source of disaster.

INTERESTS AND SECURITY

Such measures will, of course, demand the abolition of barriers hitherto considered necessary to safeguard national interests but now standing in the way of common security against unprecedented dangers. Certainly the handling of the precarious situation will demand the good will of all nations, but it must be recognized that we are dealing with what is potentially a deadly challenge to civilization itself. A better background for meeting such a situation could hardly be imagined than the earnest desire to seek a firm foundation for world security, so unanimously expressed from the side of all those nations which only through united efforts have been able to defend elementary human rights. The extent of

(caption)
A portion of the editorial page of *The Times* (London) for 11 August 1945. Besides Bohr's article, "Energy from the Atom," this page carries letters about the atomic bomb from Lord Hankey (who was Minister without Portfolio in the British War Cabinet, and chairman of the Cabinet Scientific Advisory Committee that was created in October 1940) and from John A. F. Watson, a magistrate in the London juvenile courts.

COOPERATION IN RESEARCH

In its origin, science is inseparable from the collecting and ordering of experiences, gained in the struggle for existence, which enabled our ancestors to raise mankind to its present position among the other living beings which inhabit our earth. Even in highly organized communities where, within the distribution of labor, scientific study has become an occupation by itself, the progress of science and the advance of civilization have remained most intimately interwoven. Of course, practical needs are still an impetus to scientific research, but it need hardly be stressed how often technical developments of the greatest importance for civilization have originated from studies aimed only at augmenting our knowledge and deepening our understanding. Such endeavors know no national borders, and where one scientist has left the trail another has taken it up, often in a distant part of the world. For long, scientists have considered themselves as a brotherhood working in the service of common human ideals.

In no domain of science have these lessons received stronger emphasis than in the exploration of the atom, which just now is bearing consequences of such overwhelming practical implications. As is well known, the roots of the idea of atoms as the ultimate constituents of matter go back to ancient thinkers searching for a foundation to explain the regularity which, in spite of all variability, is ever more clearly revealed by the study of natural phenomena. After the Renaissance, when science entered so fertile a period, atomic theory gradually became of the greatest importance for the physical and chemical sciences, although, until half a century ago, it was generally accepted that, due to the coarseness of our senses, any direct proof of the existence of atoms would always remain beyond human scope. Aided, however, by the refined tools of modern technique, the development of the art of experimentation has removed such limitation and even yielded detailed information about the interior structure of atoms.

DISINTEGRATION OF NUCLEI

In particular, the discovery that almost the entire mass of the atom is concentrated in a central nucleus proved to have the most far-reaching consequences. Not only did it become evident that the remarkable stability of the chemical elements is due to the immutability of the atomic nucleus when exposed to ordinary physical agencies, but a novel field of research was opened up by the study of the special conditions under which disintegrations of the nuclei themselves may be brought about. Such processes, whereby the very elements are transformed, were found to differ fundamentally in character and violence from chemical reactions, and their investigation led to a rapid succession of important discoveries through which ultimately the possibility of a large-scale release of atomic energy came into sight. This progress was achieved in the course of a few decades, and was due not least to most effective international cooperation. The world community of physicists was, so to say, welded into one team, rendering it more difficult than ever to disentangle the contributions of individual workers.

The grim realities which are being revealed to the world in these days will no doubt, in the minds of many, revive terrifying prospects forecast in fiction. With all admiration for such imagination, it is, however, most essential to appreciate the contrast between these fantasies and the actual situation confronting us. Far from offering any easy means to bring destruction forth, as it were by witchcraft, scientific insight has on the contrary made it evident that use of nuclear disintegration for devastating explosions demands most elaborate preparations, involving a profound change in the atomic composition of the materials found on earth. The astounding achievement of producing an enormous display of power on the basis of experience gained by the study of minute effects, perceptible only by the most delicate instruments, has, in fact, besides a most intensive research effort, required an immense engineering enterprise, strikingly illuminating the potentialities of modern industrial development.

Indeed, not only have we left the time far behind where each man, for self-protection, could pick up the nearest stone, but we have even reached the stage where the degree of security offered to the citizens of a nation by collective defense measures is entirely insufficient. Against the new destructive powers no defense may be possible, and the issue centers on world-wide cooperation to prevent any use of the new sources of energy which does not serve mankind as a whole. The possibility of international regulation for this purpose should be ensured by the very magnitude and the peculiar character of the efforts which will be indispensable for the production of the new formidable weapon. It is obvious, however, that no control can be effective without free access to full scientific information and the granting of the opportunity of international supervision of all undertakings which, unless regulated, might become a source of disaster.

INTERESTS AND SECURITY

Such measures will, of course, demand the abolition of barriers hitherto considered necessary to safeguard national interests but now standing in the way of common security against unprecedented dangers. Certainly the handling of the precarious situation will demand the good will of all nations, but it must be recognized that we are dealing with what is potentially a deadly challenge to civilization itself. A better background for meeting such a situation could hardly be imagined than the earnest desire to seek a firm foundation for world security, so unanimously expressed from the side of all those nations which only through united efforts have been able to defend elementary human rights. The extent of the contribution which an agreement about this vital matter would make to the removal of obstacles to mutual confidence and to the promotion of a harmonious relationship between nations can hardly be exaggerated.

In the great task lying ahead, which places upon our generation the gravest responsibility toward posterity, scientists all over the world may offer most valuable services. Not only do the bonds created through scientific intercourse form some of the firmest ties between individuals from different nations, but the whole scientific community will surely join in a vigorous effort to induce in wider circles an adequate appreciation of what is at stake and to appeal to humanity at large to heed the warning which has been sounded. It need not be added that every scientist who has taken part in laying the foundation for the new development, or has been called upon to participate in work which might have proved decisive in the struggle to preserve a state of civilization where human culture can freely develop, is prepared to assist in any way open to him in bringing about an outcome of the present crisis of humanity worthy of the ideals for which science through the ages has stood.

Niels Bohr and Nuclear Weapons

Margaret Gowing

Margaret Gowing obtained her early training at the London School of Economics and proceeded into government service. After World War II she was appointed official historian and archivist for the British Atomic Energy Authority. She has written extensively on the history of the development of atomic energy (notably in her book *Britain and Atomic Energy, 1939–1945*) and is now Professor of the History of Science at Oxford University.

It was one of the most fateful coincidences in history that the discovery of uranium fission by Otto Hahn and Fritz Strassmann and the theoretical explanation of the phenomenon by Lise Meitner and Otto Frisch were published at the beginning of the year in which World War II began. In April 1939, Frédéric Joliot-Curie's team in Paris was the first to announce experimental evidence that in fission spare neutrons are released; this opened the possibility of a nuclear chain reaction and an atomic bomb. The widespread and agitated discussion of this project diminished when, on September 1, the day Germany invaded Poland, an article by Niels Bohr and John Wheeler was published in *The Physical Review,* giving the classic interpretation of the fission process.[1] This included the important deduction (already foreshadowed by Bohr in a letter of 7 February 1939 to the same journal) that it was the rare U^{235} atom, not the U^{238} atom, that fissioned—a deduction consistent with the observation that fission was much more likely with moderated, slow neutrons than with fast ones.[2] It seemed that these slow-neutron chain reactions might produce power, but not the fantastically fast reaction necessary for a bomb.

In the subsequent two years, scientists in the belligerent countries and the nonbelligerent United States worked on bomb possibilities. The most crucial and highly organized work was done in Britain by the famous Maud Committee. The strange name of this committee was derived from a telegram sent by Meitner from Sweden in May 1940, just after Denmark had been invaded by the Germans, to the physicist Owen Richardson: "Met Niels and Margrethe recently. Both well but unhappy about events. Please inform Cockcroft and Maud Ray Kent." Cockcroft believed the last words were an anagram for "radiumtaken," and the words seemed a good code name. When Bohr arrived in England in 1943, he asked whether the message had ever reached his old governess, Maud Ray, who lived in Kent.

The Maud Committee was composed of British and refugee scientists. A paper of April 1940 by Frisch and Rudolf Peierls — both refugees at Birmingham University — had stimulated the formation of the committee, by showing that a small lump of U^{235} would give the fast reaction necessary for a bomb and by proposing an industrial method for separating the U^{235}. When France fell, two members of Joliot-Curie's team fled to Britain, and a slow-neutron team developed round them. Two members of the team soon suggested that element 94 (plutonium), whose discovery had been predicted by Edwin Macmillan and Philip Abelson in May 1940, would also be an efficient superexplosive with a small critical mass. Unknown to the British, scientists in Berkeley, California, in March 1941 demonstrated experimentally that this was so.

Several groups of American scientists were indeed working on various aspects of "the uranium problem," but in a diffuse, leisurely way. It did not become urgent to them until in July 1941 they were shown the British Maud Report, which demonstrated coherently and cogently why and how an atomic bomb was possible. Their government set up, even before Pearl Harbor ended U.S. neutrality, what was soon to become the huge Manhattan Project.

Meanwhile, German nuclear scientists had been contemplating the possibilities of an atomic bomb. In the early autumn of 1941, Fritz Houtermans wrote a paper which calculated the critical mass of U^{235} but proposed, as a route to the bomb, using the new fissile element that might be produced in a slow-neutron reactor. There was every reason to fear this competition, but in fact the German project later floundered in its science and in its organization.

After the invasion of Denmark, Bohr was preoccupied with the grave problems of his country and his institute, including as it did refugees from Nazism; he faced these problems with dignity, courage, and deep patriotism. Atomic bombs were not in the forefront of his mind. He had realized, in the wake of the paper written with Wheeler, that an explosion could indeed be achieved with a sufficiently large amount of U^{235}. He had explained this in a lecture in Birmingham just before, and in Denmark just after, the outbreak of war.[3] But he did not think that with current technical facilities it would be possible to separate enough U^{235}. He did not on these occasions consider the slow-neutron possibilities. He was therefore the more deeply disturbed by a visit from Werner Heisenberg in October 1941. Robert Jungk in his book *Brighter Than a Thousand Suns* calls this visit "a little-known peace feeler. By the expedient of a silent agreement between German and Allied atomic experts, the production of a morally objectionable weapon was to be prevented." This suggestion of German moral scruples is supported in the book by a letter to Jungk from Heisenberg about this visit to Bohr. Aage Bohr, who was close to his father in all nuclear weapons affairs, has written that Heisenberg put no proposal to Niels for a physicists' agreement to refrain from developing nuclear weapons, but that he did leave the strong impression that the Germans attributed great military importance to atomic energy.

Early in 1943 a message reached Niels Bohr from James Chadwick, who was the informal scientific leader of the British atomic project (code name: "Tube Alloys"). It was written, as a token of its authenticity, on Liverpool University notepaper. It came in great secrecy through British and Danish intelligence, in a microdot inserted in a bunch of keys. Chadwick wrote that he had heard "you have considered coming to this country if the opportunity should offer. I need not tell you how delighted I myself should be to see you again . . . There is no scientist in the world who would be more acceptable both to our university people and to the general public . . . I have in mind a particular problem in which your assistance would be of the greatest help. Darwin and Appleton are also interested in this problem and I know they too would be very glad of your help and advice.[4]

Bohr still felt that it was his duty to remain in Denmark, but replied to Chadwick that he might leave if he felt he could be of real help in other ways. He added that he did not think this probable. "Above all, I have to the best of my judgment convinced myself that, in spite of all future prospects, any immediate use of the latest marvelous discoveries of atomic physics is impracticable." Two months later he sent another message to Chadwick, reporting the rumors he had heard of German preparations for producing metallic uranium and heavy water in order to make atomic bombs. However, he was still skeptical: he took it for granted that it was impossible to separate U^{235} in sufficient quantity, and he doubted the possibility of using slow-neutron reactions for this purpose. Copies of the messages exchanged by Bohr and Chadwick were buried in the garden at Carlsberg, to be found after the war.

In September 1943, when Bohr and his family were known to be in danger of immediate arrest, their escape to Sweden was arranged and the British atomic directorate made it possible for Niels, with Aage as his assistant, to go to Britain. Niels departed on 6 October 1943, and survived a hazardous journey in an unarmed converted bomber. Aage arrived a week later.

When Niels arrived in England he was immediately told everything about the progress of the British and American projects: it now seemed almost certain that the Americans would produce nuclear weapons within a year or two. Father and son were received most warmly by the scientists, by the administrators of the Tube Alloys project, and by the minister in charge — Sir John Anderson, later Lord Waverley, who was chancellor of the exchequer and who would henceforth be a close friend of the Bohr family. The Bohrs had arrived at an important moment for the British project. The Maud Report had pushed the floundering American project off the ground, and the overconfident British had preferred atomic cooperation with the Americans to the full integration between the projects which Roosevelt suggested in the autumn of 1941. But the American project soon far outstripped the British and neither needed nor wanted British help. The British, already highly mobilized and unable to build huge atomic plants, became desperate: they could not proceed on their own and were cut off from American knowledge. It was only after a great struggle that Churchill, in August 1943, persuaded Roosevelt to sign the Quebec Agreement, which enabled British scientists to participate in some parts of the American project — most notably at Los Alamos, where the bombs were to be fabricated.

Bohr, so welcome to the British for his own sake, was also, as a member of their team in America, a trump card for them in implementing the Quebec Agreement. Bohr promised that he would not allow himself to be drawn into the American orbit, that he would assist the common effort and also do everything he could to make the association between America and Britain a real partnership. He and Aage arrived in the United States early in December 1943 under the aliases Nicholas and James Baker, or, affectionately to colleagues, Uncle Nick and Jim. They were not attached to any specific team, but Bohr's main scientific contribution was to work at Los Alamos, where he found many of his former students. There he stimulated and liberated scientific ideas, giving rise to theoretical and experimental activities which cleared up unanswered questions — for example, on the

I once asked him how much he was troubled by his false name; did he always remember to sign himself Nicholas Baker rather than Niels Bohr? His reply was, "What is the difference? My signature is quite illegible; it could be either."

O. R. Frisch,
What Little I Remember, 1979

Niels Bohr with Aage sometime after World War II.

velocity selector, the bomb assembly, and the design of the initiator. Old and new friendships flourished here.

When Bohr saw the vast Manhattan Project he was fascinated by the extraordinary edifice, built as it was on theoretical foundations he had laid. But he was vastly more impressed by the implications of the new weapon for the future of the world. He had the reputation for being the most unworldly of scientists, but the unworldliness was purely behavioral. His knowledge of philosophy, history, and politics was profound and had been deepened by his ties with the refugees from Nazism at his institute in the 1930s and by the German occupation. His exceptional imagination and intuition marked not only his science but his view of world politics. He immediately realized when he saw the Manhattan Project that it was only a beginning and that the progress of the work would disclose new and more fearsome possibilities. At Los Alamos scientists were already foreseeing a hydrogen bomb.

Bohr was concerned privately with the question of how soon the weapon would be ready for use and what role it might play in the Second World War, but he took no part in discussions about whether it should be dropped. He looked rather to the years after the war and the terrifying prospect of international competition in atomic weapons. After his first visit to Los Alamos, he wrote to Anderson that future effective control would involve not only the most intricate technical and administrative problems but also concessions over exchange of information and openness about industrial efforts and military preparations — developments that were hardly conceivable in terms of prewar international relationships. Bohr felt that the invention of atomic bombs was so climacteric that it would facilitate a whole new approach to these relationships.

Before long his thoughts crystallized into a fairly precise proposition. At a time of euphoria about brave Russian allies, Bohr believed that there would be tension between the West and Russia after the war and that confidence and cooperation might be promoted by telling Russia about the bomb before it was used. Conversely, he believed that it would be disastrous if Russia should learn on its own about the bomb. Knowing very well the competence of Russian physicists, Bohr felt certain, like most other scientists when they thought about it, that the margin of time before the Russians made a bomb themselves would be very small. This conviction was strengthened when, in London, in April 1944, he received a letter from Peter Kapitza, written six months earlier when Bohr had escaped to Sweden, inviting him to settle in Russia. This reinforced Bohr's belief that the Russians were aware of the American project. He sent back a warm, innocuous reply to Kapitza and showed the correspondence to the British authorities.

The political implications of the bomb had become Bohr's prime concern. He spent much of his time writing "political" memoranda and haunting the offices and anterooms of those who had political power or

During his visit in Los Alamos he was in a relaxed mood. He could see that the work was going well and the war was turning in our favour. The radio reports were confusing, but Bohr always listened. "We must hear all the rumours before they are denied," he joked.

O. R. Frisch,
What Little I Remember, 1979

access to it. His discursive talk and his low, indistinct voice were not easy to follow, but he made important converts: Lord Halifax and Sir Ronald Campbell, respectively ambassador and minister at the British Embassy in Washington and, most significantly, Sir John Anderson and Lord Cherwell (the scientist who was Churchill's personal advisor) and field marshal Jan Smuts. Halifax told Bohr that because of America's preponderant share in the project, any initiative would have to come from President Roosevelt. It seemed fortunate, therefore, that Bohr was able to resume a prewar friendship with Felix Frankfurter, a Supreme Court judge and a friend of Roosevelt's. Frankfurter already knew about the bomb and communicated Bohr's ideas and hopes to the president, who said the whole thing "worried him to death" and that he was most eager to explore it with Churchill.

In March 1944 Anderson wrote a long memorandum to Churchill, summarizing the situation. It seemed certain that the Americans would develop a bomb first, but it was foolish to suppose that Russia would not put forward a great effort once they had expelled the Germans. Moreover, the scale of effort needed would decrease and would become feasible for other countries. There were two possible outcomes: a particularly vicious arms race in which, at best, the United States and Britain would for a time enjoy a precarious and uneasy advantage; or some form of international control. If it was decided to work for international control, there was much to be said for communicating to Russia in the near future the bare fact that the Americans expected by a given date to have this devastating weapon and for inviting them to collaborate in preparing such a control scheme. If the Russians were told nothing, they would learn sooner or later what was afoot and might then be less disposed to cooperate. There was little risk that Russia, if she chose to be uncooperative, would be much helped by such a communication. Cherwell added his plea: "I must confess that I think plans and preparations for the postwar world and even the peace conference are utterly illusory, so long as this crucial factor is left out of account." Churchill, however, disagreed profoundly and constantly reiterated his conviction that the project must be kept as secret as possible.

Pressed by Smuts, Cherwell, and Sir Henry Dale, president of the Royal Society, Churchill saw Bohr on 16 May 1944. The meeting was a failure. Dale had expressed fears that Bohr's "mild, philosophical vagueness of expression and his inarticulate whisper" might prevent a "desperately preoccupied Prime Minister" from understanding him, and so it proved. The main point was never reached. "We did not speak the same language," said Bohr afterward. Later Churchill told Cherwell, "I did not like the man when you showed him to me, with his hair all over his head."

However, during the summer Churchill realized that he must discuss the long-term problem of the atomic bomb with Roosevelt when next they met, as they did in September 1944. Prior to the meeting, Frankfurter had sent a seven-page memorandum by Bohr to Roosevelt which outlined the scientific basis of the project, Bohr's own feelings on seeing the project after his escape from Denmark, Kapitza's approach to him, his belief that the project offered an opportunity for a new spirit and new hope in international relations, and his fears of a nuclear arms race between Russia and the West. On August 26 Roosevelt spoke with Bohr in complete privacy, in an interview lasting one and a half hours.

Bohr reiterated his belief that there was a great opportunity to better world relations, provided it was seized now rather than later. He expanded on his reasons for urging overtures to Russia and on his arguments against those who said that the West would lose thereby. He said it must be assumed that the Russians knew great efforts were being made in the United States to make a bomb; that the Russians themselves were studying the matter and would be free to mount a full effort at the end of the German war; that the Russians would probably obtain the Germans' secrets at the end of the war. If the United States and Britain said nothing before a bomb was used, they would, urged Bohr, arouse Russian suspicions and create a greater risk of fateful competition in atomic weapons. They would lose the opportunity of using an approach to Russia in order to establish confidence. Bohr emphasized that it was not necessary to begin by giving the Russians detailed information about the bomb. The approach should be general, and if the Russians responded in a cooperative spirit the way would be open for frank discussions. If not, the West would know where they stood. Bohr believed that one possible method of approach might be through preliminary and noncommittal contact between scientists.

The president could not have been friendlier to Bohr or more open and frank in his discussions of the political problems raised by the bomb. He said that an approach to Russia must be tried and that it would open a new era of human history. Stalin, he believed, was enough of a realist to understand the implications of this scientific and technological revolution. Bohr was sufficiently encouraged by his talk with Roosevelt to have a shot at a draft letter to Kapitza on the lines discussed, and held himself ready to go to Russia.

Bohr's high hopes were to be rudely dashed. In September 1944 Churchill and Roosevelt met at the second Quebec Conference and at Roosevelt's Hyde Park home and discussed the atomic bomb, with results very different from those foreshadowed during Bohr's interview with Roosevelt. On September 18 they signed an *aide mémoire* which included a paragraph saying that inquiries were to be made about Bohr and steps taken to ensure that he leaked no information, particularly to the Russians.

This agreement, besides turning down Bohr's proposal for an approach to Russia, put his own good faith and honor in question. Churchill expressed these doubts about Bohr even more forcefully to Lord Cherwell: "The President and I are much worried about Professor Bohr. How did he come into the business? He is a great advocate of publicity. He made an unauthorised disclosure to Chief Justice Frankfurter, who startled the President by telling him he knew all the details. He said he is in close correspondence with a Russian professor, an old friend of his in Russia to whom he has written about the matter and may be writing still. The Russian professor has urged him to go to Russia in order to discuss matters. What is all this about? It seems to me Bohr ought to be confined or at any rate made to see that he is very near the edge of mortal crimes."

Bohr's friends Cherwell, Anderson, Halifax, and Campbell rushed to defend Bohr and to say that Churchill was talking nonsense. They felt strongly "that the great P. J. [Panjandrum] was barking up an imaginary tree." Cherwell sent a strong reply to Churchill describing how Bohr had come into the business, giving an account of the Bohr-Frankfurter talks, the story of the approach by Kapitza, and the reply that had been agreed on by British Intelligence. "I have always found Bohr most discreet and conscious of his obligations to England to which he owes a great deal, and only the very strongest evidence would induce me to believe that he had done anything improper in this matter. I do not know whether you realise that the possibilities of a super weapon on Tube Alloys lines have been publicly discussed for at least six or seven years. The things that matter are which processes are proving successful, what the main stages are, and what stage has been reached. Most of the rest is published every silly season in most newspapers." Cherwell repeated these views to Roosevelt in the presence of Vannevar Bush, who agreed with them. Churchill accepted Cherwell's opinion about Bohr and the matter was dropped. Bohr, when he heard of the misunderstanding, was distressed; he might have been deeply offended, but his sense of humor was always stronger than his pride.

We do not know the reasons for Roosevelt's volte-face. As for Churchill, he believed passionately in the desirability and possibility of keeping atomic weapons secret. At home he kept the matter secret from the War Cabinet (including Clement Attlee, who in July 1945 became prime minister) and his defense advisors, and he refused to impart any information to the French, to whom the British had atomic obligations. He wrote, "You may be quite sure that any power that gets hold of the secret will try to make the article and that this touches the existence of human society. The matter is one out of all relation to anything else that exists in the world and I could not think of participating in any disclosure to third or fourth parties at the present time. I do not believe there is anyone in the world who can possibly have reached the position now occupied by us and the United States."

Meanwhile Bohr found himself exercising a restraining hand on Einstein, who in December 1944 sent him a *cri de coeur* about the prospect of a postwar arms race. The politicians, Einstein said, did not appreciate the threat. In all the principal countries influential scientists had the ear of political leaders — he mentioned Bohr himself, Compton (no initial, but presumably A. H. Compton), Cherwell, Kapitza, and Joffé. These men should come together to bring pressure to bear on their political leaders to strive for the internationalization of military power. "Don't be impossible" wrote Einstein to Bohr, "but wait a few days until you have accustomed yourself to these strange thoughts." Bohr went to see Einstein and explained to him that it would be quite illegitimate and might have the most deplorable consequences if anyone who was brought into confidence about the bomb should take the initiative into his own hands. Bohr assured Einstein that the attention of responsible statesmen in Britain and the United States had been called to the implications of the bomb. Einstein thereupon agreed to abstain from action and to impress on his friends the undesirability of doing anything that might complicate the delicate task of statesmen.

Bohr, conscious that time was running out, became increasingly convinced that postponement of any discussion with Russia until a bomb was demonstrated might give the appearance of an attempt at coercion in which no great nation could be expected to acquiesce. He emphasized yet again that Russia would soon learn, at the least, about the German work. Anderson, Halifax and, indeed, Anthony Eden as British foreign secretary realized that the important questions Bohr had raised would have to be faced sooner or later. In April 1945 Halifax and Frankfurter walked through Rock Creek Park in Washington discussing how to get Bohr's proposals properly considered. As they ended their walk they heard all the bells in Washington tolling. Roosevelt was dead.

In Washington, Bush and James Bryant Conant had been pressing on Henry Stimson, U.S. secretary of war, views not dissimilar to Bohr's and in May 1945 Stimson chaired an interim committee which *inter alia* discussed disclosure to Russia and possible forms of international control. Members of the committee were torn between, on the one hand, a desire for scientific openness and a conviction that the business could not remain secret for long, and, on the other hand, anxieties over deteriorating Russian behavior. The anxieties won, and the committee decided early in June 1945 that no information should be revealed to Russia or anyone else until the first bomb had been dropped on Japan.[5]

On July 24, eight days after the atomic bomb test at Alamogordo and thirteen days before a bomb was dropped on Hiroshima, President Truman told Stalin simply that the United States had a new weapon of unusual destructive force. Bohr's wartime pleas had failed. In books written in the 1960s and later, these pleas were seen as the remarkable intuition of a remarkable scientist. A leading historian of international relations, however, attacked them: "The concept of 'international control' in the minds of Bohr and others was essentially a cop-out, a flight into higher mysticism away from the unpleasant and unacceptable world of politics."[6] Such strictures were inappropriate to Bohr's essentially practical proposal. He knew that Russian physicists were extremely capable and that once a bomb had been dropped there could be no secret. To inform Russia officially would therefore carry little risk and might conceivably bring benefits. Not to inform Russia would bring little benefit and would intensify suspicions.

Bohr's idealism, that is, was set in a very practical framework of limited objectives as he looked to a future when all civilized life might be destroyed in a flash.

If Russia had been officially told about the bomb during the war, the revelation might have made no difference. The Soviets had already begun their own project and knew a great deal about the Manhattan Project from spies, notably Klaus Fuchs. The fact that they were told virtually nothing by the Allies guaranteed that attempts made just after the war to establish international control, which might have failed anyway, were doomed.

Bohr, as I have noted, made no representation in advance about the use of the atomic bomb against Japan, and he did not argue about past events once the war was over. He privately deplored the spirit in which the bomb had been used and the opportunities that had been lost,[7] but he neither made nor joined any written protestations. His thoughts were on the future and the postwar world. With his inbred and unquenchable optimism he was convinced that, although atomic bombs had introduced unprecedented threats to the world, they also provided a unique opportunity to develop a new approach to international relationships.

In the spring of 1945 Bohr wrote another memorandum intended for Roosevelt but given to Vannevar Bush after Roosevelt's death. It looked beyond the question of informing the Russians about the bomb during the war. Bohr warned that the American-British effort, immense though it was, had proved far smaller than might have have been anticipated and that any information, however scanty, that might have leaked from it would have greatly stimulated efforts elsewhere. Probably within the very near future, means would be found to "simplify the methods of production of the active substances and intensify their effects to an extent which may permit any nation possessing great industrial resources to command powers of destruction surpassing all previous imagination. Humanity will therefore be confronted with dangers of unprecedented character unless in due time measures can be taken to forestall a disastrous competition in such formidable armaments and to establish an international control of the manufacture and use of the powerful materials."[8]

Extraordinary measures would be necessary to counter secret preparations for the mastery of the new means of destruction. Not only must there be universal access to full information about scientific discoveries, but every major technical enterprise, industrial as well as military, must be open to international control. The special character of the production of the active materials, and the peculiar conditions governing their use as dangerous explosives, would, said Bohr, greatly facilitate such control and should ensure its efficiency, provided the right of supervision was guaranteed. Detailed proposals for the establishment of an effective control would have to be worked out with the assistance of scientists and technologists appointed by governments, and a standing expert committee of an international security organization could be charged with keeping account of new scientific and technical developments and with recommending appropriate adjustments of the control measures.

On recommendations from the technical committee, the organization would be able to judge the conditions under which industrial exploitation of atomic energy sources could be permitted, with adequate safeguards to prevent any assembly of active material for an explosive. All material prepared for armaments might ultimately be entrusted to the security organization, to be held in readiness for eventual policing purposes. The prewar bonds between scientists of different nations would be especially valuable in creating controls.

Elements of Bohr's ideas would eventually be found in the early postwar proposals for atomic energy control discussed at the United Nations Commission on the subject and, later, in the nonproliferation safeguards to be operated by the International Atomic Energy Agency. But the United Nations proposals came to nothing, and the nonproliferation arrangements did not apply to the existing "atomic powers." On-site inspection, which Bohr regarded as essential to "openness" and which has been a feature of all attempts to control nuclear weapons and installations, has been unacceptable to the Soviet Union. Since Russian archives are not open, we do not know whether Bohr's plan for openness might have stood a chance if it had been proposed to the Russians, along with the information that a bomb was being made, before the war ended. If there had been any such possibility, the subsequent nuclear arms race might at the very least have been contained at a lower level. However, once the immediate opportunity had passed, the bombs had been dropped on Japan, and the United Nations atomic energy commission had failed, openness as a policy became more rather than less unthinkable on both sides.

Nevertheless, Bohr continued his campaign on every possible occasion. He increasingly believed that the free access to information which was necessary for common security would have far-reaching effects in other areas of life and would bring a decisive change in international affairs. In 1948 he talked and wrote to General George Marshall, U.S. secretary of state, urging that an American initiative toward openness might still be effective and stressing that it would not entail any a priori commitment to disarmament. He was conscious of the growing danger of bacteriological and biochemical, as well as atomic, warfare. The darker the international outlook grew, the more Bohr was convinced that a great issue "suited to invoke the highest aspirations of mankind" must be raised. Openness, with free access to information about all aspects of life in every country, was to him this issue. The initative should be taken, he pleaded, even if the chances of obtaining agreement were slight, because an offer of openness would strengthen the moral position of all supporters of genuine international cooperation; the opposition of those who refused to join would amount to a confession that they lacked confidence in their own cause. His efforts culminated in his open letter to the United Nations in June 1950, which incorporated the memoranda he had written to Roosevelt and Marshall. All his memoranda and letters on this subject had been written with the same examination and reexamination of every word and every nuance that marked his scientific papers. He believed that "one should not express a thought [in a form] clearer than the thought is." The result was a density in his prose which perhaps made it opaque to many readers.

The Cold War was now rapidly intensifying, and the Korean War broke out at the same time; as Leo Szilard wrote on receiving a copy of Bohr's open letter, "unfortunately world events are moving a little too fast." As for the content of the letter, Oppenheimer was deeply pessimistic that any man in a position of political responsibility would ever take the proposal of openness as a basis for action. The letter indeed evoked very little public reaction outside Scandinavia, and some of the public comment was critical. In Britain even the liberal *Manchester Guardian* wrote an unsympathetic leading article emphasizing the need "to keep our feet on the ground." Rudolf Peierls wrote an eloquent reply, arguing, "Let us also try to keep our head out of the sand," and stressing that Bohr had not implied that the United States should forthwith publish complete blueprints of their atomic energy installations. Bohr himself remained dedicated to his main theme of openness, so much so that he would not weaken it by joining the other "peace" moves and appeals from men such as Einstein and Bertrand Russell.[9] In 1956 Bohr wrote another open letter to the secretary of the United Nations, Dag Hammarskjold.

The time and the temper had been hopelessly unfavorable to Bohr's initiative. Although it failed, subsequent events have shown that he was ahead of his time in political insight, as in so many of his scientific ideas. We do not have an open world in 1985 but East-West contacts have been extended more widely than most people expected in the dark days of 1950 and include some scientific areas that were once unthinkable, such as high-energy accelerators and thermonuclear fusion. In nuclear weapons there is still no on-site inspection, but verification by national technical means such as satellites has been accepted in the various arms control treaties. On-site inspection has been agreed on, subject to treaty, for the destruction of stocks of chemical weapons.

In today's situation of nuclear stalemate, when the danger lies in upsetting an equilibrium, the information available about the number and types of weapons on both sides has become greater than anyone in 1950 believed possible. The revolution in communications technology which Bohr foresaw has greatly added to the information that countries have about each other. If such enforced openness diminishes rather than increases suspicion, Bohr's convictions may still prove to have been clearmindedly visionary, rather than impracticably idealistic. This may, just conceivably, happen in the lifetime of his own grandchildren.

I must apologize for at least one deficiency. Some of my sentences are short, and this conflicts with Bohr's way of using any language.

Edward Teller, "Niels Bohr and the Idea of Complementarity," 1969

Meetings in Wartime and After

R. V. Jones

R. V. Jones, after obtaining his Ph.D. in physics at Oxford, became a scientific officer with the British Air Ministry, and in 1941 was appointed Assistant Director of Intelligence for the Air Staff. In this capacity he was centrally involved with wartime intelligence work on German radar and guided missiles, vividly described in his book *Most Secret War* (published in the United States as *The Wizard War*). In 1946 he became Professor of Physics at the University of Aberdeen. He is now retired.

The problem of whether Germany would attempt to develop a nuclear bomb during World War II naturally preoccupied those of us in Britain and the United States who were aware of the feasibility. We had recognized the oblique warning in a 1939 paper by Siegfried Flügge, which indicated the lines of German nuclear thought;[1] and we had positive evidence in 1941 and 1942, from a section of the Norwegian resistance led by the physical chemist Leif Tronstad, that the Germans had demanded large supplies of heavy water from the Norsk Hydro plant at Rjukan. The task of watching these developments fell to my section of British Intelligence, and one of my colleagues, Eric Welsh, suggested that it would be a notable coup if we could make contact with Niels Bohr in Copenhagen, and persuade him to escape from the Nazis and join the Allied effort.

The first step was to secure the help of someone whom Bohr would know and trust, and to ask this person to write a message inviting Bohr to Britain. We decided on James Chadwick, who hesitated until Eric Welsh, speaking in the name of Britain, convinced him that it was his duty to help. Then we had to get the message to Bohr: it was sent early in 1943 as microdots concealed in holes bored into the handles of two ordinary-looking keys which were taken to Bohr by Captain Gyth, a member of the Danish General Staff, with which we were in contact.

We had presented Bohr with one of the most difficult choices an individual can face when his country falls into enemy hands: Should he stay and fight it out, either overtly or covertly, or should he escape and take up the fight from outside? Bohr's reply came back to us by the same route as our message had gone to him. We might well have guessed it, for in a 1941 article on Danish culture he had quoted Hans Christian Andersen: "In Denmark I was born and there my home is . . . From there my world

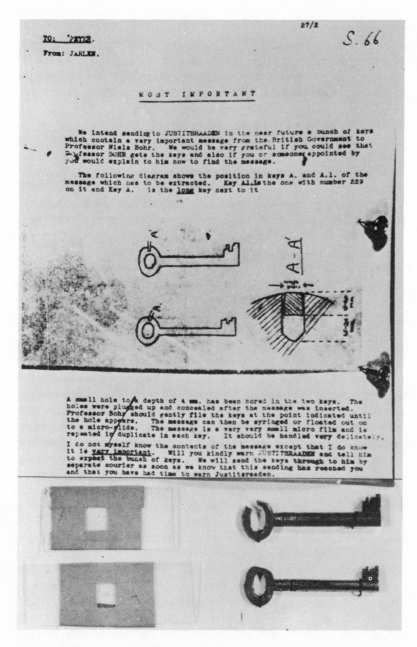

The message in the keys. Above is a facsimile of the explanatory letter that preceded them. Below is a photograph of the keys themselves and the microdot messages (barely visible as specks in the centers of the squares).

begins." He had to stay: "I feel it to be my duty in our desperate situation to help resist the threat against the freedom of our institutions and to assist in the protection of the exiled scientists who have sought refuge here." He doubted, moreover, whether he could be of great help in the West: "Above all I have to the best of my judgment convinced myself that, in spite of all future prospects, any immediate use of the latest marvelous discoveries of atomic physics is impracticable." "However," he added, "there may, and perhaps in the near future, come a moment when things look different and where I, if not in other ways, might be able modestly to assist in the restoration of international collaboration in human progress. At that moment I shall make an effort to join my friends."

Two months later Bohr heard rumors of the German demand for increased heavy-water production and for metallic uranium, and so he wrote again to Chadwick qualifying his previous disbelief in any immediate exploitation of nuclear energy — although he remained skeptical, if only because of the seeming impossibility of separating enough uranium 235. This was despite a visit that Heisenberg had paid him in Copenhagen in October 1941 which, from the questions that Heisenberg asked, Bohr subsequently interpreted as indicating that the Germans were seriously considering making a nuclear bomb.

Bohr's decision to stay in Denmark was upset in September 1943 when he and his mathematician brother, Harald, received warnings that they were to be arrested by the Germans. Bohr and his wife were therefore taken to Sweden by the Danish resistance, and their sons a few weeks later. When the news reached us, Lord Cherwell (F. A. Lindemann) sent a telegram to Bohr, inviting him to Britain, and Bohr immediately accepted. But the journey from Stockholm to Britain on 6 October 1943 was even more hazardous than the crossing from Denmark to Sweden, for courier aircraft across the North Sea were hunted by the Luftwaffe. The only reasonably safe way was by Mosquito aircraft which, with a crew of two, had not been designed to carry passengers. The only available accommodation was in the bomb bay which, like the rest of the aircraft, was unpressurized. Both crew and passengers had therefore to wear flying helmets with oxygen masks, which were to be turned on at cruising altitude on an order from the pilot over the intercommunication system through headphones in the helmets. Only after the aircraft had taken off did Bohr find that his helmet was much too tight for his large head and that he could not keep the headphones on, with the result that he failed to hear the order to switch on his oxygen and he lost consciousness. The pilot, hearing nothing from his passenger, could only hope that he had heard the order and that the absence of reply indicated nothing more serious than a microphone failure. As soon as possible the pilot dropped to low altitude to minimize oxygen starvation, but still there was no response from his passenger. I was to have gone up to the Scottish airfield at which the aircraft was due to land, but duties in connection with the prospective V-weapons campaign prevented me, and Welsh went instead. Welsh told me that he had never seen an aircraft pull up so sharply on landing, or a pilot jump out so quickly and rush to open the bomb bay. Fortunately he found Bohr alive and once again conscious, but still defeated by the helmet. He was brought to London, and a week later was followed by his physicist son, Aage.

On this first wartime visit Bohr was treated, as always, with respect as

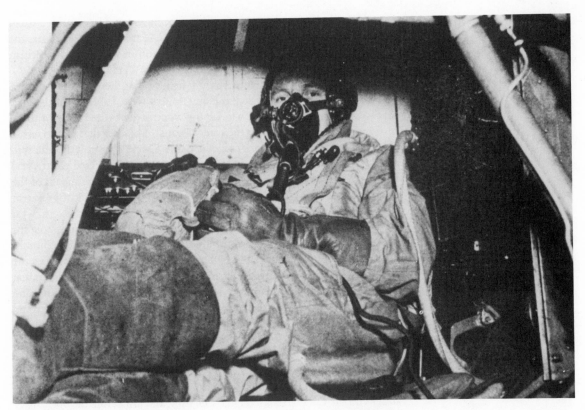

The bomb bay of a converted Mosquito bomber of the type used to transport Niels Bohr from Sweden to Britain. Throughout the Second World War, British Overseas Airways Corporation, using civilian aircraft and civilian crew, maintained flights on the route from Sweden to Britain. The aircraft, which at times made several flights a day, carried mail, news, freight (particularly ball bearings), diplomatic personnel, and "very important people." Weather, navigation problems, and hostile action combined to make the flights extremely hazardous, even after the arrival of the fast, high-flying Mosquito Airliner. In July 1943 two Mosquito bombers were converted to carry a passenger, who flew cramped and isolated in the padded bomb bay and in contact with the crew only through headphones. It was in this way that Bohr made his three-hour flight to Britain. The Mosquitoes made just over one thousand flights and flew some three quarters of a million miles. Four aircraft were lost with crew and passengers, including the flight on 25 October 1943, just after those taken by Bohr and his son Aage.

deep as it was affectionate; but the unusual circumstances led to his being kept not only from the public eye but also from the eyes of most of his former scientific colleagues, except for a very few such as Chadwick who were closely associated with the nuclear energy project, and one or two of us in British Intelligence. Bohr was installed in an apartment in St. James's Court in Westminster, and given an office in the Tube Alloys (nuclear energy) project in Old Queen Street. Stewart Menzies, head of the British

Secret Service, gave a dinner in his honor at the Savoy Hotel, with a small but cordial party that included Cherwell, F. C. Frank, and myself. After the war, Bohr told me that he had had no hesitation in working with British Intelligence because he had found it was run by a gentleman!

At the beginning, Bohr could hardly believe that the development of a nuclear bomb was so advanced; but any doubts were dispelled on the visit that was then arranged for him to see the work in the United States, and his mind immediately turned to the international problems that would ensue after the war if nuclear bombs could be made. During his visit, he renewed acquaintance with Felix Frankfurter, a justice of the Supreme Court, and he saw in this contact a chance to bring his ideas about the impact of the nuclear bomb on postwar international relations to the notice of President Roosevelt.

On his return to England in April 1944, Bohr saw a possible role for himself in bringing these same ideas to the notice also of Winston Churchill and perhaps in acting as a benevolent and enlightening link between the two leaders. I came into his discussions at this point because I was told that he and Aage were living in lonely seclusion again in St. James's Court, and that he would welcome any chance to talk. So my colleague F. C. Frank and I would take what time we could afford out of our office to call on him. Could any physicist in 1944 imagine a more frustrating situation? Here was an invitation to talk to Niels Bohr, virtually alone, for all the hours in the day (and night); yet we had to weigh almost every moment, for this was the vital time before the landings in Normandy, when we were directing the attacks on the German coastal radar defenses, and just before the opening of the V-weapons campaign, in which we were responsible for intelligence concerning the weapons. But we were fascinated by every hour that we could spend with Bohr in a marvelous series of tutorials in physics. We learned, for example, of his original difficulty in conceiving his electron-orbital model of the atom, and then in getting it accepted, because electrons in orbit were subject to centripetal acceleration, and accelerating charges should radiate electromagnetic energy. He simply had had to assume that in this particular case radiation did not take place. It is of some pedagogic interest that by the time students such as myself had come to study atomic physics in the late twenties, the objection had been so glossed over as never to be mentioned.

F. C. Frank and I were especially grateful for the simple elegance of Bohr's liquid-drop model, after the obscure formulations of other theorists concerning nuclear behavior. I was fascinated, too, by his work on complementarity and indeterminacy, to which I had been inclined in my undergraduate years by my own professor F. A. Lindemann. In London, Bohr treated us to a blow-by-blow account of his great battle with Einstein. He also recalled the day when the young Heisenberg returned to Copenhagen from Göttingen saying, "It looks as though we shall have to learn about matrices!" Again I wondered retrospectively where the imaginative idea

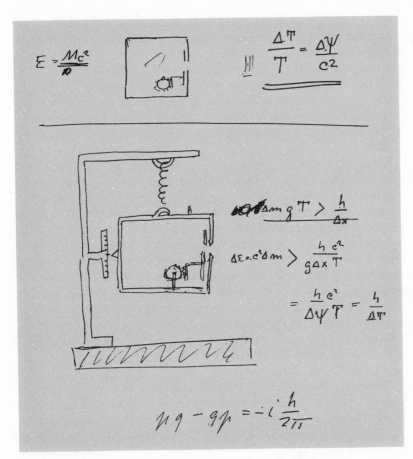

Sketch made by Niels Bohr in London, May 1944, to illustrate for R. V. Jones and F. C. Frank his account of his debate with Einstein on the uncertainty principle at the Sixth Solvay Conference in 1930. See "The Bohr-Einstein Dialogue."

came from of bringing this particular discipline into physics. I later commented on it to Max Born, who said, "That's simple. Heisenberg had come back to Göttingen for a visit, and told me about his problem in Copenhagen. I happened to know about noncommutative algebra, and gave him the idea."

Frequently Bohr became so absorbed in what he had to tell us that he would start to light his pipe, but would then put down the burning match while he went on talking, and the pipe would still not be lit, even after ten or more matches had expired and been placed in a mounting pyramid. And he would get up and repeatedly circle his desk, like one of his electrons in its orbit. Mischievously, I tested his absorption by circling the same desk, in a contra-orbit, so we crossed twice every revolution — but he never noticed. Our meetings with him were delicious mixtures of gravity and humor, the gravity being largely concerned with the postwar state of the world if the nuclear bomb materialized. Bohr was convinced that, whatever immediate

283

advantage the bomb might give the West, the Russians must sooner or later master the secrets of its design and manufacture, and with their apprehensive approach to international affairs, a very dangerous era of mutual mistrust must follow. Therefore, it might be best to dispel their mistrust for all time by showing them that they had nothing to fear from the West by giving them the secrets of the bomb as the ultimate gesture of trusting goodwill. He told us that he already had a contact with Roosevelt, and he wished that he could speak with Churchill. I offered to see what I could do, because I could see the possibility of further benefit if the two could meet.

What I had in mind was the prospect of convincing Churchill of the magnitude and imminence of the nuclear problem. I wondered whether he had fully appreciated its significance when, at the Quebec Conference of August 1943, he had signed away British postwar commercial rights to nuclear developments, with such rights to be dealt with "on terms to be specified by the President of the United States to the Prime Minister of Great Britain." I thought that if Bohr could meet him, Churchill would at least find that the world's leading nuclear physicist was so convinced that nuclear energy would be shortly released as to be deeply worried about its international implications. I also thought that Churchill might not have signed away British rights so lightly had his friend and advisor Lord Cherwell been himself more convinced of the nuclear prospect. I had even reproached Cherwell for, as I believed, his not having put the case more strongly to Churchill.

Courtesy demanded that I make the approach to Churchill through Cherwell, and I wondered whether the latter would agree. Nevertheless he did, and an appointment was made for him to take Bohr to Churchill a few days later. Much gratified, I told Bohr the good news, and he told me he would very much like my help in composing what he intended to say. He said that he knew his pronunciation of English was not good. Once he had had to speak at a dinner in London, and had had enough confidence in his pronunciation to tell "a little joke" about a Frenchman who claimed that to speak English well "was not so much a matter of the vockerbewlerie as of the assent." But when Bohr had told the story at the dinner, nobody had laughed. "Then," he said, "I realised how var' var' bad I spoke English."

Bohr proposed to write out what he intended to say, and to ask me to put it into good English, which he would then learn by heart. I did my best with his first draft, and then he modified it where he thought that I had not caught his exact meaning, and we continued successive stages of this process over the next three days or so, when Bohr declared himself satisfied. At the appointed hour, after lunch on 16 May 1944, I took him to the Cabinet Office for Cherwell to escort him over to 10 Downing Street. Leaving them, I returned to the invasion preparations. These took me to the Air Ministry building on King Charles Street, and I was making my way back to my own office again at about five o'clock along Old Queen Street when I was surprised to see Bohr coming in the opposite direction. His eyes were turned heavenward, and he walked clean past me in a daze. Turning to overtake him, I said, "Hallo, Professor Bohr, how did you get on?" All he could say was, "It was terrible. He scolded us like two schoolboys!"

Later he told me what had gone wrong. He and Cherwell were waiting in the room when Churchill came in and immediately upbraided Cherwell: "I know why you have fixed this meeting. You want to reproach me about the Quebec Agreement." Now, Cherwell had previously defended the agreement to me, but either he had since changed his mind or his defense had been purely out of loyalty to Churchill. If the latter, it was cruel that he was now being attacked; but more serious was the effect on Bohr, because his set opening speech was thereby put out of place. Bohr did his best to improvise, and no doubt suffered from his usual anxiety to be precise, but Churchill was in no mood to make allowances, and he lost patience with Bohr, telling him that if Bohr was worried about postwar problems, "I cannot see what you are worrying about. After all, this new bomb is just going to be bigger than our present bombs. It involves no difference in the principles of war; and as for any postwar problems, there are none that cannot be amicably settled between me and my friend President Roosevelt." The episode left Bohr much shaken, and my good intentions shattered.

We wondered what to do next. Bohr decided to submit his ideas in writing, no doubt based on his lost speech. In our subsequent discussions in the Secret Intelligence Service, Stewart Menzies had the idea of arranging for Bohr to meet General Jan Smuts, for whom Churchill had great respect, so that Smuts could add his weight.

I did not see Bohr again during the war; but in 1947, after I had gone to the University of Aberdeen, I was delighted by a letter from him inviting me to the next of his informal conferences in Copenhagen. I had heard of these scintillating conferences before the war, never even hoping to be invited. But Bohr was evidently grateful for what little I had been able to do during the war, and so I found myself accepted into the marvelous company of his friends — Wolfgang Pauli, Lise Meitner, Otto Stern, Victor Weisskopf, Hendrik Casimir, Hendrik Kramers, and many others among them. There was only one other Englishman present, Cecil Powell from Bristol. Although I myself could contribute nothing, Powell was on the crest of his wave with his newly identified pi and mu mesons. It was good to be present as he gave his news to the totally absorbed audience.

Niels Bohr had opened the conference by saying that some of his friends had been away from physics for most of the war, and so he would make a few introductory remarks to remind us of the state that nuclear physics had reached. I felt that this was particularly for my personal benefit, for I was the only member of the audience who had been so much out of the mainstream of physics. Much of what he said covered the ground of our 1944 tutorials, and he became so involved that all thought of tea was forgotten. Two and a half hours later he looked at his watch and said, "Oh dear! It was tea an hour ago!" He spoke so quietly that he had been difficult to follow, and the audience broke into laughter when he asked the first speaker after tea to speak up in order to be heard, at which Pauli bellowed out, "After *you*

Certainly it would be hard to imagine two great men more completely different, but opposites sometimes get on well. This time they did not. To one who had no means of assessing his greatness, Bohr's soft voice, indistinct pronunciation, and involved sentences, carefully qualified as to exclude possibilities which it was often unduly flattering to suppose one had considered, did not bring conviction.

G. P. Thomson,
Niels Bohr Memorial Lecture, 1964

everybody seems loud!" Such was the happy informality of a Copenhagen conference.

I was at a subsequent conference in 1951, at which the German physicists were now present, among them Heisenberg, Houtermans, and Kopfermann, and a stronger British contingent including C. G. Darwin, Dirac, G. P. Thomson, and J. D. Cockcroft. One of the outcomes of that conference was the proposal to form CERN.

The following year, Bohr came to stay with us in Aberdeen to receive an honorary degree. He asked to see my work in the laboratory, and when I said that I doubted whether my kind of instrumental physics would interest him he told me that when he was a research student he was the one in the laboratory to whom everyone brought their galvanometers and other instruments to repair. He was still preoccupied with the international situation, and we spent till two or three o'clock every morning talking it over. "Do you mind," he would ask my wife, "if your husband stays up so that we can have another little talk?"

One of his many comments that I can recall from these chats concerned the education of young physicists, and at what point they should begin to specialize in physics. Despite the fact that many of the greatest advances in theoretical physics had been made by young people, from which it appeared that the educational problem was to get them to the frontiers of existing knowledge at an early stage in their lives before their imagination was tired by the long journey through knowledge already won, there were those such as Heisenberg who had brilliantly come to physics after an education that was predominantly and classically in the humanities. Such examples were at

Bohr loaded with academic honors from American universities during a lecture tour after the war. The cartoon, by Bo Bojesen, appeared in the Copenhagen newspaper *Politiken* with the caption: "And now our distinguished guest will repeat his famous lecture on chain reactions."

the time being pressed by those who were resisting any expansion of physics in schools and universities. I therefore asked Bohr: Should one specialize early in physics or not? "You should not specialize," he replied, and then — with a smile and after a pause — "in too many other things!"

Fortunately our Scottish weather was kind during that 1952 visit, and we were able to show him Deeside, where he was as excited as a schoolboy at seeing salmon leaping at the Falls of Feugh. For the rest of the day he kept discreetly inquiring about the details of our return route, and it was clear that he wanted to watch the salmon again but was afraid that we might think him childish if he said so. We were astonished at his nimbleness — he was then sixty-seven, but he leaped over burns in a way that made me look clumsy. We had forgotten about his athletic youth.

A lasting memory from this same trip was a visit to an Aberdeenshire farm, whose owner, Archie Reed, and his wife were among our good friends. Archie knew and loved horses, and one of my treasures is the memory of Bohr leaning over a farm gate while Archie gave him a lecture, with a good Buchan accent, on how to judge the points of a horse. Bohr listened with all the respect due from a great expert in one field to one whom he recognized as equally expert in another. "Var' fine people," he said, recalling the incident to me years afterward.

Bohr was able to put his new knowledge to use a few days later, when, at a party at my house, he was suddenly called upon to play the part of a horse. Since we were going off to a formal dinner, we were all in full evening dress with decorations. Sir Henry Tizard, who was also one of our guests, inveigled Bohr into pretending that they were a pair of horses so they could trot round the room driven by my twenty-three-month-old daughter. The reins for Tizard were the crimson sash of his Order of the Bath; those for Bohr, the light blue of his Order of the Elephant. Never, since I had first read Bohr's name in my school textbook or had encountered Tizard in air defense, had I thought that I should see such a happy and human spectacle in my own house.

In 1959 my wife and I visited Denmark, and Bohr led me to his private retreat on the grounds of his country house at Tisvilde, on the forested shore of Zealand. During an hour or two's talk, we went over old times — the Mosquito flight, the Churchill interview and what had gone wrong with it, the open letter that Bohr had written to the United Nations in 1950, Archie Reed and his horses, the salmon, and so on. He was happier about the international situation now that the United States and the Soviet Union were together showing signs of responsibility; he felt that, after years of darkness, there was indeed hope for the world. I did not see him again, but I like to think that his life finished on this note of optimism.

In a heroic age of physics Niels Bohr was one of the very foremost figures, and his ideas stand among the turning points of human thought. In all respects the magnitude and insight of his mind were matched by the greatness and humanity of his character. What good fortune it was to have known him!

287

Open Letter to the United Nations

On 9 June 1950, Niels Bohr issued a public statement with the above title. It was the product of long thought and effort that began as soon as he learned of the wartime project in the United States to construct an atomic bomb. It represents Bohr's most complete and important statement on the problems engendered by man's discovery and exploitation of nuclear energy.

Limitations of space have prevented our reprinting the Open Letter in its entirety. The abridged version below, prepared by Dr. Erik Rüdinger at the Niels Bohr Institute, seeks to make the flow of Bohr's arguments as clear as possible, interrupted only by brief explanations of the material left out.[1]

A.P.F.

I address myself to the organization, founded for the purpose to further cooperation between nations on all problems of common concern, with some considerations regarding the adjustment of international relations required by modern development of science and technology. At the same time as this development holds out such great promises for the improvement of human welfare it has, in placing formidable means of destruction in the hands of man, presented our whole civilization with a most serious challenge.

My association with the American-British atomic energy project during the war gave me the opportunity of submitting to the governments concerned views regarding the hopes and the dangers which the accomplishment of the project might imply as to the mutual relations between nations. While possibilities still existed of immediate results of the negotiations within the United Nations on an arrangement of the use of atomic energy

Niels Bohr in his study at the Carlsberg House of Honor.

guaranteeing common security, I have been reluctant in taking part in the public debate on this question. In the present critical situation, however, I have felt that an account of my views and experiences may perhaps contribute to renewed discussion about these matters so deeply influencing international relationship.

.   .   .

For the modern rapid development of science and in particular for the adventurous exploration of the properties and structure of the atom, international cooperation of an unprecedented extension and intensity has been of decisive importance. The fruitfulness of the exchange of experiences and ideas between scientists from all parts of the world was a great source of encouragement to every participant and strengthened the hope that an ever closer contact between nations would enable them to work together on the progress of civilization in all its aspects.

Yet, no one confronted with the divergent cultural traditions and social organization of the various countries could fail to be deeply impressed by the difficulties in finding a common approach to many human problems. The growing tension preceding the Second World War accentuated these difficulties and created many barriers to free intercourse between nations. Nevertheless, international scientific cooperation continued as a decisive factor in the development which, shortly before the outbreak of the war, raised the prospect of releasing atomic energy on a vast scale.

.   .   .

Everyone associated with the atomic energy project was, of course, conscious of the serious problems which would confront humanity once the enterprise was accomplished. Quite apart from the role atomic weapons might come to play in the war, it was clear that permanent grave dangers to world security would ensue unless measures to prevent abuse of the new formidable means of destruction could be universally agreed upon and carried out.

As regards this crucial problem, it appeared to me that the very necessity of a concerted effort to forestall such ominous threats to civilization would offer quite unique opportunities to bridge international divergencies. Above all, early consultations between the nations allied in the war about the best ways jointly to obtain future security might contribute decisively to that atmosphere of mutual confidence which would be essential for cooperation on the many other matters of common concern.

In the beginning of 1944, I was given the opportunity to bring such views to the attention of the American and British governments. It may be in the interest of international understanding to record some of the ideas which at that time were the object of serious deliberation. For this purpose, I may quote from a memorandum which I submitted to President Roosevelt as a basis for a long conversation which he granted me in August 1944:

.   .   .

"It certainly surpasses the imagination of anyone to survey the consequences of the project in years to come, where in the long run the enormous energy sources which will be available may be expected to revolutionize industry and transport. The fact of immediate preponderance is, however, that a weapon of an unparalleled power is being created which will completely change all future conditions of warfare.

"Quite apart from the question of how soon the weapon will be ready for use and what role it may play in the present war, this situation raises a number of problems which call for most urgent attention. Unless, indeed, some agreement about the control of the use of the new active materials can be obtained in due time, any temporary advantage, however great, may be outweighed by a perpetual menace to human security.

• • •

"The present moment where almost all nations are entangled in a deadly struggle for freedom and humanity might at first sight seem most unsuited for any committing arrangement concerning the project. Not only have the aggressive powers still great military strength, although their original plans of world domination have been frustrated and it seems certain that they must ultimately surrender, but even when this happens, the nations united against aggression may face grave causes of disagreement due to conflicting attitudes toward social and economic problems.

"By a closer consideration, however, it would appear that the potentialities of the project as a means of inspiring confidence just under these circumstances acquire most actual importance. Moreover the momentary situation would in various respects seem to afford quite unique possibilities which might be forfeited by a postponement awaiting the further development of the war situation and the final completion of the new weapon."

[Bohr further developed these arguments in a second memorandum of March 1945, also quoted in the open letter:]

"Above all, it should be appreciated that we are faced only with the beginning of a development and that, probably within the very near future, means will be found to simplify the methods of production of the active substances and intensify their effects to an extent which may permit any nation possessing great industrial resources to command powers of destruction surpassing all previous imagination.

"Humanity will, therefore, be confronted with dangers of unprecedented character unless, in due time, measures can be taken to forestall a disastrous competition in such formidable armaments and to establish an international control of the manufacture and use of the powerful materials.

• • •

291

"As argued in the memorandum, it would seem most fortunate that the measures demanded for coping with the new situation, brought about by the advance of science and confronting mankind at a crucial moment of world affairs, fit in so well with the expectations for a future intimate international cooperation which have found unanimous expression from all sides within the nations united against aggression.

"Moreover, the very novelty of the situation should offer a unique opportunity of appealing to an unprejudiced attitude, and it would even appear that an understanding about this vital matter might contribute most favorably toward the settlement of other problems where history and traditions have fostered divergent viewpoints.

"With regard to such wider prospects, it would in particular seem that the free access to information, necessary for common security, should have far-reaching effects in removing obstacles barring mutual knowledge about spiritual and material aspects of life in the various countries, without which respect and goodwill between nations can hardly endure.

. . .

"All such opportunities may, however, be forfeited if an initiative is not taken while the matter can be raised in a spirit of friendly advice. In fact, a postponement to await further developments might, especially if preparations for competitive efforts in the meantime have reached an advanced stage, give the approach the appearance of an attempt at coercion in which no great nation can be expected to acquiesce.

"Indeed, it need hardly be stressed how fortunate in every respect it would be if, at the same time as the world will know of the formidable destructive power which has come into human hands, it could be told that the great scientific and technical advance has been helpful in creating a solid foundation for a future peaceful cooperation between nations."

Looking back on those days, I find it difficult to convey with sufficient vividness the fervent hopes that the progress of science might initiate a new era of harmonious cooperation between nations, and the anxieties lest any opportunity to promote such a development be forfeited.

Until the end of the war I endeavored by every way open to a scientist to stress the importance of appreciating the full political implications of the project and to advocate that, before there could be any question of use of atomic weapons, international cooperation be initiated on the elimination of the new menaces to world security.

I left America in June 1945, before the final test of the atomic bomb, and remained in England, until the official announcement in August 1945 that the weapon had been used. Soon thereafter I returned to Denmark and have since had no connection with any secret, military or industrial, project in the field of atomic energy.

[Bohr described the difficult political situation in the postwar years:]

Without free access to all information of importance for the interrelations between nations, a real improvement of world affairs seemed hardly imaginable. It is true that some degree of mutual openness was envisaged as an integral part of any international arrangement regarding atomic energy, but it grew ever more apparent that, in order to pave the way for agreement about such arrangements, a decisive initial step toward openness had to be made.

The ideal of an open world, with common knowledge about social conditions and technical enterprises, including military preparations, in every country, might seem a far remote possibility in the prevailing world situation. Still, not only will such relationship between nations obviously be required for genuine cooperation on progress of civilization, but even a common declaration of adherence to such a course would create a most favorable background for concerted efforts to promote universal security. Moreover, it appeared to me that the countries which had pioneered in the new technical development might, due to their possibilities of offering valuable information, be in a special position to take the initiative by a direct proposal of full mutual openness.

I thought it appropriate to bring these views to the attention of the American government without raising the delicate matter publicly.

[Bohr quoted a memorandum of March 1948 addressed to the secretary of state. The quotation ends with the following passage:

"Under the circumstances it would appear that most careful consideration should be given to the consequences which might ensue from an offer, extended at a well-timed occasion, of immediate measures toward openness on a mutual basis. Such measures should in some suitable manner grant access to information, of any kind desired, about conditions and developments in the various countries and would thereby allow the partners to form proper judgment of the actual situation confronting them.

"An initiative along such lines might seem beyond the scope of conventional diplomatic caution; yet it must be viewed against the background that, if the proposals should meet with consent, a radical improvement of world affairs would have been brought about, with entirely new opportunities for cooperation in confidence and for reaching agreement on effective measures to eliminate common dangers.

"Nor should the difficulties in obtaining consent be an argument against taking the initiative since, irrespective of the immediate response, the very existence of an offer of the kind in question should deeply affect the situation in a most promising direction. In fact, a demonstration would have been given to the world of preparedness to

293

live together with all others under conditions where mutual relationships and common destiny would be shaped only by honest conviction and good example.

"Such a stand would, more than anything else, appeal to people all over the world fighting for fundamental human rights, and would greatly strengthen the moral position of all supporters of genuine international cooperation. At the same time, those reluctant to enter on the course proposed would have been brought into a position difficult to maintain since such opposition would amount to a confession of lack of confidence in the strength of their own cause when laid open to the world.

"Altogether, it would appear that, by making the demand for openness a paramount issue, quite new possibilities would be created, which, if purposefully followed up, might bring humanity a long way forward toward the realization of that cooperation on the progress of civilization which is more urgent and, notwithstanding present obstacles, may still be within nearer reach than ever before."

.   .   .

Within the last years, worldwide political developments have increased the tension between nations, and at the same time the perspectives that great countries may compete about the possession of means of annihilating populations of large areas and even making parts of the earth temporarily uninhabitable have caused widespread confusion and alarm.

As there can hardly be question for humanity of renouncing the prospects of improving the material conditions for civilization by atomic energy sources, a radical adjustment of international relationship is evidently indispensable if civilization shall survive. Here, the crucial point is that any guarantee that the progress of science is used only to the benefit of mankind presupposes the same attitude as is required for cooperation between nations in all domains of culture.

Also in other fields of science recent progress has confronted us with a situation similar to that created by the development of atomic physics. Even medical science, which holds out such bright promise for the health of people all over the world, has created means of extinguishing life on a terrifying scale which imply grave menaces to civilization, unless universal confidence and responsibility can be firmly established.

The situation calls for the most unprejudiced attitude toward all questions of international relations. Indeed, proper appreciation of the duties and responsibilities implied in world citizenship is in our time more necessary than ever before. On the one hand, the progress of science and technology has tied the fate of all nations inseparably together; on the other hand, it is on a most different cultural background that vigorous endeavors for national self-assertion and social development are being made in the various parts of our globe.

An open world where each nation can assert itself solely by the extent to which it can contribute to the common culture and is able to help others with experience and resources must be the goal to be put above everything else. Still, example in such respects can be effective only if isolation is abandoned and free discussion of cultural and social developments permitted across all boundaries.

Within any community it is only possible for the citizens to strive together for common welfare on a basis of public knowledge of the general conditions in the country. Likewise, real cooperation between nations on problems of common concern presupposes free access to all information of importance for their relations. Any argument for upholding barriers for information and intercourse, based on concern for national ideals or interests, must be weighed against the beneficial effects of common enlightenment and the relieved tension resulting from openness.

In the search for a harmonious relationship between the life of the individual and the organization of the community, there have always been and will ever remain many problems to ponder and principles for which to strive. However, to make it possible for nations to benefit from the experience of others and to avoid mutual misunderstanding of intentions, free access to information and unhampered opportunity for exchange of ideas must be granted everywhere.

In this connection it has to be recognized that abolition of barriers would imply greater modifications in administrative practices in countries where new social structures are being built up in temporary seclusion than in countries with long traditions in governmental organization and international contacts. Common readiness to assist all peoples in overcoming difficulties of such kind is, therefore, most urgently required.

The development of technology has now reached a stage where the facilities for communication have provided the means for making all mankind a cooperating unit, and where at the same time fatal consequences to civilization may ensue unless international divergencies are considered as issues to be settled by consultation based on free access to all relevant information.

The very fact that knowledge is in itself the basis for civilization points directly to openness as the way to overcome the present crisis. Whatever judicial and administrative international authorities may eventually have to be created in order to stabilize world affairs, it must be realized that full mutual openness, only, can effectively promote confidence and guarantee common security.

Any widening of the borders of our knowledge imposes an increased responsibility on individuals and nations through the possibilities it gives for shaping the conditions of human life. The forceful admonition in this respect which we have received in our time cannot be left unheeded and should hardly fail in resulting in common understanding of the seriousness of the challenge with which our whole civilization is faced. It is just on this background that quite unique opportunities exist today for furthering cooperation between nations on the progress of human culture in all its aspects.

I turn to the United Nations with these considerations in the hope that they may contribute to the search for a realistic approach to the grave and urgent problems confronting humanity. The arguments presented suggest that every initiative from any side toward the removal of obstacles for free mutual information and intercourse would be of the greatest importance in breaking the present deadlock and encouraging others to take steps in the same direction. The efforts of all supporters of international cooperation, individuals as well as nations, will be needed to create in all countries an opinion to voice, with ever increasing clarity and strength, the demand for an open world.

Copenhagen, June 9th, 1950

Niels Bohr

VI. PHILOSOPHICAL IDEAS

The Philosophy of Niels Bohr

Aage Petersen

Attempting to describe Niels Bohr's philosophy puts me in a situation very much like that of the young man in one of Bohr's favorite stories.[1] In an isolated village there was a small Jewish community. A famous rabbi once came to the neighboring city to speak and, as the people of the village were eager to learn what the great teacher would say, they sent a young man to listen. When he returned he said, "The rabbi spoke three times. The first talk was brilliant — clear and simple. I understood every word. The second was even better — deep and subtle. I didn't understand much, but the rabbi understood all of it. The third was by far the finest — a great and unforgettable experience. I understood nothing, and the rabbi himself didn't understand much either."

For me, Niels Bohr's philosophy also fell into three parts: one which I thought I grasped, one which I did not understand but which I felt was clear to Bohr, and one which Bohr himself saw only dimly. Thus, my description can be only a weak reflection of his wonderfully rich thought, and I am sure he would have expressed many points differently. Yet I hope I can give a feeling of the breadth and depth of his philosophy, and perhaps also a glimpse of that intellectual harmony in which he lived.

Bohr never referred to his philosophy as his own. He used to speak of it as a general lesson to be drawn from quantum mechanics. Yet when, shortly after the development of quantum mechanics, he told his old friend Edgar Rubin, a psychologist, of this general lesson, Rubin replied, "Yes, it's very interesting, but you must admit that you said just the same thing twenty years ago." Can this really have been the case?

The route along which Bohr's philosophical ideas developed was as remarkable as the ideas themselves. Even as a child, Bohr was considered the thinker of the family, and his father listened closely to his views on funda-

Aage Petersen worked closely with Niels Bohr in Copenhagen, but subsequently moved to the United States. He is Professor of Physics at Yeshiva University, New York City; his particular interest has been the philosophical aspects of quantum theory.

mental problems. Bohr has said that as far back as he could remember he liked to dream of great interrelationships. His philosophical attitude seems to have been shaped early with very little influence from outside, and he spent much time developing it. Shortly before his death, Bohr spoke of his youthful philosophical work. When asked what place this work then had in his existence, he replied, "It was, in a way, my life!"

Before the end of his university studies, he had come so far in his thought on philosophical problems that he planned to write a book. He felt that he had a point of view that was sufficiently clear and complete to be published. But he did not write it. Instead he took up other work, and his philosophical writing was put off for twenty years. He began to do experiments on surface tension for a competition paper by the Royal Danish Academy. He built the apparatus himself, and was especially fascinated by glass blowing. Commenting on this transition from philosopher to experimental physicist, he said, "I was not a daydreamer; I was willing to do hard work." He continued on in physics and wrote a doctoral thesis on the electron theory of metals. After the thesis he became interested in the problems of atomic constitution, and from 1913 to 1927 led the development of quantum physics. This development, in which all sides of Bohr's intellect came to fruition, brought him back to philosophy.

For Bohr, the new theory was not only a wonderful piece of physics; it was also a philosophical treasure chamber which contained, in a new form, just those thoughts he had dreamed about in his early youth. He no longer regretted that he had not written the epistemological book he had planned earlier, because he felt that he could now express himself far more clearly. Moreover, in 1927, Bohr was no longer a young student, but one of the world's leading physicists. Because of his enormous authority as a scientist, he was in a unique position for philosophical innovation. But even though he could now use the quantal description as a medium for expressing his ideas, his philosophical message was still hard to understand. From his essays one gets a strong impression that it was also difficult to present.

In his enthusiasm for the new prospects now open, Bohr planned to start a journal for the philosophical investigations that quantum physics suggested. This plan, too, was never realized. Again other problems demanded his working power. But during the following thirty-five years he published a series of articles in which his philosophical viewpoint was developed. Almost all of these articles originated from lectures and speeches that he gave on various occasions. In all of them Bohr played the same theme again and again with slight variations, in a continual attempt to make the attitude, argument, and terminology clearer. If one said to Bohr that his articles were all very similar, he would smile and tell a little story about a Greek philosopher of the Sophist school. This philosopher had been away from Athens for a long time, and when he returned he was surprised to see Socrates engaged in discussion at the usual place. "Are you," he asked, "still standing here, Socrates, saying the same things about the same things?"

Socrates replied, "Don't you ever say the same things about the same things?"

In trying to survey what he called "our situation," Bohr did not proceed in the same way as the philosophers who had formed the Western philosophical tradition. He did not give new answers to old questions. His philosophy cannot be described in terms of the usual philosophical "isms" or schools. To Bohr, philosophical problems were neither about existence or reality, nor about the structure and limitations of human reason. They were communication problems. They dealt with the general conditions for conceptual communication.

When asked what he meant by that, Bohr would say, "What is it that we human beings ultimately depend on? We depend on our words. We are suspended in language. Our task is to communicate experience and ideas to others. We must strive continually to extend the scope of our description, but in such a way that our messages do not thereby lose their objective or unambiguous character."

The general conditions for the use of language include a law requiring a proper balance between content and form in conceptual communication. When we describe and order experience, we must use a system of concepts. No experience can be understood or communicated without being fixed in a logical frame. The frame — that is, the way we characterize and combine experience — determines what we can talk about and what relationships we can express. We must always be prepared to find that a conceptual framework is too narrow to contain the content we want to press into it. In such a situation we are confronted with a logical disharmony, because we try to speak about something for which our conceptual system has no room. And in efforts to restore harmony, even the frames that are apparently the most solid, those defining our elementary concepts, may prove to be blinders that conceal more fundamental relationships. Yet logical possibilities for extending or generalizing any frame lie like seeds in the presuppositions for using our concepts. The extension enables us to talk about new things and to express new kinds of regularities. "More and more deeply explored presuppositions may reveal relationships of greater and greater scope."

These fundamental aspects of description problems have been illuminated especially by mathematics. Deductive reasoning has taught us the significance of the conceptual framework. We have also learned to prove that certain problems — for example, the trisection of an angle — cannot be solved within a given framework. In addition, mathematics has shown us the wealth of unsuspected possibilities for conceptual extension or generalization that are latent in the way we use our simplest words.

Even in his school years, Bohr was intensely interested in the foundations of the mathematical approach and its relation to the general conditions for the use of language. Especially, he pondered the remarkable limitation of ordinary numbers which had been discovered by ancient Greek mathematicians when they had tried to express the length of the diagonal in the unit

[Such] quantitative analysis [is] characteristic of the exact sciences, whose task, according to the program of Galileo, is to base all description on well-defined measurements. Notwithstanding the help which mathematics has always offered for such a task, it must be realized that the very definition of mathematical symbols and operations rests on simple logical use of common language. Indeed, mathematics is not to be regarded as a special branch of knowledge based on the accumulation of experience, but rather as a refinement of general language, supplementing it with appropriate tools to represent relations for which ordinary verbal communication is imprecise or too cumbersome.

Niels Bohr,
Rutherford Memorial Lecture, 1958

square and which is believed to have been a major stimulus to the development of the axiomatic method. Ordinary numbers and ordinary arithmetical operations form a system within which many numerical problems can be asked and answered. We depend on the number system when we express numerical relationships. Yet there are numerical problems which, so to say, fall outside the scope of this system. In the system of integers and fractions, one cannot express the length of the diagonal of the unit square. But the system can be extended or generalized, and in the extended system there is a number, $\sqrt{2}$, which when multiplied by itself gives 2.

As far as I can see, the doctrine that we are, philosophically speaking, suspended in language, that we depend on our conceptual framework for unambiguous communication, and that the scope of the frame may be extended by generalization in the way illustrated in mathematics, forms the general basis of Bohr's philosophy. In his writings he never gave a detailed exposition of this view. Nor did he discuss its relation to other conceptions of the philosophical status of language. He considered it completely obvious and was surprised that others found it so difficult to understand.

Traditional philosophy has accustomed us to regard language as something secondary, and reality as something primary. Bohr considered this attitude toward the relation between language and reality inappropriate. When one said to him that it cannot be language which is fundamental, but that it must be reality which, so to speak, lies beneath language, and of which language is a picture, he would reply, "We are suspended in language in such a way that we cannot say what is up and what is down. The word 'reality' is also a word, a word which we must learn to use correctly." Bohr was not puzzled by ontological problems or by questions as to how concepts are related to reality. Such questions seemed sterile to him. He saw the problem of knowledge in a different light.

The chief characteristic of the sort of description we seek both in science and in practical life is objectivity. In Bohr's usage, an objective message was an unambiguous message, one that could not be misunderstood. If our communications are to be understood, their content must be clearly delineated. There must be, so to speak, a partition between the subject which communicates and the object which is the content of the communication. This partition is indispensable in every objective description, and Bohr saw in it the core of the problem of knowledge.

We may get an idea of the significance of the subject-object partition by considering the problem of describing our own thinking activity. When we think, we confront an objective content with a thinking subject. But the subject, our own ego, can also be made a part of the content of consciousness. In introspection we make that which is usually the subject, and therefore outside the description, a part of the object about which we communicate. Yet the very delineation of this extended content of consciousness is performed by a new subject. We can thus move the partition between actor and spectator or between stage and audience, and we can therefore, in a certain sense, talk about ourselves. But even a message about ourselves requires, if it is to be unambiguous, a subject and a partition, and the meaning of the message depends on where the partition is placed.

Thus, our situation is characterized by the fact that, on the one hand, we separate subject and object, while, on the other hand, we ourselves belong to that about which we are talking. In Bohr's opinion, the problems in epistemology originate primarily because we do not master the dialectics of the movable subject-object partition. The difficulty of delineating clearly the content of our messages is the chief source of ambiguity and paradox in conceptual communication.

In early youth, Bohr thought that he had found a way to handle the dialectics of the movable subject-object partition. When he heard lectures on the theory of complex functions, it struck him that there was a remarkable logical similarity between the problems of introspection and those of multivalued functions. He became convinced that the ingenious geometric device for eliminating the ambiguity of such functions, mapping the various functional values on different sheets of a so-called *Riemann surface,* could be exploited for clarifying paradoxes connected with the subject-object partition. This discovery presumably was one of his chief motives for writing a book on epistemology. He set to work to apply Riemann's idea to the problem of free will. In the meantime, the quantal description was developed, and it took over the role of the Riemann surface. Bohr's philosophical ideas were not originally inspired by physics, but the characteristics of the new theory fitted his philosophy wonderfully well.

Quantum physics is the temporary climax of a long development which started at about the same time that the Greek mathematicians discovered that $\sqrt{2}$ cannot be expressed as a ratio between two integers. The aim of this development had been to understand why things in nature possess the properties they do. One school of Greek thinkers found the logical core of this problem in the concept of atomicity. In other words, nature's stability and specific forms originate from the fact that there is something in nature that is indivisible. The Greek atomists assumed that there is a limitation in the divisibility of matter, that all matter is built up of immutable particles called atoms.

Around 1900 it was discovered that atomicity in nature is not restricted to matter, but that there is in nature another feature of indivisibility — an indivisibility of physical processes. Considered from the standpoint of ordinary physics, this discovery, like the Greeks' discovery of a length that cannot be measured by ordinary numbers, was a shock. It was soon realized that the problem of encompassing indivisible processes was related logically to the problem of inexpressible numbers. One saw that just as it had been necessary to extend the number concept, it was necessary to extend or generalize ordinary physics. This program, embodied in Bohr's correspondence argument, was finally carried through in 1925 by the creation of quantum mechanics.

The Danish theologian Søren Kierkegaard concluded there to be two kinds of truth: objective and subjective truth. When the truth appeared from an objective point of view to be paradoxical, it was an indication, he said, that one should be seeking a more subjective truth, one involving one's own participation. According to this particular strand of theological thought, one finds it necessary, as in modern physics, to take a step back from the objects of one's enquiry — whether they be God and Jesus, or light and matter — and be content to speak only of one's interactions with those objects. As a postscript, it is necessary to note that Bohr was an avid reader of Kierkegaard. Could it be that twentieth-century physics owes a modest debt to a nineteenth-century theologian's contemplation of a fourth-century Christian creed?

Russell Stannard, *The Times,* London, 3 December 1983

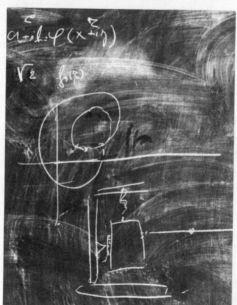

Niels Bohr and his blackboards. Discussions at the blackboard were an important part of Bohr's method of working. Clockwise, from upper left: Bohr as a young man after giving a lecture on his early quantum theory; lecturing at Princeton in 1948; talking about diffraction and interference; working on a problem with his son Aage; his last blackboard, recording a discussion that took place the day before his death. The lower sketch on this last blackboard is of the famous photon box that featured in his discussion with Einstein; the upper sketch shows a contour in the complex plane, used by Bohr to illustrate his comparison of ambiguity in language with the mathematics of multiform functions displayed on the different planes of a Riemann surface.

The quantal description, which encompasses individual physical processes, the regularities mainly responsible for nature's stability and specific forms, is thus an extended or generalized mode of description. It contains relationships that cannot be formulated within the narrow framework of ordinary physics. Quantum physics showed that even the causal mode of description is only one very special way of tying together natural phenomena, and thus lawfulness in nature is not equivalent to mechanical causality. We have had to learn that the causal mode of description is a limiting case that encompasses only the simplest features of nature.

By providing a clear *physical* example of a generalized conceptual framework, the quantal description supported and sharpened Bohr's view of language. The tenets of this view are particularly visible in his investigation of problems concerning the measurability of physical quantities. Basic to this investigation is the idea that although physics is an evolutionary enterprise with regard to its data and the algorithms correlating them, the account of the handling and functioning of measuring instruments must always be expressed in "ordinary plain language suitably supplemented by the terminology of classical physics." In that sense, the language of Newton and Maxwell will remain the language of physics. In support of this far-reaching idea, Bohr offered the "purely logical argument" that "by the very word 'experiment,' we refer to a situation where we can tell others what we have done and what we have learned."

When asked whether the algorithm of quantum mechanics could be considered as somehow mirroring an underlying quantum world, Bohr would answer, "There is no quantum world. There is only an abstract quantum physical description. It is wrong to think that the task of physics is to find out how nature is. Physics concerns what we can say about nature." Bohr felt that every step in the development of physics has strengthened the view that the problem of establishing an unambiguous description of nature has only one solution. He regarded all attempts to replace our elementary concepts or to introduce a new logic to account for the peculiarities of quantum phenomena as not merely unnecessary but also incompatible with our most fundamental conditions, since we are suspended in a unique language. "Of course," he once said jokingly, "it may be that when, in a thousand years, the electronic computers begin to talk, they will speak a language completely different from ours and lock us all up in asylums because they cannot communicate with us. But our problem is not that we do not have adequate concepts. What we may lack is a sufficient understanding of the unambiguous applicability of the concepts we have."

Quantum physics also throws light on Bohr's thoughts about the subject-object partition. How can the dialectics of introspection come into physics? In physics, if anywhere, we keep ourselves, as observers, outside descriptions. Quantum physics, however, made us see the significance of a feature of physical description that had so far been given little attention. In physics one distinguishes between the system investigated and the measur-

305

ing tools used for the investigation. This distinction between system and measuring tool lies in the general concept of experiment that so deeply characterizes physics. The conditions of observation defined by the experimental arrangement specify the physical situation in which the observed system finds itself. They are therefore essential in making communications about the system unambiguous. We can move the partition between measured system and measuring tool. For instance, when we observe an object in a microscope, we consider the light used to localize the object as a part of the measuring tool. But there is nothing to prevent us from making this light an object of investigation by introducing new measuring tools to define and obtain the information we want about the new system.

In the language of physics there are various sets of concepts such as space and time, and so-called dynamic concepts like momentum and energy. Corresponding to these different sets of concepts are different types of measuring tools. For example, to determine the position of an object, one must use rulers firmly attached together to form a reference frame. On the other hand, to measure an object's momentum one may let it collide with a freely movable test body of known mass, and then measure the resultant velocity of the test body.

In classical physics, the measuring tools that correspond to elementary physical concepts may, despite their dissimilarity, be used in combination to investigate a system. The findings provided by each arrangement simply supplement each other, and it is just this combination of dissimilar information that is needed in order to give a causal description of the behavior of the system.

In quantum physics we use the same concepts and thus the same measuring tools, but here the dissimilarity between the measuring tools becomes crucially important. Here we cannot use different types of instruments in combination. We cannot combine the information about the system that we get from one type of instrument with the information we get from another. Therefore, a quantum physical phenomenon is characterized by the type of measuring instrument we use. Two phenomena obtained by observing the same system with two different types of instruments are mutually exclusive. Bohr called this logical relation of exclusion *complementarity*.

The logic of quantum physics is related to the logic of introspection because in both physics and psychology we use the concept of observation. Quantum physics illuminates the dialectics of introspection because the physicist's partition between system and measuring tool corresponds closely to the epistemologist's partition between object and subject. The description of indivisible processes has taught the quantum physicist how to handle his partition. The epistemologist can learn the art from him. Like the Riemann surface with its separate sheets, the quantal description with its complementarity is a prototype of a logical device for combining experiences whose definition must include the circumstances under which they were obtained. In all areas where such experiences arise, we meet contrasts similar to those in quantum phenomena. But because quantum phenomena are so simple, they could point the way to complementarity.

Guided by the insight into description problems provided by quantum physics, Bohr again set out to investigate general epistemological problems and paradoxes. As before, these investigations took the form of tracing and analyzing logical analogies. It was shown that the conditions of description in a variety of fields are structurally similar to those in quantum physics, and that the relation of complementarity is suited to eliminate ambiguity in the comprehension of experience in these fields. Since some of the areas in question are closer to the sphere of common knowledge than are quantum phenomena, the logical analogies also served to throw light back on the situation in quantum physics. Let us consider a few examples that illustrate in a particularly striking way Bohr's general philosophical views.

Biology was the first domain outside physics that Bohr considered. He had been interested in fundamental biological problems since childhood. Like his father, he was especially concerned with the relation between physics and biology. With his deep understanding of what physics is, it was clear to Bohr that, if there is to be order in nature, the laws of physics cannot be broken. All talk about a special "life force," or the notion that the existence of organisms contradicts the principles of thermodynamics, he considered experimentally unfounded and epistemologically incorrect. On the other hand, he stuck to the view that biology is a topic different from physics in principle as well as in practice. Thus, he thought there was a core of truth in the vitalist attitude.

Vitalists have always been confronted with the difficult task of defining precisely the limitation of physics in biological phenomena. Bohr's philosophical viewpoint had prepared him for dealing with this task. In quantum physics he investigated the conditions under which one can employ the various elementary physical concepts meaningfully. Here the problem is to inquire into the conditions for doing physics. In order to see how there can be room for natural phenomena that are not fully describable in physical terms, we must specify what it is we do when we do physics.

In science one tries, whenever possible, to obtain knowledge through experiments. Part of the art of experimentation consists in being able precisely to define and to control the experimental conditions. Physics is a science where experience can be gained under conditions that in principle can be specified precisely and controlled completely. That is why the laws of physics are precise, or exact. The statement that physics is concerned with a world governed by exact laws is only one side of the story. Physics is concerned just as much with an experimenter who can carry his individual experiments to an arbitrary degree of precision. It is the freedom of the experimenter to choose and control the experimental conditions that permits the physical description to be stringent and complete.

Bohr's idea was that biological phenomena occur under conditions in

Complementarity is no system, no doctrine with ready-made precepts. There is no *via regia* to it: no formal definition of it can even be found in Bohr's writings, and this worries many people. The French are shocked by this breach of the Cartesian rules: they blame Bohr for indulging in "clair-obscur" and shrouding himself in "les brumes du Nord." The Germans in their thoroughness have been at work distinguishing several forms of complementarity and studying in hundreds of pages their relations to Kant. Pragmatic Americans have dissected complementarity with the scalpel of symbolic logic and undertaken to define this gentle art of the correct use of words without any words at all.

Léon Rosenfeld, "Niels Bohr's Contribution to Epistemology," 1963

307

which physical concepts can be applied to a certain extent, but that, if the phenomenon is to remain biological, the conditions cannot be controlled in the detail necessary to make the physical description complete. Just as an atom displays chemical properties only under conditions that exclude investigation of the position of its electrons, the phenomena of life occur only under conditions that exclude an exhaustive analysis of the single atoms in the organism. To carry through such an analysis, we would have to use experimental tools that would kill the organism, and we would no longer be dealing with a biological situation or a biological phenomenon. Thus, the experimenter in biology does not have the same possibilities of definition and control as in physics. Organisms, Bohr said, are not the results of experiments we perform ourselves. Rather, they have to be considered as results of nature's own experiments — if we can talk about experiments in a situation where no tools are used.

In principle, we cannot determine the physical state of a living organism. Hence, the word "life" refers not to a special quality or force that may permeate some physical systems, but rather, like "quantum of action," to a relationship of exclusion between conditions of observation or between the conceptual tools these conditions define. Thus, physical concepts are not sufficient to describe biological regularities. In order to obtain an adequate description of biological regularities, we must also use words that do not belong to physics, words like "purpose," "self-preservation," and "self-adaptation," which refer to the organism as a whole. The approach that uses these concepts, the vitalist (or finalist) approach, is complementary to the mechanistic approach.

I must confess that this attitude to fundamental problems in biology has never become completely clear to me. The crucial question is, of course, what we may understand by "typically biological conditions of observation" and the corresponding "typically biological concepts."

The instinctive behavior of animals is another example of how possibilities in nature can be used to preserve and reproduce life. Bohr was very interested in the wonders of instinct, and his view on the relationship between instinctive behavior and conscious thinking illuminates his general attitude in a particularly striking way. One of his favorite examples was the salmon's fantastic pathfinding ability. A salmon is born in a lake in the mountains. When some months old, it swims down brooks and rivers to the ocean, where it lives until the time comes for it to reproduce, and then it begins the long journey back. It jumps with unbelievable force over waterfalls, and even though the streams and brooks divide many times, it can still find its way back to the pool in which it was born.

How can it do that? To those who say, "It must have a sixth sense!" Bohr would reply, "It is easy enough to count from five to six, but it is not so easy to say what the physical basis for the functioning of such a sixth sense would be." Bohr thought that the salmon can do it because it does not know how it does it. It has only one task to perform, and it does not solve it by selecting among alternatives. The salmon is not in the same situation as someone who decides to buy something in a certain shop in a certain street. That person could have decided to buy something else in another shop somewhere else. The salmon is not confronted with alternatives: it makes no choices. It has not, so to speak, divided up the world. To Bohr, it was the fact that we have divided up the world, that we communicate with each other by means of concepts, that separates us from animals. It is primarily the ability to think in concepts, or the possession of a language, that makes us human beings. Bohr even suggested that the adjective "human" should be reserved for only those characteristics not directly connected with genetic inheritance. As long as a child has not yet learned to use concepts, it cannot, strictly speaking, be regarded as a human being. Yet, of course, it is fundamentally different from animals because it possesses the organic possibilities of receiving through education a culture that will enable it eventually to take its place in some human society.

A child does not become a human being until it has learned to talk. It can support a culture not only because its physiological background is sufficiently complex, but also because it possesses a means of communication through which a culture can be implanted. I think it was Bohr's opinion that if any other animal were to possess the anatomical apparatus for talking and thus the ability to receive a culture, it would not be possible beforehand to set limits for the extent to which culture could be supported by its physiological background. As he said, "The latent possibilities in any living organism are not easily fathomed." He did not think that there is any connection between a man's genes and his so-called spiritual faculties such as intelligence and morality or, in general, between his biological makeup and his capacity to acquire a culture. Bohr thought that cultures were like flowers in a field. The same field could have grown another sort of flower, and the same flower could have grown in another field. The variability of cultures indicates the numerous possibilities for social life. In the relations between cultures as well as in the relations between the individual and society, he found new complementary features.

By dividing up the world, we suppress our inherited instinctive behavior. But in return, we call into play a new kind of behavior based on conceptual communication. Ordinary language is a perfectly adequate tool to describe our conscious life. Bohr pointed out that, while the causal mode of description was indispensable as a starting point for the ordering of physical phenomena, the conceptual framework in which we communicate our states of mind has been since the origin of language a complementary mode of description. He liked to compare the relation between situations in which we express thoughts and those in which we express feelings with that obtaining between quantum phenomena described by space-time coordination and conservation laws, respectively. The richness and contrasts of quantum phenomena and psychological experiences are due to the fact that they are sensitive to the placing of the partitions between system and

I remember especially how, at my last stay with [Rutherford] a few weeks before his death, he was fascinated by the complementary approach to biological and social problems and how eagerly he discussed the possibility of obtaining experimental evidence on the origin of national traditions and prejudices by such unconventional procedures as the interchange of newborn children between nations.

Niels Bohr,
Rutherford Memorial Lecture, 1958

instrument, and between conscious content and "the background that we loosely refer to as 'ourselves.'" Only two sets of concepts are used to account for quantum phenomena, whereas a personal pronoun may be connected with a multitude of verbs referring to different contents of consciousness.

Bohr's general attitude was epistemologically oriented to an unusual degree. For him, the primary task was to obtain a survey of our situation based on objective description. Yet he did not want to exclude any side of existence, and he felt that from the viewpoint of complementarity one could understand that there is room for all features of our situation. Art shows us harmonies beyond the scope of objective description. Bohr considered poetry, painting, and music to be means of expression where a freer and freer display of fantasy is made possible by a greater and greater relaxation of definition. An object formed by nature and not by a human hand cannot be called a work of art, since art is a *human* activity.

Bohr wanted to understand existence through insight into the conditions of human life. It is by understanding our conditions that we can overcome disharmony. It was Buddha's insight into man's situation which gave him the ability to console others. Bohr was an optimist. "Nobody can deny," he said, "that we have a feeling of being able to make the best of circumstances." His view of life is beautifully illustrated by a little story he liked very much. Three philosophers came together to taste vinegar, the Chinese symbol for the spirit of life. First Confucius drank of it. "It is sour," he said. Next, Buddha pronounced the vinegar bitter. Then Lao-tze tasted it and exclaimed, "It is fresh!"[2]

Light and Life

In August 1932 Bohr gave an address with the above title at the opening meeting of the International Congress on Light Therapy in Copenhagen.[1] His theme was to explore the extent to which the principles of quantum physics could explain the phenomenon of life. At the time, he considered that there were limitations to this possibility, related in basic ways to his ideas of complementarity, and these limitations are discussed at length in the body of the address.

Thirty years later, at the inauguration of the new Institute of Genetics at the University of Cologne (directed by Bohr's longtime friend and colleague Max Delbrück), in a lecture entitled "Light and Life Revisited" Bohr reexamined what he had called in his earlier lecture "the impossibility of a physical or chemical explanation of the peculiar functions characteristic of life."[2] Influenced perhaps by the developments in molecular biology in the intervening decades, Bohr modified his earlier position and expressed the now generally accepted view that there is not "any limitation in the application to biology of the well-established principles of atomic physics." It was Bohr's last major address; he died less than six months later.

Below is reprinted almost the full text of the 1932 lecture, followed by the brief passage in his 1962 address in which he states his revised opinion on these matters.

A.P.F.

As a physicist whose studies are limited to the properties of inanimate bodies, it is not without hesitation that I have accepted the kind invitation to address this assembly of scientists met together to forward our knowledge of the beneficial effects of light in the cure of diseases. Unable as I am

311

to contribute to this beautiful branch of science that is so important for the welfare of mankind, I could at most comment on the purely inorganic light phenomena which have exerted a special attraction on physicists through-out the ages, not least owing to the fact that light is our principal tool of observation. I have thought, however, that on this occasion it might perhaps be of interest in such a comment to enter on the problem of how far the results reached in the more limited domain of physics may influence our views as regards the position of living organisms within the general edifice of natural science. Notwithstanding the subtle character of the riddles of life, this problem has presented itself at every stage of science, the very essence of scientific explanation being the analysis of more complex phenomena into simpler ones. At the moment it is the essential limitation of the mechanical description of natural phenomena revealed by the recent development of atomic theory which has lent new interest to the old problem. This development originated just in the closer study of the interaction between light and material bodies, which presents features that defeat certain demands hitherto considered as indispensable in a physical explanation. As I shall endeavor to show, the efforts of physicists to master this situation resemble in some way the attitude toward the aspects of life always taken more or less intuitively by biologists. Still, I wish to stress at once that it is only in this formal respect that light, which is perhaps the least complex of all physical phenomena, exhibits an analogy to life which shows a diversity beyond the grasp of scientific analysis.

From a physical standpoint, light may be defined as transmission of energy between material bodies at a distance. As is well known, such effects find a simple explanation within the electromagnetic theory, which may be regarded as a rational extension of classical mechanics suited to alleviate the contrast between action at a distance and at contact. According to this theory, light is described as coupled electric and magnetic oscillations differing from ordinary electromagnetic waves of radio transmission only by the greater frequency of vibration and the smaller wavelength. In fact, the practically rectilinear propagation of light, on which rests the location of bodies by direct vision or by suitable optical instruments, depends entirely on the smallness of the wavelength compared with the dimensions of the bodies concerned and of the instruments. At the same time, the wave character of light propagation not only forms the basis for our account of color phenomena, which in spectroscopy have yielded such important information on the constitution of material bodies, but is also essential for every refined analysis of optical phenomena. As a typical example, I need only mention the interference patterns which appear when light from one source can travel to a screen along two different paths. Here we find that the effects which would be produced by the separate light beams are strengthened at such points of the screen where the phases of the two wave-trains coincide, that is, where the electric and magnetic oscillations in the two beams have the same directions, while the effects are weakened and may even disappear at points where these oscillations have opposite directions and where the wave-trains are said to be out of phase with one another. These interference patterns offer so thorough a test of the wave picture of light propagation that this picture cannot be considered as a hypothesis in the usual sense of this word, but may rather be regarded as the adequate account of the phenomena observed.

Still, as you all know, the problem of the nature of light has been subjected to renewed discussion in recent years, on account of the discovery of an essential feature of atomicity in the mechanism of energy transmission, which is quite unintelligible from the point of view of the electromagnetic theory. In fact, any energy transfer by light can be traced down to individual processes in each of which a so-called light quantum is exchanged whose energy is equal to the product of the frequency of the electromagnetic oscillations and the universal quantum of action, or Planck's constant. The obvious contrast between this atomicity of the light effect and the continuity of the energy transfer in the electromagnetic theory presents us with a dilemma of a character hitherto unknown in physics. Thus, in spite of its obvious insufficiency, there can be no question of replacing the wave picture of light propagation by some other picture leaning on ordinary mechanical ideas. Especially, it should be emphasized that light quanta cannot be regarded as particles to which a well-defined path in the sense of ordinary mechanics can be ascribed. Just as an interference pattern would completely disappear if, in order to make sure that the light energy traveled only along one of the two paths between the source and the screen, we would stop one of the beams by a nontransparent body, so is it impossible in any phenomenon for which the wave constitution of light is essential to trace the path of the individual light quanta without essentially disturbing the phenomenon under investigation. Indeed, the spatial continuity of our picture of light propagation and the atomicity of the light effects are complementary aspects in the sense that they account for equally important features of the light phenomena which can never be brought into direct contradiction with one another, since their closer analysis in mechanical terms demands mutually exclusive experimental arrangements. At the same time, this very situation forces us to renounce a complete causal account of the light phenomena and to be content with probability laws based on the fact that the electromagnetic description of energy transfer remains valid in a statistical sense. This forms a typical application of the so-called correspondence argument, which expresses the endeavor of utilizing to the utmost extent the concepts of the classical theories of mechanics and electrodynamics, in spite of the contrast between these theories and the quantum of action.

At first, this situation may appear very uncomfortable, but, as has often happened in science when new discoveries have led to the recognition of an essential limitation of concepts hitherto considered as indispensable, we are rewarded by getting a wider view and a greater power to correlate phenomena which before might even have appeared as contradictory. Indeed, the limitation of classical mechanics symbolized by the quantum of action has offered a clue to our understanding of the intrinsic stability of atoms on which the mechanical description of natural phenomena is essentially based. Of course, it has always been a fundamental feature of the atomic theory that the indivisibility of atoms cannot be understood in mechanical terms, and this situation remained practically unchanged even after the indivisibility of atoms was replaced by that of the elementary electric particles, electrons and protons, of which atoms and molecules are built up. What I am referring to is not the problem of the intrinsic stability of these elementary particles but that of the atomic structures composed of them. If we attack this problem from the point of view of mechanics or of electromagnetic theory, we find no sufficient basis on which to account for the specific properties of the elements and not even for the existence of rigid bodies on which all measurements used for ordering phenomena in space and time ultimately rest. These difficulties are now overcome by the recognition that any well-defined change in an atom is an individual process consisting in a complete transition of the atom from one of its so-called stationary states to another. Moreover, since just one light quantum is exchanged in a transition process by which light is emitted or absorbed by an atom, we are able by means of spectroscopic observations to measure directly the energy of each of these stationary states. The information thus derived has also been most instructively corroborated by the study of the energy exchanges which take place in atomic collisions and in chemical reactions.

In recent years a remarkable development of atomic mechanics along the lines of the correspondence argument has taken place, affording us proper methods of calculating the energies of the stationary states of atoms and the probabilities of transition processes, thus making our account of atomic properties as comprehensive as the coordination of astronomical experience by Newtonian mechanics. Notwithstanding the greater complexity of the general problems of atomic mechanics, the lesson taught us by the analysis of the simpler light effects has been most important for this development. Thus, the unambiguous use of the concept of stationary states stands in a similar relation of complementarity to a mechanical analysis of intra-atomic motions as do light quanta to the electromagnetic theory of radiation. Indeed, any attempt to trace the detailed course of a transition process would involve an uncontrollable exchange of energy between the atom and the measuring instruments, which would completely disturb the very energy balance we set out to investigate. The causal mechanical coordination of experience can be accomplished only in cases where the action involved is large compared with the quantum and where, therefore, a subdivision of the phenomena is possible. If this condition is not fulfilled, the action of the measuring instruments on the object under investigation cannot be disregarded and will entail a mutual exclusion of the various

The recollection came to my mind of an afternoon in March 1931 when in a brilliant improvisation Bohr, put on the subject by Delbrück, rediscovered at the blackboard the explanation of the stability of Saturn's rings which Maxwell had given in his classic paper. Thrilled by the simplicity of the solution, he feelingly expressed his admiration for nature's workings on the cosmic scale. Then he added, with a twinkle in his eye: "The stability of atoms is also wonderful."

Léon Rosenfeld
On the Constitution of Atoms and Molecules, 1963

kinds of information required for a complete mechanical description of the usual type. This apparent incompleteness of the mechanical analysis of atomic phenomena issues ultimately from the ignorance of the reaction of the object on the measuring instruments inherent in any measurement. Just as the general concept of relativity expresses the essential dependence of any phenomenon on the frame of reference used for its coordination in space and time, the notion of complementarity serves to symbolize the fundamental limitation, met with in atomic physics, of the objective existence of phenomena independent of the means of their observation.

This revision of the foundations of mechanics, extending to the very idea of physical explanation, not only is essential for the full appreciation of the situation in atomic theory but also creates a new background for the discussion of the problems of life in their relation to physics. In no way does this mean that in atomic phenomena we meet with features which show a closer resemblance to the properties of living organisms than do ordinary physical effects. At first sight, the essentially statistical character of atomic mechanics might even seem to conflict with the marvelously refined organization of living beings. We must keep in mind, however, that just this complementary mode of description leaves room for regularities in atomic processes foreign to mechanics but as essential for our account of the behavior of living organisms as for the explanation of the specific properties of inorganic matter. Thus, in the carbon assimilation of plants, on which so largely depends also the nourishment of animals, we are dealing with a phenomenon for the understanding of which the individuality of photochemical processes is clearly essential. Likewise, the nonmechanical stability of atomic structures is markedly exhibited in the characteristic properties of such highly complicated chemical combinations as chlorophyll or hemoglobin, which play a fundamental part in the mechanism of plant assimilation and animal respiration. Still, analogies from ordinary chemical experience, like the ancient comparison of life with fire, will of course yield no more satisfactory explanation of living organisms than will their resemblance with such purely mechanical contrivances as a clockwork. Indeed, the essential characteristics of living beings must be sought in a peculiar organization in which features that may be analyzed by usual mechanics are interwoven with typically atomistic features to an extent unparalleled in inanimate matter.

An instructive illustration of the degree to which this organization is developed is exhibited by the construction and function of the eye, for the exploration of which the simplicity of light phenomena has again been most helpful. I need not here go into details but shall just remind you how ophthalmology has revealed to us the ideal properties of the human eye as an optical instrument. Indeed, the limit imposed on the image formation by the unavoidable interference effects coincides practically with the size of such partitions of the retina, which have separate nervous connection with the brain. Moreover, since the absorption of a single light quantum by each

A part of the development and impact of science is probably determined. If a sufficient number of semi-intelligent people (like ourselves) look at the world, they are bound to arrive at some conclusions. In exceptional cases, however, there are developments that I am not sure were bound to occur. Bohr's idea of complementarity is an example.

Edward Teller, "Niels Bohr and the Idea of Complementarity," 1969 of these retinal partitions is sufficient for a sight impression, the sensitiveness of the eye may be said to have reached the limit set by the atomic character of the light processes. The efficiency of the eye in both of these respects is actually the same as that obtained in a good telescope or microscope connected with a suitable amplifier so as to make the individual processes observable. It is true that it is possible by such instruments to essentially increase our powers of observation, but, due to the limits imposed by the fundamental properties of the light phenomena, no instrument is imaginable which is more efficient for its purpose than the eye. Now, this ideal refinement of the eye, recognized through the recent development of physics, suggests that also other organs, whether they serve for the reception of information from the surroundings or for the reaction to sense impressions, will exhibit a similar adaptation to their purpose, and that also here the feature of individuality symbolized by the quantum of action is of decisive importance in connection with some amplifying mechanism. That it has been possible to trace this limit in the eye but not, so far, in any other organ is due simply to the extreme simplicity of the light phenomena to which we have referred before.

The recognition of the essential importance of atomistic features in the mechanism of living organisms is in no way sufficient, however, for a comprehensive explanation of biological phenomena. The question at issue, therefore, is whether some fundamental traits are still missing in the analysis of natural phenomena before we can reach an understanding of life on the basis of physical experience. Notwithstanding the fact that the multifarious biological phenomena are practically inexhaustible, an answer to this question can hardly be given without an examination of the meaning to be given to physical explanation still more penetrating than that to which the discovery of the quantum of action has already forced us. On the one hand, the wonderful features which are constantly revealed in physiological investigations and which differ so markedly from what is known of inorganic matter have led biologists to the belief that no proper understanding of the essential aspects of life is possible in purely physical terms. On the other hand, the view known as vitalism can hardly be given an unambiguous expression by the assumption that a peculiar vital force, unknown to physics, governs all organic life. Indeed, I think we all agree with Newton that the ultimate basis of science is the expectation that nature will exhibit the same effects under the same conditions. If, therefore, we were able to push the analysis of the mechanism of living organisms as far as that of atomic phenomena, we should not expect to find any features foreign to inorganic matter. In this dilemma it must be kept in mind, however, that the conditions in biological and physical research are not directly comparable, since the necessity of keeping the object of investigation alive imposes a restriction of the former which finds no counterpart in the latter. Thus, we should doubtless kill an animal if we tried to carry the investigation of its organs so far that we could tell the part played by the single atoms in vital functions. In every experiment on living organisms there must remain some uncertainty as regards the physical conditions to which they are subjected, and the idea suggests itself that the minimal freedom we must allow the organism will be just large enough to permit it, so to say, to hide its ultimate secrets from us. On this view, the very existence of life must in biology be considered as an elementary fact, just as in atomic physics the existence of the quantum of action has to be taken as a basic fact that cannot be derived from ordinary mechanical physics. Indeed, the essential nonanalyzability of atomic stability in mechanical terms presents a close analogy to the impossibility of a physical or chemical explanation of the peculiar functions characteristic of life.

In tracing this analogy, however, we must remember that the problems present essentially different aspects in atomic physics and in biology. While in the former field we are primarily interested in the behavior of matter in its simplest forms, the complexity of the material systems with which we are concerned in biology is of a fundamental nature, since even the most primitive organisms contain large numbers of atoms. It is true that the wide field of application of ordinary mechanics, including our account of the measuring instruments used in atomic physics, rests just on the possibility of largely disregarding the complementarity of the description entailed by the quantum of action in cases where we are dealing with bodies containing a great number of atoms. Notwithstanding the essential importance of the atomistic features, it is typical of biological research, however, that we can never control the external conditions to which any separate atom is subjected to the extent possible in the fundamental experiments of atomic physics. In fact, we cannot even tell which particular atoms really belong to a living organism, since any vital function is accompanied by an exchange of material through which atoms are constantly taken up into and expelled from the organization which constitutes the living being. Indeed, this exchange of matter extends to all parts of a living organism to a degree which prevents a sharp distinction on an atomic scale between those features of its mechanism which can be unambiguously accounted for on usual mechanics and those for which a regard of the quantum of action is decisive. This fundamental difference between physical and biological research implies that no well-defined limit can be drawn for the applicability of physical ideas to the problems of life which corresponds to the distinction between the field of causal mechanical description and proper quantum phenomena in atomic mechanics. This apparent limitation of the analogy in question is rooted in the very definitions of the words "life" and "mechanics," which are ultimately a matter of convenience. On the one hand, the question of a limitation of physics in biology would lose any meaning if, instead of distinguishing between living organisms and inanimate bodies, we extended the idea of life to all natural phenomena. On the other hand, if, in accordance with common language, we were to reserve the word "mechanics" for the unambiguous causal description of natural phe-

317

nomena, such a term as "atomic mechanics" would become meaningless. I shall not enter further into such purely terminological points but only add that the essence of the analogy being considered is the obvious exclusiveness between such typical aspects of life as the self-preservation and the self-generation of individuals, on the one hand, and the subdivision necessary for any physical analysis, on the other hand. Owing to this essential feature of complementarity, the concept of purpose, which is foreign to mechanical analysis, finds a certain field of application in biology. Indeed, in this sense teleological argumentation may be regarded as a legitimate feature of physiological description which takes due regard of the characteristics of life in a way analogous to the recognition of the quantum of action in the correspondence argument of atomic physics.

In discussing the applicability of purely physical ideas to living organisms we have, of course, treated life just as any other phenomenon of the material world. I need hardly emphasize, however, that this attitude, which is characteristic of biological research, involves no disregard of the psychological aspect of life. On the contrary, the recognition of the limitation of mechanical concepts in atomic physics would rather seem suited to conciliate the apparently contrasting viewpoints of physiology and psychology. Indeed, the necessity of considering the interaction between the measuring instruments and the object under investigation in atomic mechanics exhibits a close analogy to the peculiar difficulties in psychological analysis arising from the fact that the mental content is invariably altered when the attention is concentrated on any special feature of it. It will carry us too far from our subject to enlarge upon this analogy which offers an essential clarification of the psychophysical parallelism. However, I should like to emphasize that considerations of the kind here mentioned are entirely opposed to any attempt at seeking new possibilities for a spiritual influence on the behavior of matter in the statistical description of atomic phenomena. For instance, it is impossible, from our standpoint, to attach an unambiguous meaning to the view sometimes expressed that the probability of the occurrence of certain atomic processes in the body might be under the direct influence of the will. In fact, according to the generalized interpretation of the psychophysical parallelism, the freedom of the will is to be considered as a feature of conscious life which corresponds to functions of the organism that not only evade a causal mechanical description but resist even a physical analysis carried to the extent required for an unambiguous application of the statistical laws of atomic mechanics. Without entering into metaphysical speculations, I may perhaps add that an analysis of the very concept of explanation would, naturally, begin and end with a renunciation as to explaining our own conscious activity.

In conclusion, I need hardly emphasize that with none of my remarks have I intended to express any kind of skepticism as to the future development of physical and biological sciences. Such skepticism would, indeed, be far from the mind of physicists at the present time, when just the recognition of the limited character of our most fundamental concepts has resulted in such a remarkable development of our science. Nor has the renunciation of an explanation of life impeded the wonderful progress which has taken place in all branches of biology, including those which have proved so beneficial in the art of medicine.

[In his 1962 address, Bohr returned to what for him was the central question. After a passage in which he discussed the application of basic thermodynamic principles to biological systems, he said:]

Notwithstanding such general considerations, it appeared for a long time that the regulatory functions in living organisms, disclosed especially by studies of cell physiology and embryology, exhibited a fineness so unfamiliar to ordinary physical and chemical experience as to point to the existence of fundamental biological laws without counterpart in the properties of inanimate matter studied under simple reproducible experimental conditions. Stressing the difficulties of keeping the organisms alive under conditions which aim at a full atomic account I therefore suggested that the very existence of life might be taken as a basic fact in biology in the same sense as the quantum of action has to be regarded in atomic physics as a fundamental element irreducible to classical physical concepts.

In reconsidering the conjecture from our present standpoint, it must be kept in mind that the task of biology cannot be that of accounting for the fate of each of the innumerable atoms permanently or temporarily included in a living organism. In the study of regulatory biological mechanisms the situation is rather that no sharp distinction can be made between the detailed construction of these mechanisms and the functions they fulfil in upholding the life of the whole organism. Indeed, many terms used in practical physiology reflect a procedure of research in which, starting from the recognition of the functional role of the parts of the organism, one aims at a physical and chemical account of their finer structures and of the processes in which they are involved. Surely, as long as for practical or epistemological reasons one speaks of life, such teleological terms will be used in complementing the terminology of molecular biology. This circumstance, however, does not in itself imply any limitation in the application to biology of the well-established principles of atomic physics.

In the last resort, it is a matter of how one makes headway in biology. I think that the feeling of wonder which physicists had thirty years ago has taken a new turn. Life will always be a wonder, but what changes is the balance between the feeling of wonder and the courage to understand.

Complementarity as a Way of Life

R. V. Jones

R. V. Jones, who described his meetings with Bohr in Part V of this volume, here discusses some social and human applications of Bohr's ideas of complementarity.

Physicists have had many successes through recognizing analogies between phenomena in new fields of experience and those in the well-beaten fields which have yielded to experience and thought in the past. Niels Bohr's model of the atom as a miniature solar system, and his model of the nucleus as a liquid drop, are outstanding successful examples; and with such successes in mind it is tempting to contemplate whether there could be even broader analogies between the phenomena of physics and those of human society.

Heisenberg's formulation of the uncertainty relation early in 1927 crystallized in Bohr's mind some thoughts on the association of wave and particle properties that he had been entertaining for some time. And Bohr first used the term "complementary" to describe these properties in a lecture at the momentous conference in Como in September 1927. Bohr had been much interested in William James's *The Principles of Psychology* (1884), and one of his earliest papers (1929) on complementarity draws a psychological analogy: "The necessity of taking recourse to a complementary, or reciprocal, mode of description is perhaps familiar to us from psychological problems."[1] And Max Jammer could well be right in suggesting that Bohr was inspired by another passage from James: "In certain persons the total consciousness may be split into two parts which coexist but mutually ignore each other, and share the objects of knowledge between them. More remarkable still, they are complementary."[2] Following Bohr, most physicists have grown accustomed to having to accept the reconciliation of seemingly irreconcilable concepts, to think constructively forward from the reconciliation, and to regard it as a recognition of the ultimately inevitable breakdown of any single model based on earlier experience.

It is interesting that Bohr, such a master of analogies and models, should have come to the concept of complementarity by an analogy with psychology. In turn, we may ask whether, reciprocally, complementarity can be applied analogically to other human problems. Bohr himself saw an analogy in the development of thought: "In particular, the apparent contrast between the continuous onward flow of associative thinking and the preservation of the unity of the personality exhibits a suggestive analogy with the relation between the wave description of the motions of material particles, governed by the superposition principle, and their indestructible individuality."[3]

Early in the eighteenth century a philosophical statesman, the Marquess of Halifax, had seen a duality in the object of constitutional law to keep the balance "between the excess of unbounded power and the extravagance of liberty not enough restrained" (*The Character of a Trimmer*, 1717). This has something in common with the Golden Mean of Aristotle, who argued that any virtue was the mean of two vices—for example, bravery is the mean between rashness and cowardice. Hegel went further: "Every truth, every reality, is the unification of two contradictory elements or partial aspects which are not merely contrary like black and white but contradictory, like same and different." And this had a great influence on Marx, who, in the preface to the second edition of *Das Kapital*, claimed to out-Hegel Hegel: "In Hegel's writings, dialectic stands on its head . . . My own dialectical method is not only fundamentally different from the Hegelian method, but is its direct opposite."

One of Hegel's points was that quantity and quality, seemingly distinct, are related to the extent that a change in quantity may produce a change of quality. (Coal, for example, which can be ignited only with some difficulty when in large lumps, becomes a dangerous explosive when in the form of dust, largely because of the changed ratio of surface area to volume.) And Marx's colleague Engels, when pursuing Hegel's argument, cited a vivid point made by Napoleon concerning his military actions against the Mameluke horsemen in Egypt. Individually they were so good, Napoleon said, that two Mamelukes could overcome three of his cavalrymen—but a force of a thousand cavalrymen could defeat fifteen hundred Mamelukes. The qualitative outcome of an action, victory or defeat, could therefore depend merely on a change of quantity, or scale. On the small scale, individual horsemanship was the key factor; on the large, it was the disciplined application of force at the right place and time. Both factors were present on both scales, but the balance between them changed. Once again, we are concerned with the complementary nature of individual and community behavior.

My own contact with the military field leads me to point to other analogies with complementarity, many of which apply not only to military matters but to community affairs generally. Obedience, for example, is a civil virtue as well as a military one; but even in the military field, is it

Bohr was always as eager to learn as to teach, however, and when much later, some time in the early thirties, his old friend the psychologist Rubin called his attention to William James's *Principles of Psychology*, he joyfully recognized in this great book a general attitude akin to his own: he was particularly enthusiastic about James's brilliant description of the stream of consciousness.

Léon Rosenfeld, "Niels Bohr's Contribution to Epistemology," 1963

321

always the best course? Sometimes the man on the spot is the first to be aware of some change in the situation from that which his commander expected. Should he therefore disobey his orders and act as he thinks best? This possibility was at one time recognized in the old Austrian army, where an officer who had been vindicated by success when he had disobeyed orders on the battlefield could be awarded the Order of Maria Teresa — but the penalty for failure was to be shot.

Finding the best balance between obedience and initiative is rarely an easy problem, but it is only one of several of its kind that are likely to be encountered in any command structure, military or civil. One of these concerns the balance between tradition and innovation. There have been many military actions where regimental tradition has caused men to hold out in the face of frightening odds, and many others where armies have been defeated because they were slow to modernize their weapons and methods. This is also true of states: Japan, for example, advanced spectacularly through adopting Western methods and technology, both civil and military, whereas ancient China, with a state brilliantly run for many centuries by an outstanding civil service full of traditional wisdom, and with a nationwide respect for authority, could not stand up against Western military technology until Mao's revolution. The Western advance had been largely a result of the challenge to authority in the fifteenth and sixteenth centuries typified by the Protestant movement.

Yet too much protest and individualism can result in anarchy, and it has been a fortunate thing for science that we have on the whole struck a successful balance. No worker in science, however eminent, is above challenge not only from his accepted peers but also from his juniors, when he makes a statement in science; but we can have reasonably ordered laboratories only if there is some degree of authority in their running.

Again, the running of an organization requires a degree of balance between individual and committee rule. Sometimes it is best that one form of rule predominate; sometimes, the other. Committees tend to act at the rate of their most timorous members, particularly when a danger has to be faced; at other times they may talk themselves into an unrealistic viewpoint, much as a crowd can mutually interact to produce crazy behavior that may be no more harmful than a fiesta but which may be as dangerous as a riot. In the military field the choice often comes down to accepting either a single commander or a council of war, and much has been said in favor of both. A council of war very nearly stopped the Greeks from fighting the decisive battle of Marathon; Frederick the Great would not allow his commanders to hold councils of war because he believed that these encouraged indecision; and General Robert Clive said that "he had never called but one council of war, and that, if he had taken the advice of that council, the British would never have been masters of Bengal." Of course, what we do not know is how many disastrous battles have been avoided through good advice given by councils of war. One, at least,

Poltava, was lost by Charles XII of Sweden fighting deep in Russia against the advice of his senior officers.

One of the most fascinating balances of all that has to be struck between conflicting factors in command concerns the extent to which the commander or leader should involve himself in "front-line" or "sharp-end" action. This was a problem that repeatedly dogged Winston Churchill, whose inclination was always toward front-line involvement. For example, he had seen the reluctance of the British admirals in 1917 to introduce the convoy system, of which he wrote, "The firmly inculcated doctrine that an Admiral's opinion was more likely to be right than a Captain's, and a Captain's than a Commander's, did not hold good when questions entirely novel in character, requiring keen and bold minds unhampered by long routine, were under debate." So when new conditions of warfare had to be faced, it was important for the high command to draw on the up-to-the moment experience of the men directly confronting the enemy. And there was danger that this experience might be attenuated if it was transmitted up the normal hierarchy of command. At the same time, Churchill saw the need for some degree of detachment for those who lead. In an essay on Moses (1931) he wrote, "Every prophet has to come from civilisation, but every prophet has to go into the wilderness. He must have a strong impression of a complex society and all that it has to give, and then he must serve periods of isolation and meditation. This is the process by which psychic dynamite is made." So withdrawal and involvement are complementary factors in creative leadership. Clerk Maxwell had a kindred thought. In a letter of 21 April 1862 to a friend, he wrote, "I hope you enjoy the absence of pupils . . . The total oblivion of them for definite intervals is a necessary condition for doing them justice at the proper time."

Yet another aspect of complementarity concerns the balance between long-term and short-term interests. I myself encountered this in World War II regarding intelligence: How should we dispose our cryptographic effort in Britain? There was a temptation to devote it to those ciphers whose solution would affect operations immediately; but if we carried this too far we would have no effort available for ciphers whose solution was concerned with, for example, developments at Peenemünde, which might become of intense operational importance in a few months' time. So a balance had to be struck, and anyone planning either his own future or that of an organization to which he belongs is likely sooner or later to have to strike such a balance. By looking exclusively to the future he may lose the present, and by thinking only of the present he may sacrifice the future.

The creative individual is, in a sense, complementary to the society in which he lives, rather as the soloist in a concerto. Both the basic ideas of science and the key inventions of mankind have generally been conceived in the minds of individuals, while the effort to gain the data on which the ideas and inventions have been based, and the subsequent effort to turn them to good account, have required the contributions of many besides the

Bohr had great expectations about the future role of complementarity. He upheld them with unshakable optimism, never discouraged by the scant respect he got from our unphilosophical age. On one of those unforgettable strolls during which Bohr would so candidly disclose his innermost thoughts, we came to consider that what many people nowadays sought in religion was a guidance and consolation that science could not offer. Thereupon Bohr declared, with intense animation, that he saw the day when complementarity would be taught in the schools and become part of general education: and better than any religion, he added, a sense of complementarity would afford people the guidance they needed.

Léon Rosenfeld, "Niels Bohr's Contribution to Epistemology," 1963 inventor and the originator of ideas. So the individual and the community are necessary to one another, and both have to be given their due weights in a successful society.

Politicians would do well to heed the lesson of complementarity, regarding the balance of interests between the community and the individual. If only they could see that these interests are complementary, some, at least, of the more bitter struggles between socialism and individualism might be avoided. Each viewpoint represents an aspect of truth, and neither can completely deny the other, although the optimum balance between them will change with circumstances.

Max Born put it well in his Joule memorial lecture of 1950:[4] "Complete freedom of the individual in economic behaviour is incompatible with the existence of an orderly state, and the totalitarian state incompatible with the development of the individuum. There must exist a relation between the latitudes of freedom Δf and of regulation Δr of the type $\Delta f \cdot \Delta r \approx p$, which allows a reasonable compromise. But what is the 'political constant' p? I must leave this to a future quantum theory of human affairs. The world, which is so ready to learn the means of mass destruction from physics, would do better to accept the message of reconciliation contained in the philosophy of complementarity."

The Complementarity Principle and Eastern Philosophy

D. S. Kothari

The principle of complementarity, which we owe principally to Niels Bohr, is perhaps the most significant and revolutionary concept of modern physics.[1] The complementarity approach can enable people to see that seemingly irreconcilable points of view need not be contradictory. These, on deeper understanding, may be found to be complementary and mutually illuminating—the two opposing contradictory aspects being parts of a "totality," seen from different perspectives. It allows the possibility of accommodating widely divergent human experiences into an underlying harmony, and bringing to light new social and ethical vistas for exploration and for alleviation of human suffering. Bohr fervently hoped that one day complementarity would be an integral part of everyone's education and would provide guidance in the problems and challenges of life.

Hideki Yukawa was once asked whether young physicists in Japan, like most young physicists in the West, found it difficult to comprehend the idea of complementarity. He replied that Bohr's complementarity always appeared to them as quite evident: "You see, we in Japan have not been corrupted by Aristotle."

The core of the profound ethical and spiritual insights propounded in the Upanishads, Buddhism, and Jainism rests essentially on the complementarity approach to the problems of life and existence, though the formulations vary. Sri Aurobindo, perhaps the greatest exponent of the Upanishadic thought in our times, writes in his commentary on the Isha Upanishad:

> The principle it follows throughout is the uncompromising reconciliation of uncompromising extremes . . . The pairs of opposites successively taken up by the Upanishad and resolved are, in the order of their

D. S. Kothari, Professor Emeritus at the University of Delhi, worked at Cambridge under Rutherford and R. H. Fowler, and also spent some time at the Niels Bohr Institute. He was a pioneer in applying nuclear and quantum principles to the theory of white dwarf stars around 1930. From 1961 to 1973 he was Chairman of the University Grants Commission of India. He is Chancellor of Jawaharlal Nehru University, Delhi.

succession: (1) The conscious Lord and phenomenal Nature; (2) Renunciation and Enjoyment; (3) Action in Nature and Freedom in the Soul; (4) The One stable Brahman and the multiple Movement; (5) Being and Becoming; (6) The Active Lord and the indifferent Akshara Brahman; (7) Vidya (Knowledge) and Avidya (Ignorance); (8) Birth and Non-Birth; (9) Works and Knowledge.[2]

The Jain formulation of the complementarity approach is based on the Syādvāda dialectic (*Syād* means "may be").[3] The Syādvāda logic is indispensable for the theory and practice of *ahimsa* (nonviolence) in thought, word, and deed. Syādvāda and *ahimsa* go integrally together. Syādvāda asserts that the knowledge of reality is possible only by denying the absolutistic attitude. What is new is the fact that relativity and quantum mechanics embody the same line of thought as one finds in the Syādvāda logic. Further, the Syādvāda approach enriches our understanding of complementarity in physics. As pointed out by P. C. Mahalanobis[4] and J. B. S. Haldane,[5] the foundations of the theory of probability are also in keeping with the Syādvāda logic.

The recognition that in atomic phenomena we are concerned with an application of complementarity which can be precisely formulated provides a basic motivation for eventually discovering deeper and richer levels of complementarity encompassing both matter and mind. Bohr concludes his essay "Causality and Complementarity" as follows:

In general philosophical perspective, it is significant that, as regards analysis and synthesis in other fields of knowledge, we are confronted with situations reminding us of the situation in quantum physics. Thus, the integrity of living organisms and the characteristics of conscious individuals and human cultures present features of wholeness, the account of which implies a typical complementary mode of description. Owing to the diversified use of the rich vocabulary available for communication of experience in those wider fields, and above all to the varying interpretations, in philosophical literature, of the concept of causality, the aim of such comparisons has sometimes been misunderstood. However, the gradual development of an appropriate terminology for the description of the simpler situation in physical science indicates that we are not dealing with more or less vague analogies, but with clear examples of logical relations which, in different contexts, are met with in wider fields.[6]

Bohr's first and continuing preoccupation with philosophical problems related to the use of language for unambiguously describing our experiences. A fundamental difficulty in this regard arises from the inescapable fact that man is both *actor* and *spectator* in the universe. Thus, when I am "seeing" a thing, I am also "acting": my choice to see the particular thing is an "act" on my part. We often use the *same* word to describe both a state of our consciousness and the associated accompanying behavior of the body. How to avoid the ambiguity? Bohr drew attention to the beautiful analogy between the concept of multiform functions and the concept of a Riemann surface: the different values of a multiform function are distributed on different planes of a Riemann surface. Similarly, we may say that the different meanings of the same word belong to different "planes of objectivity."[7]

Bohr used to tell how the ancient Indian thinkers had emphasized the futility of our ever understanding the "meaning of existence." And he would add that the one certain thing is that a statement like "existence is meaningless" is itself devoid of any meaning.

As lucidly pointed out by Heisenberg, the concepts of ordinary or natural language have undergone changes due to developments of modern science.[8] Further changes are to be anticipated as a result of continuing advancements. The ambiguities and contradictions faced in science have been attributed to the use of the terminology of natural language. Contradictions are inherent in natural language, as well as in precise scientific language. The role of the complementarity approach and of Syādvāda logic is to give a less ambiguous meaning to the terminology of natural language and to provide greater insight into the relationship between human mind and reality.

Consider the following idealized situation, or "thought experiment," discussed by Heisenberg.[9] There is an atom in a closed box that is divided by a partition into two equal compartments. The partition has a very small hole so that the atom can pass through it. The hole can be closed by a shutter, if desired. According to classical logic, the atom will be either in the left compartment (L) or in the right compartment (R). There is no third possibility. But quantum physics forces us to admit other possibilities to explain adequately the results of experiments. If we use the words "box" and "atom" at all, then there is no escape whatsoever from admitting that in some strange way, which totally defies description in words, *the same atom is, at the same time, in both compartments* (when the hole is open). Such a situation cannot be expressed properly in ordinary language — it is inexpressible (except mathematically). As we shall see, it is *avayakta* in the terminology of Syādvāda. It is an idea crazy beyond words. But there is no escape; for, totally unlike large objects, particles at the atomic level exhibit a

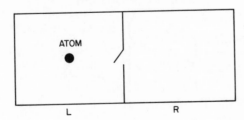

wave aspect as well as a particle aspect. These two aspects, which are contradictory and mutually exclusive in the everyday domain, are complementary in atomic phenomena.

Bohr's famous analysis of a two-slit interference experiment made this complementarity quantitative.[10] The figure here shows a slight variant of the thought experiment involved. A plate P receives the photons. If, as in (a) in the figure, the plate is rigidly fixed, the interference pattern is built up by the arrival of many photons. But with a very weak beam, in which photons cross the apparatus one at a time, and with P suspended so that it can recoil along the y direction, as in (b), one might try to infer whether an individual photon came through hole A or hole B by measuring the transverse momentum $\pm h\nu\theta/c$ transferred to P. This, however, will, by the Heisenberg indeterminacy principle, make the y coordinate of P uncertain by an amount $\Delta y > h/(h\nu\theta/c)$; that is, $\Delta y > \lambda/\theta$. But for observation of interference fringes, it is necessary that this uncertainty be less than the fringe spacing, which requires Δy to be *less* than λ/θ. For interference fringes to be produced, photons must in some sense go through both holes, but this can happen only if we forgo any attempt to observe them. It is because of this mutual exclusiveness of the two setups (a) and (b) that the particle and the wave aspects are complementary and not contradictory.

A similar situation would apply if one observed Xrays scattered from the atom in the two-compartment box. One could either locate the atom as being in one compartment or the other, or one could observe an interference pattern arising from its partial presence in both compartments. Moreover, choosing at a given instant ("now") to make one type of observation or the other would seem to imply that this decision influenced the state of the atom at an earlier time (earlier by the transit time of Xrays from the atom to the plate).[11] This looks utterly strange. The lesson is that the behavior of "small objects" is not visualizable. It is not describable in ordinary language. Nevertheless, it is real. As Wheeler has remarked:

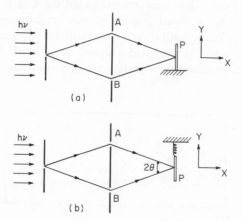

(a)

(b)

"There is no more remarkable feature of the quantum world [characterized by the Planck constant] than the strange coupling it brings about between future and past.[12] Every observation which implies a freedom of choice (that is, free will) between mutually exclusive alternatives is, in a sense, a participation in genesis (giving a new meaning to our being "actors" and "spectators" in the drama of existence). Perhaps, as we probe deeper in our understanding of nature, other levels of complementarity may be discovered.[13]

Let us now return to the Syādvāda formulation as applied to the wave-particle duality.[14] According to the Syādvāda scheme, every fact of reality leads to seven ways or modes of description. These are combinations of affirmation, negation, and inexpressibility — namely, (1) Existence, (2) Nonexistence, (3) Occurrence of Existence and Nonexistence, (4) Inexpressibility or Indeterminateness, (5) Inexpressibility as qualified by Existence, (6) Inexpressibility as qualified by Nonexistence, and (7) Inexpressibility as qualified by both Existence and Nonexistence.

The fourth mode — the inexpressibility known as *avayakta* — is the key element of the Syādvāda dialectic. This is especially well brought out by the foregoing discussion of the wave-particle duality in modern physics. As mentioned earlier, Mahalanobis and Haldane have discussed the significance of Syādvāda for the foundations of modern statistics.

The physical example of the atom and the box can be presented diagrammatically and compared with the seven modes of Syādvāda, as shown in the table on the following page. The quantum-mechanical description in the usual notation appears in the middle column. The atom when *observed* is either in the state $|L\rangle$ or $|R\rangle$. The *superposed state* $|P\rangle = |L\rangle + |R\rangle$ is not directly observable using the type of apparatus for observing $|L\rangle$ and $|R\rangle$ states.

Take any meaningful statement. Call it *A*. It may describe a fact of experience. It could be a proposition of logic or mathematics. The Syādvāda dialectic demands that in the very nature of things the negative of the given statement is also correct under certain conditions. Denote by *not-A* the negative statement of *A*. The conditions under which the two statements *A* and *not-A* are correct cannot, of course, be the same; in general, the respective conditions are mutually exclusive. Given a statement *A*, it may not be at all easy to discover the conditions or situations under which *not-A* holds. It may even appear at the time impossible. But faith in Syādvāda should encourage one to continue the search. For example, in Euclidean geometry the sum of the three angles of a triangle is equal to the sum of two right angles. The negation of this theorem is a new geometry in which the sum of the three angles of a triangle is not equal to the sum of two right angles. Not until two thousand years after Euclid was non-Euclidean geometry discovered, in the nineteenth century; Einstein's theory of general relativity is based on this geometry.

For special relativity theory, the Syādvāda approach is directly applicable.

An old legend describes a dialogue between Abraham and Jehovah. Jehovah chides Abraham, "You would not even exist if it were not for me!" "Yes, Lord, that I know," Abraham replies, "but also You would not be known if it were not for me." In our time the participants in the dialogue have changed. They are the universe and man.

John A. Wheeler, *Quantum Theory and Measurement*, 1983

329

Seven modes of *Syādvāda*, illustrated by the example of an atom in a box with two compartments.

Atom in a box	Quantum-mechanical representation (in the usual notation)	Syādvāda mode of description
1. Atom in left compartment (L)	System in state $\lvert L\rangle$	Existence (atom in L)
2. Atom in right compartment (R)	System in state $\lvert R\rangle$	Nonexistence (in L)
3. Cases (1) and (2), *at different times;* or two similar boxes at the same time	Mixture of $\lvert L\rangle$ and $\lvert R\rangle$ represented by $\lvert L\rangle\langle L\rvert + \lvert R\rangle\langle R\rvert$	Existence (in L) and Nonexistence (in L)
4. Atom in *both* compartments, at *the same time*; this wave aspect is nonvisualizable	System in a state which is superposition of $\lvert L\rangle$ and $\lvert R\rangle$ $\lvert P\rangle = \lvert L\rangle + \lvert R\rangle$	*Avayakta* (Inexpressibility)
5. (4) and (1) at different times; or two boxes at the same time, one box for (4) and another box for (1)	Mixture $\lvert P\rangle\langle P\rvert + \lvert L\rangle\langle L\rvert$	*Avayakta* and Existence (in L)
6. (4) and (2), at different times; or two boxes at the same time	Mixture $\lvert P\rangle\langle P\rvert + \lvert R\rangle\langle R\rvert$	*Avayakta* and Nonexistence (in L)
7. (4) and (3), at different times; or three boxes at the same time	Mixture $\lvert P\rangle\langle P\rvert + \lvert R\rangle\langle R\rvert + \lvert L\rangle\langle L\rvert$	*Avayakta* and Existence and Nonexistence

An object traveling with any velocity is at rest with respect to an observer traveling with the object. Syādvāda logic implies the existence of the negation of this proposition. Thus, according to Syādvāda, there must exist an entity such that to imagine an observer traveling with it must imply a logical contradiction. Syādvāda associates this with light, whose existence is the foundation of the relativity theory.

When we know that both *A* and *not-A* exist, we are ready to move on to a deeper layer or a new plane of reality corresponding to the simultaneous existence of both *A* and its negation. The new plane cannot be described in terms of the conceptual framework which described *A* and *not-A*. Syādvāda logic, indispensable for ethical and spiritual quest and for *ahimsa*, is also of the greatest value for the advancement of natural science.

For the quest of truth — scientific, moral, and spiritual — what is most important is the Syādvāda or the complementarity approach. The precise definitions and number of alternative modes are less important.

Complementarity and Marxism-Leninism

Loren Graham

Loren Graham is Professor in the
Program for Science, Technology,
and Society at the Massachusetts
Institute of Technology. He is a
specialist on science and philosophy
in the Soviet Union.

Physicists and philosophers of science do not agree on a single definition of
"complementarity." One way of defining the principle would be to say that
contradictory properties of a microbody are reconciled by granting these
individual properties existence only at separate moments of measurement.
Another formulation, which evades the question of "existence" of properties but which nonetheless is commonly given, is to say that the quantum
description of phenomena divides into two mutually exclusive classes that
complement each other in the sense that one must combine them in order to
have a complete description in classical terms. This latter view was the one
accepted by Oppenheimer when he stated that the notion of complementarity "recognizes [that] various ways of talking about physical experience
may each have validity, and may each be necessary for the adequate description of the physical world, and may yet stand in a mutually exclusive
relationship to each other, so that to a situation to which one applies, there
may be no consistent possibility of applying the other."[1] It must also be
added that even such early leaders in quantum mechanics as Niels Bohr and
Wolfgang Pauli did not entirely agree in their definitions of complementarity. The essential problem was the perennial one in the history of
science: that of giving a verbal interpretation to a mathematical relationship.

Prior to World War II the views of Soviet physicists on quantum mechanics were quite similar to those of advanced scientists elsewhere. Russian
physics was in many ways an extension of central and west European
physics. The work of such men as Bohr and Heisenberg influenced scientists in the Soviet Union as it did everywhere. Indeed, Soviet physicists
spoke of the "Russian branch" *(filial)* of the Copenhagen School, composed
of a group of highly talented theoretical physicists, including Matvei P.

Bronshtein, Lev D. Landau, Igor E. Tamm, and Vladimir A. Fock. And yet behind this exterior of agreement with scientists everywhere on quantum mechanics (or, more accurately, disagreements similar to those everywhere), as early as the 1920s certain Soviet physicists had become aware that dialectical materialism might some day be interpreted in a way that could influence their research. Lenin had, after all, devoted an entire book, *Materialism and Empirio-Criticism,* to the crisis in interpretations of physics and had particularly criticized the neopositivism of Ernst Mach, out of which much of the philosophy of modern physics grew.[2] Lenin's assertion that a dialectical materialist must recognize the existence of matter separate and independent from the mind, while not inherently contradictory to quantum mechanics, could be regarded as at least uncongenial to the Copenhagen School's disinclination to comment upon matter in the absence of sensible measurement. And the extension of the concept of complementarity beyond physics to other realms, including ethical and cultural problems, by certain members of the Copenhagen School almost guaranteed some conflict with representatives of Marxism. As early as 1929 the leading Soviet philosopher at that time, Abram M. Deborin, gave a lecture entitled "Lenin and the Crisis of Contemporary Physics" to the Academy of Sciences. But the first serious Soviet critique of the customary interpretation of quantum mechanics in a physics journal, rather than a philosophy journal, appeared in 1936, written by Konstantin V. Nikol'skii. In the dispute that developed between Nikol'skii and V. A. Fock, a leading interpreter of quantum mechanics in the Soviet Union until his death in 1974 and originally an adherent of the Copenhagen School, Nikol'skii called the Copenhagen interpretation "idealistic" and Machist," two appellations that were to be frequently used after World War II by Soviet Marxist critics.

Fock's interpretation in 1936 of the physical significance of the wave function was essentially the same as that of the Copenhagen School, which combined Born's emphasis on the mathematical description of man's knowledge of the microworld with its own emphasis on the role of measurement. Fock stated in an introduction to a Russian translation of the 1935 debate of Einstein, Podolsky, and Rosen versus Bohr: "In quantum mechanics the conception of state is merged with the conception of 'information about the state obtained as a result of a specific maximally accurate operation.' In quantum mechanics the wave function describes not the state in the usual sense, but rather this 'information about the state.'"[3] The importance of this prewar position of Fock's lies in its subtle difference from his stated views after the war, when he was placed under heavy pressure to desert the Copenhagen School. Nevertheless, Fock's change in position was small compared to the swerves that occurred in the views of several other prominent Soviet philosophers and scientists.

The debate of the 1930s did not, however, leave a permanent imprint on Soviet attitudes toward quantum mechanics. Even many philosophers accepted much of the Copenhagen view. Early in 1947 Mikhail E. Omel'ianovskii, a Ukrainian philosopher, argued a position on quantum mechanics close enough to the Copenhagen orientation to cause him intense embarrassment only a few months later. In his book, *V. I. Lenin and Twentieth-Century Physics,* Omel'ianovskii accepted much of the common interpretation of quantum mechanics.[4] He recognized and used such terms as "the uncertainty principle" and "Bohr's principle of complementarity." (A year later Omel'ianovskii's terminology became "the so-called 'principle of complementarity.'") He guarded against using these concepts in a way that might deny physical reality, as he said certain people (including Bohr on occasion) had done, but his major thesis was a defense of the surprising but necessary concepts of modern physics against adherents of the determinism of Laplace, by then clearly outdated. Buried within Omel'ianovskii's arguments, however, one may observe, at least in retrospect, the core of his own interpretation of quantum mechanics and of his later criticisms of the Copenhagen School. Although he acquiesced in the vocabulary of Copenhagen, he emphasized that the correct interpretation of quantum mechanics began with a recognition of the peculiar qualities of microparticles, not with problems of cognition: "And so we have come to the conclusion that Heisenberg's uncertainty principle, like Bohr's principle of complementarity, is a generalized expression of the facts of the dual (corpuscular and undulatory) nature of microscopic objects." Thus, the uncertainty principle was not actually an epistemological limitation or a limitation of knowledge, but a direct result of the combined wave-like and corpuscle-like nature of micro-objects, which was the *material* reason why classical concepts could not be applied to the microworld. In view of this material source of the phenomenon of canonically conjugate parameters, one could not expect ever to possess simultaneous exact values of position and momentum of elementary particles. For his recognition of the basic position of contemporary views on quantum mechanics, Omel'ianovskii was soon criticized severely, and eventually produced a second edition of his book, in which, most notably, he repudiated the principle of complementarity.

The most important event of the postwar years for Soviet scholarship was Andrei A. Zhdanov's speech on 24 June 1947, at the discussion of Georgii F. Aleksandrov's *History of Western European Philosophy,* an event well known to historians of the Soviet Union. Only near the end of that speech did Zhdanov mention specific issues in science, and less than a sentence referred directly to quantum mechanics: "The Kantian vagaries of modern bourgeois atomic physicists lead them to inferences about the electron's possessing 'free will,' to attempts to describe matter as only a certain conjunction of waves, and to other devilish tricks."[5]

Although Zhdanov's speech is now known as the beginning of the most intense ideological campaign in the history of Soviet scholarship, the Zhdanovshchina, the first few issues of the new journal *Problems of Philosophy* that appeared after the speech were surprisingly unorthodox. Evidently taking seriously the slogan of the journal—"to develop and carry further"

Beginning students sometimes find themselves explaining this principle [complementarity] with the words of the Abbé of Galiana (1728–1787): "One cannot bow in front of somebody without showing one's back to somebody else."

John A. Wheeler, address at ceremony conferring the Atoms for Peace Award on Niels Bohr, 1957

Marxist-Leninist theory — the editors promoted vital discussions of several philosophic questions. In no field was this vitality more apparent than in the philosophy of physics; the second issue of *Problems of Philosophy* contained an article by the outstanding theoretical physicist Moisei A. Markov, a specialist in the relativity theory of elementary particles, which may well still be the most outspoken presentation of the Copenhagen point of view to appear in a Soviet philosophy journal since World War II.[6]

Markov accepted modern quantum theory completely and agreed with Bohr's position in Bohr's debate with Einstein, Podolsky, and Rosen. Thus, Markov considered quantum mechanics to be complete, in the technical sense that no experiment that did not contradict it could yield results not predicted by it; and he consequently rejected all attempts to explain the behavior of microparticles on the basis of "hidden parameter" theories that would later permit restitution of the concepts of classical physics: "It is impossible to regard quantum mechanics as a classical mechanics that has been corrupted by our 'lack of knowledge.'" Such complementary functions as "momentum" and "position" simply did not have simultaneous values; to suggest that they did would mean contradicting quantum theory.

Not only was Markov's view on conjugate parameters typical of the Copenhagen School, but his approach to science bore few traits of dialectical materialism despite his initial quotations from Marxist classics. He asked that no statements be made that could not be empirically verified; he accepted relativity theory, including relativity of spatial and temporal intervals; he used the term "complementarity" without hesitation. To be sure, he affirmed that his view of science was "materialist" and criticized James Jeans and other non-Soviet commentators on science, but nowhere in his article did he make any effort to illustrate the relevance of dialectical materialism to science.

Markov maintained that "truth" is obtained from many sources. When we speak of knowledge of the microworld, which we gain with instruments, we are speaking about knowledge that has come from three sources: nature, the instrument, and man. The language we use to describe our knowledge is perforce always "macroscopic" language, since this is the only language we possess. The measuring instrument performs the role of "translating" the microphenomenon into a macrolanguage accessible to man. "We consider physical reality to be that form of reality in which reality appears in the macroinstrument." Thus, according to Markov, our concept of physical reality is subjective to the extent that it is formed in macroscopic language and is "prepared" by the act of measurement, but it is objective in the sense that physical reality in quantum mechanics is a macroscopic form of the reality of the microworld.

The role of the measuring instrument is one of the thorniest issues in quantum mechanics. Markov's view was essentially in agreement with that of the Copenhagen interpretation, according to which the wave describing a physical state spreads out over larger and larger values until a measurement is made, when a reduction of this spread (wave packet) to a sharp value occurs. Such an interpretation does indeed imply that complementary microphysical quantities have no inherent sharp values but that such values instead result from, or are "prepared by," the measurement. The most imaginative attempt, no doubt, to illustrate the striking results of some interpretations of this view of microphysical measurements was made by Schrödinger.[7] A live cat is imprisoned in a box that also contains a Geiger counter and a very weak radioactive source. If the counter fires, it actuates a relay that breaks open a container of cyanide. The source is of such a strength that there is a 50 percent chance that the Geiger counter will be triggered during the space of one hour. If the cat is sealed inside the box at a certain time — say, twelve o'clock — and then no one looks at the cat until one o'clock, are we still to assume, Schrödinger asked, that the cat is neither

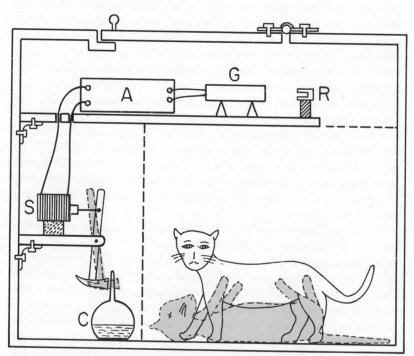

Schrödinger's cat. R indicates the radioactive source, G the Geiger counter, A the amplifier, S the solenoid relay, and C the cyanide bottle. After one hour in what Schrödinger called this "hellish contraption," there is a 50 percent chance that the Geiger counter will have been triggered by a particle from the source. In Schrödinger's words: "If at the end of an hour the cat is still living, one would say that no atom has decayed. An indication of the first decay would be the presence of equal parts of the living and the dead cat." Schrödinger was here attacking indeterminacy, which by such a situation "is transferred from the atomic to the crude macroscopic level, which can then be *decided* by direct observation. This prevents us from accepting a 'blurred model' too naively as a picture of reality."

dead nor alive until someone actually looks in the box and thus "prepares" the cat in a state of life or death?

The cat paradox raises a basic question about the difference in quantum mechanics between relational aspects and subjective aspects. It is quite possible to interpret the cat paradox from a relational standpoint in which subjective considerations play no role. The moment of interaction between the microentity and the macroentity is not the moment when the observer looks in the box, but the moment when (if indeed it happens) the Geiger counter is triggered by a particle from the radioactive source, thereby poisoning the cat. As Hilary Putnam has pointed out, contemporary physicists explain the cat paradox by saying that all macro-observables always have sharp values, and therefore the poisoning of the cat (if it occurred) was in itself a "measurement." Similarly, the moment of measurement when a photographic plate is used to record a quantum phenomenon is the moment of exposure, not the moment a human being looks at the plate. Thus, although measurement theory is still an uncertain area in the philosophy of science, measurement can be made an essential part of quantum mechanics without necessarily including subjective considerations.

I think I must accuse Bohr — though in actual fact he is one of the kindliest persons I ever came to know — of an unnecessary cruelty for his proposing to kill his victim by observation.

Erwin Schrödinger, *Science and Humanism*, 1951

Whatever Markov's views on the cat paradox, his opinion that the instrument "prepares" the state of microphysical reality, together with his acceptance of the Copenhagen interpretation in general, exposed him to criticisms from a number of quarters, ranging from dogmatic ideologues to ordinary physicists with hopes for the eventual replacement of the views of Bohr and his colleagues by an interpretation more agreeable to common-sense intuition. The Markov article very quickly became the occasion for a full-blown controversy, involving several dozen participants, on the nature of physical reality and the dialectical materialist interpretation of quantum mechanics.

The polemic began with the appearance of an article by Aleksandr A. Maksimov in the *Literary Gazette,* an unusual place for a commentary on the philosophy of science. The article, entitled "Concerning a Philosophic Centaur," contained very serious allegations against Markov.

The central point of Maksimov's article was that Markov was a supporter of the Copenhagen interpretation of quantum mechanics and was trying to whitewash this view with a few statements about its agreement with dialectical materialism. According to Maksimov, "M. A. Markov, following directly behind Bohr, asserts that in physical experiments there is a mutual influence of the microworld and the instrument, which in essence can never be overcome. However, this argument in no way touches upon the basic epistemological question: Does microreality with such properties exist before the application of an instrument by man? M. A. Markov answers that it does not exist, but is 'prepared' by the instrument."[8]

After the appearance of Maksimov's article in the *Literary Gazette,* the editors of *Problems of Philosophy* published a discussion of quantum mechanics. But the factor that determined the eventual disapproval of Mar-

There is always a certain time-lag between the views held by learned men and the views held by the general public about the views of those learned men. I do not think that fifty years is an excessive estimate for the average length of that timelag. Be that as it may, the fifty years that have just gone by — the first half of the twentieth century — have seen a development of science in general, and of physics in particular, unsurpassed in transforming our Western outlook on what has often been called the Human Situation. I have little doubt that it will take another fifty years or so before the educated section of the general public will have become aware of this change . . . *We have to labour for it.* In this labour I take my share, trusting that others will take theirs. It is part of our task in life.

Erwin Schrödinger,
Science and Humanism, 1951

kov's article was a decision, beyond any doubt promoted by the Party, to replace Bonifatii M. Kedrov as editor of *Problems of Philosophy* by Dmitrii L. Chesnokov. It is clear that Maksimov's attack on Markov played an important role in Kedrov's downfall. Maksimov's *Literary Gazette* article was a clear criticism of Kedrov, and in a statement that appeared after Kedrov's dismissal, Maksimov commented: "Only a decisive rejection of the idealistic inventions of N. Bohr and M. A. Markov, only a decisive repudiation of the position taken on this question by the editorial board of the journal *Problems of Philosophy* [in no. 1, 1948] can lead our philosophical organ out of this blind alley into which it attempted to lure several sections of our intelligentsia, those inclined to waver on the basic questions of Marxist-Leninist ideology."[9] In a note in the third issue of 1948 the reformed editorial board of *Problems of Philosophy* admitted that the journal had not taken the correct position on quantum mechanics and particularly on Markov's article, which had "weakened the position of materialism." The article had contained "serious mistakes of a philosophic character" and was in essence a departure from dialectical materialism "in the direction of idealism and agnosticism."

In terms of personnel, the immediate casualty of the Markov affair was Kedrov, but in terms of the philosophy of science, the casualty was the principle of complementarity. The period from 1948 to roughly 1960 may be called, with respect to discussions of quantum mechanics in the Soviet Union, the age of the banishment of complementarity. Only a few scientists in this period, most notably Vladimir A. Fock, attempted to include complementarity as an integral part of quantum theory.

This critical attitude toward complementarity after 1948 was made clear in an article by Iakov P. Terletskii that immediately preceded the final statement on the Markov controversy by the editors of *Problems of Philosophy*. Terletskii observed that Markov's article was actually an attempt to justify the acceptance of complementarity by maintaining that as a result of the role of measuring instruments as "translators" of reality, statements about microphysics often contradict each other. Such a view, thought Terletskii, was merely a restatement of Mach's opinion that scientists must describe nature in terms of sensations. A true dialectical materialist approach, however, showed, Terletskii continued, that the principle of complementarity was in no way a basic physical principle and that quantum mechanics could very well "get along without it."

The result of the Markov affair, then, was a victory for dogmatic ideologists. Maksimov, an ideologist, had triumphed over Markov, an active theoretical physicist in the Academy of Sciences. But it also became fairly clear that Maksimov was not capable of advancing an interpretation of quantum mechanics that held any chance of official acceptance. The period after 1948 was dominated instead by physicists and a small number of philosophers with some knowledge of physics, all of whom, however, were influenced by the atmosphere created by the Markov affair. Until approximately 1958 the major interpreter of quantum mechanics was the philosopher of science Omel'ianovskii, who drew upon the theories of the physicist Dmitrii I. Blokhintsev, advocate of the "ensemble" interpretation. Also important was Fock, who termed his interpretation a recognition of the "reality of quantum states."

One of the most complete statements of Fock's interpretation of quantum mechanics appeared in a collection of articles on philosophic problems of science published in Moscow in 1959.[10] Written at a time of relative freedom from ideological restriction, it is a statement of both scientific rigor and philosophic conviction. Fock began his discussion by considering and then dismissing attempts to interpret the wave function according to classical concepts. De Broglie's and Schrödinger's attempts originally to explain the wave function as a field spread in space, similar to electromagnetic and other previously unknown fields, were examples of classical interpretations, as was also de Broglie's later view that a field acts as a carrier of the particle and controls its movement (pilot-wave theory). Bohm's "quantum potential" was essentially the same type of explanation, since it attempted to preserve the concept of trajectory. Similarly, J. P. Vigier's concept of a particle as a point or focus in a field was an attempt to preserve classical ideas in physics. All these interpretations, according to Fock, were extremely artificial and had no heuristic value; not only did they not permit the solution of problems that were previously unsolvable, but their authors did not even attempt such solutions.

Fock believed that the true significance of the wave function began to emerge in the statistical interpretation of Max Born, especially after Bohr combined this approach with his own view of the importance of the means of observation. This emphasis on measuring instruments was essential for quantum mechanics, Fock agreed, but it was exactly on this point that Bohr also slipped: "In principle it seems that it is possible to reduce a description to the indications of instruments. However, an excessive emphasis on the role of instruments is reason for reproaching Bohr for underrating the necessity for abstraction and for forgetting that the object of study is the properties of the micro-object, and not the indications of the instruments."[11] Bohr then compounded the confusion, said Fock, by utilizing inexact terminology — terminology he was forced to invent in order to cover up the discrepancy that arose when he attempted to use classical concepts outside their area of application. One of the most important of these uses of inexact terminology was his opposition of the principle of complementarity to the principle of causality. According to Fock, if one defines terms with the necessary precision, no such opposition exists. The complementarity that *does* exist in quantum mechanics is a complementarity between *classical descriptions* and causality. But this does not deny causality in general because classical descriptions of macroparticles are necessarily inappropriate for microparticles. Using classical descriptions (macrolanguage) is merely a necessary method, since we do not have a microlanguage. Realizing that a microdescription of microparticles would be different from a classical description of the same particles, we can say that on both levels (micro- and macro-) the principle of causality holds. Since we always use a macrodescription, however, we should redefine causality in such a way that it fits both levels. Our new approach, said Fock, should be to understand causality as an affirmation of the existence of laws of nature, particularly those connected with the general properties of space and time (finite velocity of action, the impossibility of influencing the past). Causal laws can, therefore, be either statistical or deterministic. The true absence of causality in nature would mean to Fock that not even probabilistic descriptions could be given; all outcomes would be equally probable. Fock concluded his remarks on causality by commenting that in his recent conversations he had found Bohr in agreement with these observations. Thus, a few redefinitions of complementarity and causality would go far toward strengthening the Copenhagen interpretation.

The interpretation of quantum mechanics is still a very open question, not only in the Soviet Union but in all countries where there is an active concern with current problems of the philosophy of science. The Soviet discussions of causality, of the influence of the observer, and of the possibility of hidden parameters were quite similar to the worldwide controversy. In the Soviet Union the main participants in the debate — Fock, Blokhintsev, and Omel'ianovskii — all had disagreements with each other, and outside the Soviet Union the interpreters of quantum mechanics also have had intense disputes.

All scientists in the course of their investigations must proceed beyond physical facts and mathematical methods; such theorization is one of the bases of scientific explanation. Choices must be made among alternative courses that are equally justifiable on the basis of the mathematical formalism and the physical facts. The choice will often be based on philosophic considerations and will often have philosophic implications. Thus, Fock in his interpretation of quantum mechanics defined "complementarity" as a "complementarity between classical descriptions of microparticles and causality." In his subsequent choice between retaining either a classical description or causality, he chose causality, and thereby lost the possibility of a classical description. He could have gone the other way. This decision inevitably involved philosophy.

The Soviet scientists and philosophers drew attention to a significant and fruitful concept when they observed that as long as even probabilistic descriptions of nature are possible, the principle of causality can be retained. To them, the nonexistence of causality in quantum mechanics would mean that all possible values of position and momentum for a micro-object would have equal probability. In such a world, a science of quantum mechanics would be impossible. If the present opinion of most scientists is confirmed and it becomes increasingly clear that causality must be interpreted probabilistically if it is to be retained at all, the resulting discussions could lead to refreshing developments in the age-old debates over determinism and free will, particularly in the Marxist framework in which freedom is seen as knowledge of natural laws; Marxists could allow room for a given situation to generate a range of possible outcomes without resorting to any factors outside the natural world. But the full significance of quantum mechanics in its present form has not yet been adequately absorbed by specialists in other fields, Marxists or non-Marxists.

Whether the future of quantum mechanics will reassure the probabilists or the determinists will depend on science. In the meantime, Soviet philosophers and scientists have found an interpretation — or rather, several interpretations — that makes the world seem more intelligible to them and that could handle either eventuality.[12]

VII. EPILOGUE

A Glimpse of the Other Side

Philip Morrison

At the end of this commemorative volume, it is fitting to ask what limits can be placed on the utility of Niels Bohr's legacy to physicists, and to a wider circle still.

Let me begin with some anecdotal history, to establish my stance for the reader. I am not one of the generation of the Golden Age, that group of physicists who found themselves between the wars in Copenhagen, at Blegdamsvej, enchanted by the subtle quantum music of the north. (I did not visit Bohr's institute until 1959, and at that time Bohr himself was not in the city.) But I am one of the immediate generational successors to those heroic physicists; my own teachers, Robert Oppenheimer and Hans Bethe chief among them, shared the Copenhagen spirit, even if perhaps they were not steadily under its spell. During 1937, my first eye-opening year as a naive graduate student at Berkeley, Bohr himself came to that campus to give a series of lectures for the learned public. Of course, he attended Oppenheimer's series of theoretical seminars at which I was still an awed and generally uncomprehending kid in the back row.

I took careful notes at Bohr's public lectures, a bit intimidated by their content—but I wonder if I ever studied them with care? My ideas of quantum mechanics came mainly from the problems and style set by Oppenheimer, and from long hours of study of the famous Pauli article.[1] However, I have one strong conviction that does date back to Bohr's Berkeley lectures: the experimental arrangement, the classical limit that makes the inevitable cut away from description through the quantum state, is best represented by the pseudorealism of Bohr, those wonderful drawings of slits milled out of heavy brass, held to an optical bench by serious nuts and bolts. Measurement theory for me will always have that decisive background.

Philip Morrison, Institute Professor at the Massachusetts Institute of Technology, is a theoretical physicist with special interests in astrophysics and cosmology. He has also been very active for many years in the area of science education, both for schools and for the general public. His books *My Father's Watch* (with Donald Holcomb) and *Powers of Ten* (with Phylis Morrison) are expressions of this interest.

Epilogue

While Gamow and Landau were at the institute, the three of us often went to the movies together, and we had great preference for bad films. Sometimes we could entice Bohr to come with us to see a Western or a gangster film we had selected. His comments were always remarkable because he used to introduce some of his ideas on observations and measurements. Once, after a thoroughly stupid Tom Mix film, his verdict went about as follows: "I did not like that picture; it was too improbable. That the scoundrel runs off with the beautiful girl is logical; it always happens. That the bridge collapses under their carriage is unlikely, but I am willing to accept it. That the heroine remains suspended in mid-air over a precipice is even more unlikely, but again I accept it. I am even willing to accept that at that very moment Tom Mix is coming by on his horse. But that at that very moment there should be a fellow with a motion picture camera to film the whole business — that is more than I am willing to believe."

H. B. G. Casimir, "Recollections from the Years 1929–1931," 1967

It was at Los Alamos that I came into more direct contact with Bohr. Again I was no intimate; but since I was close to Oppenheimer and to Berkeley classmates (such as Robert Wilson and Bob Christy) I found myself quite often in a circle around the Bohrs, Niels and Aage, both at work and at evening gatherings. My political stance about the bomb and its international meaning owed much to Bohr's ideas, filtered for most of us through Oppenheimer. Bohr's public views persisted a little time into the postwar world, and the idea of an international agency charged not only with the policing task but also with the positive development of nuclear science and technology caught me up for a while. It is surely a conception that goes back to the interaction of Bohr and Oppenheimer. The chill of the Cold War caused me, an old campus radical, to shiver rather early. The idealism of the war's end withered in the frosty political context of the late forties, and soon the urgent claims of national security, East and West, put a stop to Bohr's "open world," the hope for a universal nuclear community.

One Sunday afternoon in the late summer of 1944, many newcomers to the mesa country made the modest drive over the Jemez Mountains from Los Alamos to Jemez Pueblo, to witness the marvelous ancient dances held there. Bohr came too, and I watched him within that scene of unpredictable contrasts with the sense of wonder so often induced by the omens and incongruities of wartime Los Alamos. It was not long afterward that my first few hours of personal contact with Bohr took place. Somehow it fell to me to offer to accompany Bohr to a Sunday evening film at the post theater. It was a large-scale Western, perhaps with John Wayne or some similar star of gun and saddle. I undertook to interpret the peculiar genre of the Hollywood Western to Bohr the European, and we sat together as I told him how to watch for white hats and black, how to part the good guys from the bad. He seemed to enjoy the film, and to appreciate my whispered interpretations. It may have been a decade before I read recollections by others of their years in Copenhagen, and realized that Bohr possessed substantial critical expertise on Westerns, arising out of long study. My help had been as superfluous as it was ingenuous, but his warmth and insight led him to keep his secret: subtle as always.

A GENERATION OF NATURAL SCIENCE SINCE BOHR

Niels Bohr died in 1962. It is not hard, in retrospect at least, to list the chief discoveries in the natural sciences that have grown clearly visible in the years since Bohr's work ended.

The oldest of these — the key findings go back to the early fifties — was the rise of modern biology, especially the genetic biology of information storage and transfer, based upon the double helix and its template method of reproduction. A generation after the double helix, the science has acquired great powers, both intellectual and technological. It has opened the way to the understanding of the development of multicellular organisms, a goal still not close at hand.

Niels Bohr in about 1962, photographed by Lotte Meitner-Graf, sister-in-law of Lise Meitner and aunt by marriage of Otto Frisch.

Next, as the sixties opened, came knowledge of sea-floor spreading. Eventually it was convincingly shown that the motion of distinct plates of continental size, which together form the earth's outer crust, must be a steady feature of earth history. For the first time, geology became open to a broad theoretical account, more or less as a century ago paleontology became clarified as the consequence of Darwin's theory of evolution. Whereas previously a special history had had to be invoked for every region and for every period, even though the physical processes were roughly similar over time, now the whole globe had a unitary albeit complex history. Geologists who hardly knew that the world beyond the domain of their special interest existed now saw the earth as one interactive system.

Also in the sixties, the ideas of relativistic cosmology were extrapolated through aeons of history, back to a time when the universe appeared qualitatively different from what it is today. The discovery of quasars — discrete galaxy-like bodies bright enough to be seen at distances corre-

347

sponding to red-shifts several times greater than any known for ordinary galaxies — made plain that the universe had a long history. By the end of the decade there remained little doubt that the distant microwave background was a residue of the ubiquitous early presence of space-filling plasma — matter directly observable that showed complete uniformity and isotropy, and gave no sign of discrete condensed objects. This provided evidence for the long-predicted early hot universe, seat of the synthesis of helium in cosmic space, whereas all subsequent nucleosynthesis would be within stars and galaxies not yet gravitationally assembled out of the uniform hydrogen-helium plasma.

Bohr himself would probably have rejoiced most over the great advances in particle physics. From the end of World War II until about 1960, experiments disclosed a plethora of candidates for designation as fundamental particles. But for some time there was little understanding of these results. Then, beginning with the successes of the algebraic grouping called the Eight-fold Way, theorists rapidly developed schemes to explain the entire array of particle states we now know. That understanding rests squarely upon advances in quantum field theory — advances that have brought about the demonstrated unification of the electromagnetic and the weak nuclear interactions. Then came the recognition, perhaps less well established but compelling all the same, that the strong nuclear interactions too could be understood on the basis of quarks, which are believed to be the inner constituents of neutrons, protons, and other strongly interacting particles. Quarks now seem so real (despite the fact that, according to the theorists, they can never be found free) that their involvement at energies far beyond experimental reach, as described by the so-called grand unified theories, lays widely recognized claim to good sense. Today the edifice of particle physics is quite impressive, with only a few hints from experiment that closure of the physics of accessible energy might not be final.[2]

ASSESSING COMPLEMENTARITY

How have Bohr's general views — his principle of complementarity, his concern with correct language and the recognition of its limits — stood up throughout the decades of such scientific novelty?

The answer seems, fittingly enough, to be itself dual. The new biology, wonderfully apt, and growing by leaps and bounds in scope and depth, is positively anti-Copenhagen. We can see this with some confidence, for in his lecture "Light and Life" Bohr gave us insight into how he expected a biology of power to look. He knew, of course, that it would require molecular analysis if it were to join the rest of science; life depends on fine structure. But certainly, at the time Bohr gave the lecture, he believed that we would require a new viewpoint on chemistry, on reactions and structures. That viewpoint, he suggested, would come out of the exploitation of the complementarity between the microdescription of an organism and its living function.

Before Bohr died, he came to realize that his initial view was wrong. Something must be added to chemistry. That novelty lies in two directions. Theoretically, we need to follow not only free energies and structures in the usual way, but we must keep in mind the category information. Replication of pattern and an intricate set of variations on that theme are key matters. Moreover, there is a structural level beyond the simple collections of sequenced monomers. Polymer chemistry is evident, but probably the work of membranes and other components of cells goes beyond that, toward surface structure, closer to solid-state physics than to test-tube reactions. Yet nothing appears that is more subtle: no new interactions, no dynamics achieved only at the expense of form. It looks as though simple close fit of enzyme molecules and substrates, sometimes with ancillary molecular partners, will account for the reaction chains and cycles. It looks, too, as though the origin of secondary and tertiary structure arises from the long sequences that make up the key polymers. Normal collisional processes allow eventual use of secondary and tertiary bonding to build the architecture beyond the long chains. Cunning evolutionary design is enough to construct such molecular mechanisms as ribosomes out of molecules specified only chemically, although at tedious length and detail. Schrödinger's metaphor of the aperiodic crystal seems a better hint than any of Bohr's more dynamic suggestions — or so we think today. Life ought to exhibit complementarity, if anything does, but that subtlety enters perhaps beyond the molecular level, in evolutionary processes and in learning systems associated with neural networks.[3]

It is almost self-evident that plate tectonics, even though we do not know much about its driving forces, is a mechanism of a fully Newtonian kind. At that scale, there is not much room for subtler views. Yet how slow we were to recognize these worldwide processes that can be described in Cartesian language.

But in the physics of particles, quantum theory — quantum field theory — conceptually complete by the 1930s, seems to have found flawless application up to energies of a hundred thousand electron masses at least. The theorists of today have been original and ingenious; their daring in extending the old ways is a marvel. Plainly, dynamical quantities go well beyond ordinary space-time, but they fit the ideas of quantum field theory like a hand in a snug glove. When Bohr and Rosenfeld made their painstaking study of the measurement physics of the vacuum fluctuations implied by quantum electrodynamics, they showed how quantum complementarity ideas held in strength, and how these ideas were required by, and in turn illuminated, the meaning of the formal theory. That all still holds true. It is even the case that the dynamics of the vacuum — a true quantum ether — is physically more and more important. Some day we will extend the wonderful complementary relationship that Bohr and Rosenfeld showed to exist between the assignment of sharp particle numbers and the description of the fields.

Today the cosmology of the early universe is thoroughly integrated with

It is of decisive importance to realize that the fruitful application of structural formulae in chemistry rests solely on the fact that the atomic nuclei are so much heavier than the electrons that, in comparison with molecular dimensions, the indeterminacy in the position of the nuclei can be largely neglected.

Niels Bohr,
Rutherford Memorial Lecture, 1958

particle physics. Not only do the conjectured high temperatures of that early state make the relationship inevitable (if the extraordinary extrapolation of the expansion holds as expected), but even more the union leads to testable explanations for the symmetries of the universe we see. Here the ground is so newly entered that we must be tentative. But I can recall from the dim past of 1937 Bohr's firm opinion, when he was asked for comment upon the then entirely geometrical relativistic cosmologies of the day, that we would understand the universe as a whole only when we understood in depth the nature of the fundamental particles; the trouble was that there was as yet no recognized context which the two subjects shared. Now we have found that context (if we are right) in the early expansion and—a straightforward interpretation of evidence now at hand—in the unexpected uniformity and isotropy of early matter as a whole.

The history of particle physics is an account of splendid successes, old and new, and of shortcomings and limitations too, to be expected in any single work of our finite species. They built a city shining upon a hill, they who half a century ago founded the quantum mechanics of particles and fields, and the design of this city is understood best through Niels Bohr.

Some Closing Reflections

A. P. French

In 1885, when Niels Bohr was born, a number of scientists (although admittedly a minority) still considered atoms to be no more than a theoretical recipe for organizing our experience. For example, the chemist August Kekulé (of benzene-ring fame) once said, "The question of whether atoms exist or not has but little significance from a chemical point of view; its discussion belongs rather to metaphysics." And Bohr was already more than ten years old when the electron was discovered, and with it the beginning of an awareness that the structure of matter must be wildly different from anything hitherto imagined. The discovery of the quantum — not to speak of the mystery of radioactivity — added further complexity to a confused and confusing situation. Most physicists did not realize it, but their traditional ways of modeling nature in classical and largely mechanical terms were doomed to failure when it came to the atom. Only someone who had the genius and the courage to introduce radically new ideas would be able to usher in a new physics that was adequate to the situation. As we know, it fell to Niels Bohr to accomplish this.

The great Rutherford had of course set the stage with his discovery of the nucleus, but it was astonishing to see how quickly Bohr built on that discovery. As we have seen, even before he began thinking about electron orbits, he had conceived the idea that the nucleus and its electric charge were the clue to the existence of physically different but chemically identical atoms — what Frederick Soddy later called "isotopes." Bohr was, in fact, too far ahead of the evidence for Rutherford, who urged caution and so, perhaps, denied Bohr the credit for being the discoverer of isotopy. But Bohr apparently felt no grudge and, although he could not have known it, glory in plenty awaited him.

It is not my purpose here to recapitulate what has been so amply pre-

A. P. French received his professional education at Cambridge University and taught there for a number of years. During World War II he was a member of the British and American nuclear bomb projects. He is now Professor of Physics at the Massachusetts Institute of Technology, with a particular interest in education and physics history.

351

sented and discussed in this book, but rather to consider what sort of a man this was who changed the face of physics — not once but twice, or perhaps more than that. He was, it is clear, a daring innovator; yet as John Heilbron has described in his article, Bohr's way was to keep all the facts in the picture, and never to discard without good reason. According to a well-known story, the young Bohr, age eleven, was given an assignment at school to draw a certain house with a fenced garden. Niels's drawing was extremely precise: gables and roof peaks receded in perfect perspective, and trees were drawn in exquisite leafy detail. When he came to the fence, Niels broke off drawing long enough to run up to count the actual pickets, lest his drawing depart in any significant way from the reality it depicted. This insistence on keeping in touch with *all* the facts of nature became a life-long characteristic. Because of it, Bohr never wandered, as Einstein did in his later years, into isolated areas of thought which, however elegant and beautiful, were not ultimately fruitful.

Nevertheless, Bohr's deep philosophical concern with the *process* of knowing and describing nature was ignored or dismissed by a large fraction of physicists. Quantum mechanics, once developed, was a supremely successful tool for analyzing atomic and nuclear phenomena. What need was there to worry about what might seem to be academic, even metaphysical questions? But for Bohr, such concerns were absolutely basic. For him the central problem was to locate the boundary separating the subject doing the thinking from the object of the thought. In his 1929 essay "The Quantum of Action and the Description of Nature," Bohr refers to the situation of someone in a darkened room trying to orient himself with the help of a stick. "When the stick is held loosely, it appears to the sense of touch to be an object. When, however, it is held firmly, we lose the sensation that it is a foreign body, and the sensation of touch becomes immediately localized at the point where the stick is touching the body under investigation." This problem of setting boundaries appears in its sharpest form, as Aage Petersen has discussed, when we think about our own thinking activity. Here subject and object appear to alternate in an infinite regression. Bohr was fascinated by the presentation of the dilemma in Paul Møller's fable *Tale of a Danish Student.* A student, contemplating his consciousness at work considering itself, is dismayed by the vision: "In short, thinking becomes dramatic and quietly acts out the most complicated plots with itself, and the spectator again and again becomes actor." Bohr would quote this sentence often in later years.

This might have seemed a topic which, however interesting as abstract philosophy, had little to do with the sharp-edged realities of quantitative physical measurements. But more and more we have come to see the profundity of Bohr's insights in such matters. He would insist that physics does not deal with nature per se, but only with what we can *say* about nature; language, and the limitations of language, are forever present. And when it comes to an actual process of measurement, and the choice of a line

Physicists from all over the world came to Copenhagen to work with Niels Bohr and spread his way of thinking, and naturally I do the same, having worked there for five years. But I think it is quite wrong to regard the Copenhagen School as an establishment set on perpetuating the views of its founder. Bohr's views have been debated and criticized for half a century, and no alternative has seriously shaken them. Newton may be accused of having suppressed the wave theory of light, simply by his great prestige and by not sharing the view of his older contemporary, the Dutchman Christian Huygens. Will the future level a similar accusation against Bohr? I doubt it; but only the future can tell.

O. R. Frisch,
What Little I Remember, 1979

of demarcation between observer and observed, then real and testable consequences emerge, as in the delayed-choice interference experiments for photons, which have shown that, in this matter at least, Bohr was right and Einstein was wrong. When Einstein declared his conviction that God does not throw dice, Bohr's response was that we can know only what observation tells us, and (as Heisenberg has quoted him) "it is not our business to prescribe to God how he should run the world."

Nothing could be further from the public stereotype of the scientist than Bohr's life and personality. Both professionally and personally, he radiated warmth and humanity. The physicists who as young men came to Copenhagen to work with Bohr speak of his having shaped their characters as well as their minds. Contrary to the image of the theorist as solitary thinker (which indeed was largely correct for Einstein), Bohr could not function without constant interaction and exchange with others. By endless, passionate, occasionally boisterous discussions, he groped his way forward to an understanding of how things must be. In the process, he made an indelible mark on scores of others. Otto Frisch has recalled the quality of those sessions: "The most personal contacts we had on those frequent occasions when Bohr had invited a number of us out to Carlsberg where, sipping our coffee after dinner, we sat close to him — some of us literally at his feet, on the floor — so as not to miss a word. There, I felt, was Socrates come to life again, tossing us challenges in his gentle way, lifting each argument onto a higher plane, drawing wisdom out of us which we didn't know was in us (and which of course wasn't). Our conversations ranged from religion to genetics, from politics to art; and when I cycled home through the streets of Copenhagen . . . I felt intoxicated with the heady spirit of Platonic dialogue." Through such experiences, Bohr instilled and transmitted a whole attitude to physics and to life that has had a lasting influence on the scientific world.

When one looks at Bohr's personal achievements in the theory of the atom, in the fundamentals of quantum mechanics, and in nuclear physics, it is plain that he did more, probably, than any other single person to create what we call modern physics. And yet there is a strong temptation to agree with Sir George Thomson's comments (quoted at the very beginning of this book): "[Bohr] led science through the most fundamental change of attitude it has made since Galileo and Newton, by the greatness of his intellect and the wisdom of his judgments. But quite apart from their unbounded admiration for his achievements, the scientists of all nations felt for him an affection which has perhaps never been equalled. What he was counted for even more than what he had done." It would, however, be pointless and inappropriate to try to analyze or assess that final judgment. We only hope that in this book we have managed to do justice to both of these aspects of a very great man.

It was no wonder that so many eminent natural scientists emerged from his classes. He was one of those rare teachers who know how to apply caution and, if necessary, force, to the task of liberating the slumbering genius in each man's mind. Like Socrates, whose expression of thought through dialogue he considered exemplary, Bohr was a midwife of ideas.

C. F. von Weizsäcker, quoted in Robert Jungk, *Brighter Than a Thousand Suns,* 1956

Notes

A SHORT BIOGRAPHY

1. Léon Rosenfeld, "Niels Bohr's Contribution to Epistemology," *Physics Today* 16, no. 10 (1963): 47–54.

2. The phrase "the Great Trilogy," as a description of the 1913 papers, was originated by Léon Rosenfeld in the introductory essay he wrote for a 1963 reprint of these papers. See Niels Bohr, *On the Constitution of Atoms and Molecules* (Copenhagen: Munksgaard, 1963; New York: W. A. Benjamin, 1963), introduction by Rosenfeld.

3. The Solvay conferences were established by the Belgian industrial chemist and social reformer Ernest Solvay (1838–1922). He had a deep interest in fundamental physics, and as a result of meeting the German physical chemist Walther Nernst in 1910 he gave his backing to the organization of an international conference which was held in Brussels, 30 October–3 November 1911. Its topic was "Radiation Theory and Quanta." In 1912 Solvay established the Instituts Internationaux de Physique et de Chimie, which made it possible for subsequent conferences to be held every few years (except when Europe was disrupted by war) from that time on. Bohr's last published address was a fine historical review, delivered at the Twelfth Solvay Conference (October 1961), of these meetings. The address is reprinted in Bohr, *Essays, 1958–1962, on Atomic Physics and Human Knowledge* (New York: Wiley-Interscience, 1963, pp. 79–100).

4. This article is reprinted in full in Part V of this volume.

5. Other biographical references: American Physical Society, "Niels Bohr Memorial Session," *Physics Today* 16, no. 10 (1963): 21–64 (includes contributions by J. Rud Nielsen, Felix Bloch, Aage Bohr, John A. Wheeler, Léon Rosenfeld, and Victor F. Weisskopf]; John D. Cockcroft, "Niels Henrik David Bohr," in *Biographical Memoirs of Fellows of the Royal Society* 9 (1963): 37–49; Ruth Moore, *Niels Bohr: The Man and the Scientist* (New York: Knopf, 1966; rpt. Cambridge, Mass.: MIT Press, 1985); Léon Rosenfeld, "Niels Bohr: Biographical Sketch," in Niels Bohr, Collected Works, vol. 1, ed. J. Rud Nielsen (Amsterdam: North-Holland, 1972), pp. xvii–xlviii; Stefan Rozental, ed., *Niels Bohr: His Life and Work as Seen by His Friends and Colleagues* (Amsterdam: North-Holland, 1967) [a treasurehouse of affectionate memoirs].

A PERSONAL MEMOIR

1. This reminiscence is based on excerpts from James Franck, "Niels Bohrs Persönlichkeit," *Naturwissenschaften* 9 (1963): 341–343. Trans. by the editors, with the assistance of Brigitte and Stephen Steadman.

2. This refers to what quickly became known as the Franck-Hertz experiment, which gave the first direct evidence of discrete energy levels in atoms. Franck and Hertz shared the Nobel Prize for physics in 1925 for this discovery.

NIELS BOHR, THE QUANTUM, AND THE WORLD

1. "Light and Life," an address given at the opening meeting of the International Congress on Light Therapy, Copenhagen, August 1932. The address is reprinted in Part VI of this book.

2. For additional reading, see: Henry A. Boorse and Lloyd Motz, eds., *The World of the Atom,* 2 vols. (New York: Basic Books, 1966) [a massive anthology of selections from the original literature, with commentaries, covering the whole history of atomic and quantum physics, from Lucretius to elementary-particle physics in the 1960s]; Wolfgang Pauli, ed., *Niels Bohr and the Development of Physics* (Oxford: Pergamon, 1955), esp. the essays by C. G. Darwin, Werner Heisenberg, Francis L. Friedman and Victor F. Weisskopf, and John A. Wheeler; Victor F. Weisskopf, *Physics in the Twentieth Century* (Cambridge, Mass.: MIT Press, 1972).

BOHR'S FIRST THEORIES OF THE ATOM

1. This account of Bohr's work abbreviates that given in J. L. Heilbron and Thomas S. Kuhn, "The Genesis of the Bohr Atom," *Historical Studies in the Physical Sciences* 1 (1969): 211–290. A lively portrait of Bohr during the period covered in this chapter can be found in Léon Rosenfeld and Erik Rüdinger, "The Decisive Years, 1911–1918," in Stefan Rozental, ed., *Niels Bohr: His Life and Work as Seen by His Friends and Colleagues* (Amsterdam: North-Holland, 1967), pp. 38–73.

2. An overview of the early history of theories of atomic structure is J. L. Heilbron, "Lectures on the History of Atomic Physics, 1900–1922," in Charles Weiner, ed., *History of Twentieth-Century Physics* (New York: Academic Press, 1977), pp. 40–108.

3. The quotations by Bohr in this article are taken from his *Collected Works,* ed. Léon Rosenfeld et al. (Amsterdam: North-Holland, 1972–).

4. From Haas's expression $h = 2\pi e(ma)^{1/2}$ it follows that $a = h^2/4\pi^2 e^2 m$, precisely the value later obtained by Bohr, from quite different premises, for the radius of the ground-state orbit in hydrogen. That the radius in both cases must be proportional to h^2/me^2 follows from the dimensions of the atomic constants. It is only a coincidence, but a striking one, that Haas's calculations, based on the Thomson model, resulted in the same numerical value as Bohr's calculations based on the Rutherford model.

5. Paul Langevin and Maurice de Broglie, eds., *La Théorie du rayonnement et des quanta* (Paris: Gauthier-Villars, 1912).

6. Rutherford was justifiably proud of the beautiful experiments by which he (with Hans Geiger) showed directly that alpha particles have charge $2e$ and (with T. Royds) proved spectroscopically that neutralized alpha particles are helium atoms. His Nobel Prize lecture in 1908 (for chemistry) was devoted to these topics.

THE THEORY OF THE PERIODIC SYSTEM

1. Louis de Broglie worked with his older brother Maurice, a famous X-ray spectroscopist, before making his own name in wave mechanics.

2. Further historical details about Bohr's theory of the periodic system can be found in Helge Kragh, "Niels Bohr's Second Atomic Theory," *Historical Studies in the Physical Sciences* 10 (1979): 123–186. The celtium-hafnium priority conflict is analyzed in Helge Kragh, "Anatomy of a Priority Conflict: The Case of Element 72," *Centaurus* 23 (1980): 275–301. For Stoner's and Pauli's contributions, see J. L. Heilbron, "The Origins of the Exclusion Principle," *Historical Studies in the Physical Sciences* 13 (1983): 261–310.

For additional information, see: Niels Bohr, *Collected Works,* vol. 4, ed. J. Rud Nielsen (Amsterdam: North-Holland, 1977); A. d'Abro, *The Rise of the New Physics,* vol. 2 (New York: Dover, 1951); Hendrik A. Kramers and H. Holst, *Das Atom und die Bohrsche Theorie seines Baues* (Berlin: Julius Springer, 1925); Edward M. MacKinnon, *Scientific Explanation and Atomic Physics* (Chicago: University of Chicago Press, 1982); Jagdish Mehra and Helmut Rechenberg, *The Historical Development of Quantum Theory,* vol. 1 (New York: Springer Verlag, 1982); Peter Robertson, *The Early Years: The Niels Bohr Institute, 1921–1930* (Copenhagen: Akademisk Forlag, 1979); Stefan Rozental, ed., *Niels Bohr: His Life and Work as Seen by His Friends and Colleagues* (Amsterdam: North-Holland, 1967).

BOHR AND RUTHERFORD

1. The Kapitza Club was an informal discussion club in Cambridge, made up mainly of young research physicists and named in honor of Peter Kapitza, who worked with Rutherford at the Cavendish Laboratory.

2. The Russell-Einstein manifesto, issued in July 1955, was a peace appeal publicly addressed to the governments of the world. Its chief architects were Bertrand Russell and Albert Einstein.

Bohr's Rutherford Memorial Lecture (1958) gives a fascinating and deeply human account of his relationship with Rutherford and of his own scientific development. It was originally published in *Proceedings of the Physical Society* (London) 78 (1961): 1083–1115; reprinted in Niels Bohr, *Essays, 1958–1962, on Atomic Physics and Human Knowledge* (New York: Wiley-Interscience, 1963).

BOHR, GÖTTINGEN, AND QUANTUM MECHANICS

1. An English translation of these lectures can be found in Niels Bohr, *Collected Works,* vol. 4, ed. J. Rud Nielsen (Amsterdam: North-Holland, 1977), pp. 341–419.

2. This is a reference to the classic work by Arnold Sommerfeld, *Atombau und Spektrallinien,* first published in 1919 (Braunschweig: Vieweg). It was soon translated into English: *Atomic Structure and Spectral Lines* (New York: Dutton, 1922), trans. Henry L. Brose. Both versions went through several editions to keep up with the development of quantum theory.

3. See also Werner Heisenberg's reminiscence, "The Beginnings of Quantum Mechanics in Göttingen," in Heisenberg, *Tradition in Science* (New York: Seabury, 1983), pp. 37–55.

THE TRILOGY

1. From Niels Bohr, *Collected Works,* vol. 1, ed. J. Rud Nielsen (Amsterdam: North-Holland, 1972), p. 559.

2. Niels Bohr, "Reminiscences of the Founder of Nuclear Science and of Some

Developments Based on His Work," *Proceedings of the Physical Society* (London) 78 (1961): 1086, 1084.

3. Ibid., p. 1092.

NOBEL PRIZE LECTURE

1. This lecture is reprinted in full in the following publications: *Nobel Lectures: Physics, 1922–1941* (Amsterdam: Elsevier, 1965), pp. 7–43; Niels Bohr, *Collected Works*, vol. 4, ed. J. Rud Nielsen (Amsterdam: North-Holland, 1977), pp. 467–482; *Nature* 112 (1923): 29–44; *Naturwissenschaften* 11 (1923): 606–624. The excerpts reproduced here are taken from *Nobel Lectures: Physics, 1922–1941*.

BOHR ON THE FOUNDATIONS OF QUANTUM THEORY

1. Niels Bohr, Hendrik A. Kramers, and John C. Slater, "The Quantum Theory of Radiation," *Philosophical Magazine* 47 (1924): 785–802; "Über die Quantentheorie der Strahlung," *Zeitschrift für Physik* 24 (1924): 69–87.

2. Paul A. Hanle, "The Schrödinger-Einstein Correspondence and the Sources of Wave Mechanics," *American Journal of Physics* 47 (1979): 644–648; Linda Wessels, "Schrödinger's Route to Wave Mechanics," *Studies in the History and Philosophy of Science* 10 (1977): 311–340.

3. Mara Beller, "Matrix Theory before Schrödinger: Philosophy, Problems, Consequences" *Isis* 74 (1983): 469–491.

4. For additional information, see: Gerald Holton, "The Roots of Complementarity," in Holton, *Thematic Origins of Scientific Thought* (Cambridge, Mass.: Harvard University Press, 1973), pp. 115–161; John Honner, "The Transcendental Philosophy of Niels Bohr," *Studies in the History and Philosophy of Science* 13 (1980): 1–29; Max Jammer, *The Philosophy of Quantum Mechanics* (New York: Wiley, 1974), esp. chs. 3–6; Martin J. Klein, "The First Phase of the Bohr-Einstein Dialogue," *Historical Studies in the Physical Sciences* 2 (1970): 1–39; Edward MacKinnon, *Scientific Explanation and Atomic Physics* (Chicago: University of Chicago Press, 1982); Niels Bohr, *Collected Works*, ed. Léon Rosenfeld et al. (Amsterdam: North-Holland, 1972–); Paul A. Schilpp, ed., *Albert Einstein: Philosopher-Scientist* (Evanston, Ill.: Library of Living Philosophers, 1949), esp. pp. 1–95.

THE BOHR-EINSTEIN DIALOGUE

1. Paul A. Schilpp, ed., *Albert Einstein: Philosopher-Scientist* (Evanston, Ill.: Library of Living Philosophers, 1949).

2. The excerpt is taken from Schilpp, *Albert Einstein*, pp. 208–235 and 240–241. Reprinted by permission of the Open Court Publishing Company, La Salle, Illinois, copyright 1949, 1951, and © 1969 by the Library of Living Philosophers, Inc.

3. Werner Heisenberg, "Über den anschaulichen Inhalt der quanten-theoretische Kinematik und Mechanik" [The descriptive content of quantum-theoretical kinematics and mechanics] *Zeitschrift für Physik* 43 (1927): 172–198.

4. Published in *Atti del Congresso Internazionale dei Fisici*, Como, September 1927; reprinted in *Nature* 121 (1928): 580–590.

5. Institut International de Physique Solvay, *Rapports et discussions du 5ᵉ Conseil* (Paris: Gauthier-Villars, 1928), pp. 253–256.

6. Ibid., pp. 248–289.

7. Albert Einstein, Boris Podolsky, and Nathan Rosen, "Can Quantum-Mechanical Description of Physical Reality Be Considered Complete?" *Physical Review* 47 (1935): 777–780.

8. Niels Bohr, "Can Quantum-Mechanical Description of Physical Reality Be Considered Complete?" *Physical Review* 48 (1935): 696–702.

9. Albert Einstein, "Physik und Realität," *Journal of the Franklin Institute* 221 (1936): 313–347; "Physics and Reality," ibid., pp. 349–382 (trans. Jean Piccard).

A BOLT FROM THE BLUE

1. Albert Einstein, Boris Podolsky, and Nathan Rosen, "Can Quantum-Mechanical Description of Physical Reality Be Considered Complete?" *Physical Review* 47 (1935): 777–780.

2. David Bohm, *Quantum Theory* (Englewood Cliffs, N.J.: Prentice-Hall, 1951), pp. 614–619.

3. Alain Aspect, Philippe Grangier, and Gérard Roger, "Experimental Tests of Realistic Local Theories via Bell's Theorem," *Physical Review Letters* 47 (1981): 460–463; idem, "Realization of E-P-R-Bohm Gedankenexperiment: A New Violation of Bell's Inequalities," *Physical Review Letters* 49 (1982): 91–94; Alain Aspect, Jean Dalibard, and Gérard Roger, "Experimental Test of Bell's Inequalities Using Time-Varying Analyzers," *Physical Review Letters* 49 (1982): 1804–1807.

4. Léon Rosenfeld, in Stefan Rozental, ed., *Niels Bohr: His Life and Work as Seen by His Friends and Colleagues* (Amsterdam: North-Holland, 1967), pp. 114–136.

5. Niels Bohr, "Can Quantum-Mechanical Description of Physical Reality Be Considered Complete?" *Physical Review* 48 (1935): 696–702.

6. John S. Bell, "On the Einstein-Podolsky-Rosen Paradox," *Physics* (New York) 1 (1964): 195–200.

7. Abraham Pais, *"Subtle is the Lord": The Science and the Life of Albert Einstein* (New York: Oxford University Press, 1982), p. 455.

8. Henry P. Stapp, "Are Superluminal Connections Necessary?" *Nuovo Cimento* 40B (1977): 191–204.

9. Albert Einstein, "Physik und Realität," *Journal of the Franklin Institute* 221 (1936): 313–347; "Physics and Reality," ibid., pp. 349–382 (trans. Jean Piccard).

10. See the detailed but quite elementary discussion of this example in N. David Mermin, "Bringing Home the Atomic World: Quantum Mysteries for Anybody," *American Journal of Physics* 49 (1981): 940–943.

11. Albert Einstein, "Quanten-Mechanik und Wirklichkeit," *Dialectica* 2 (1948): 320–323; reprinted (in English) as "Quantum Mechanics and Reality," in *The Born-Einstein Letters*, pp. 168–173.

DELAYED-CHOICE EXPERIMENTS

1. Niels Bohr, "Discussion with Einstein on Epistemological Problems in Atomic Physics," in Paul A. Schilpp, ed., *Albert Einstein: Philosopher-Scientist* (Evanston, Ill.: Library of Living Philosophers, 1949), p. 222.

2. Warner A. Miller and John A. Wheeler, "Delayed-Choice Experiments and Bohr's Elementary Quantum Phenomenon," in S. Kamefuchi, ed., *Proceedings of the International Symposium on Foundations of Quantum Mechanics* (Tokyo: Physical

Society of Japan, 1983), pp. 140–151; John A. Wheeler, "The 'Past' and the 'Delayed-Choice' Double-Slit Experiment," in A. Ransom Marlow, ed., *Mathematical Foundations of Quantum Theory* (New York: Academic Press, 1978), pp. 9–48.

3. Carroll O. Alley, A. Jakubowicz, C. A. Steggerda, and W. C. Wickes, "A Delayed Random Choice Quantum Mechanics Experiment with Light Quanta," in Kamefuchi, *Proceedings*, pp. 158–164.

4. Warner A. Miller, "A Proposed Delayed-Choice Experiment Using Ultra-Cold Neutrons," in Kamefuchi, *Proceedings*, pp. 153–157.

5. John A. Wheeler, "Beyond the Black Hole," in Harry Woolf, ed., *Some Strangeness in the Proportion* (Reading, Mass.: Addison-Wesley, 1980), p. 356.

6. Daniel M. Greenberger, Michael A. Horne, Clifford G. Shull, and Anton Zeilinger, "Delayed-Choice Experiments with the Neutron Interferometer," in Kamefuchi, *Proceedings*, pp. 294–299.

7. John A. Wheeler, "Genesis and Observership," in Robert E. Butts and Jaakko Hintikka, eds., *Foundational Problems in the Special Sciences* (Dordrecht, Netherlands: D. Reidel, 1977), pp. 3–33; idem, "The 'Past' and the 'Delayed-Choice' Double-Slit Experiment;" idem, "Delayed-Choice Experiments and Bohr's Elementary Quantum Phenomenon," in Kamefuchi, *Proceedings*, pp. 147–149.

8. John A. Wheeler, "Law without Law," in John A. Wheeler and Wojciech H. Zurek, eds., *Quantum Theory and Measurement*, (Princeton: Princeton University Press, 1983), p. 194.

9. Readers wishing to pursue the discussions opened by the Bohr-Einstein dialogue and to look through "the magic window of physics" into the "participatory universe," the "self-excited universe," or the "many worlds" will find a rich collection of articles (including the complete text of Bohr's "Discussion with Einstein") in Wheeler and Zurek, *Quantum Theory and Measurement*.

ON BOHR'S VIEWS CONCERNING THE QUANTUM THEORY

1. This essay is reprinted from Ted Bastin, ed., *Quantum Theory and Beyond* (Cambridge: Cambridge University Press, 1971). It was originally presented as an address to a colloquium entitled "Quantum Theory and Beyond," held at Cambridge, England, in July 1968.

2. Niels Bohr, *Atomic Theory and the Description of Nature* (Cambridge: Cambridge University Press, 1934); idem, *Atomic Physics and Human Knowledge* (New York: Wiley, 1958).

3. Albert Einstein, Boris Podolsky, and Nathan Rosen, "Can Quantum-Mechanical Description of Physical Reality Be Considered Complete?" *Physical Review* 47 (1935): 777–780.

4. Niels Bohr, "Can Quantum-Mechanical Description of Physical Reality Be Considered Complete?" *Physical Review* 48 (1935): 696–702.

5. David Bohm, "On the Role of Hidden Variables in the Fundamental Structure of Physics," in Bastin, *Quantum Theory and Beyond*, pp. 95–116.

6. Ibid.

7. Ibid.

8. A later discussion of the ideas treated in this essay can be found in David Bohm, *Wholeness and the Implicate Order* (London: Routledge and Kegan Paul, 1980).

WAVES AND PARTICLES: 1923–1924

1. This essay has been excerpted from John C. Slater, *Solid-State and Molecular Theory: A Scientific Biography* (New York: Wiley-Interscience, 1975), ch. 2.

REMINISCENCES FROM 1926 AND 1927

1. This essay has been excerpted from Werner Heisenberg, *Physics and Beyond* (New York: Harper and Row, 1971), trans. Arnold J. Pomerans from Heisenberg, *Der Teil und das Ganze* [The part and the whole] (Munich: Piper, 1969; Deutscher Taschenbuch Verlag, 1973).

2. For Heisenberg's own presentation of basic quantum mechanics, see his classic book *The Physical Principles of the Quantum Theory*, trans. Carl Eckart and Frank C. Hoyt (Chicago: University of Chicago Press, 1930). The following books deal more generally with the development of quantum theory: William H. Cropper, *The Quantum Physicists* (London: Oxford University Press, 1970) [a lively mixture of biography and physics]; B. d'Espagnat, ed., *Foundations of Quantum Mechanics* (New York: Academic Press, 1971); William C. Price, Seymour S. Chissick, and T. Ravensdale, eds., *Wave Mechanics: The First Fifty Years* (London: Butterworth, 1973); Jagdish Mehra and Helmut Rechenberg, *The Historical Development of Quantum Theory* (New York: Springer Verlag, 1982–) [a massively detailed record].

AT THE NIELS BOHR INSTITUTE IN 1929

1. Max Born, "Zur Quantenmechanik der Stossvorgänge" (The Quantum Mechanics of Collision Processes), *Zeitschrift für Physik* 37 (1926): 863–867.

2. See the detailed discussion in the next chapter.

NIELS BOHR AND THE PHYSICS OF SIMPLE PHENOMENA

1. For a later discussion of this problem, see Jørgen Kalckar, "On the Measurability of the Spin and Magnetic Moment of the Free Electron," *Nuovo Cimento* 8A (1972): 759–776.

2. The actual formula is

$$U = -\frac{23}{4\pi}\,\hbar c \cdot \frac{\alpha_1(0)\,\alpha_2(0)}{R^7},$$

where $\alpha_1(0)$ and $\alpha_2(0)$ are the static polarizabilities of the interacting atoms.

A REMINISCENCE FROM 1932

1. This article is based on excerpts from the essay, "Bohr und Heisenberg: Eine Erinnerung aus dem Jahr 1932," in C. F. von Weizsäcker, *Wahrnehmung der Neuzeit* [Observation of the New Age] (Munich: Carl Hauser, 1983). The editors are grateful to Nicholas Kemmer for help with the translation.

2. A complete English translation can be found in George Gamow, *Thirty Years That Shook Physics* (New York: Doubleday, 1966), pp. 165–218.

THE COMO LECTURE

1. Niels Bohr, "The Quantum Postulate and the Recent Development of Atomic Theory," *Nature* 121 (1928): 580–590.

2. Idem, "The Quantum Postulate and the Recent Development of Atomic Theory," in *Atti del Congresso Internazionale dei Fisici* (Bologna: Nicola Zanichelli, 1928), pp. 565–588.

NIELS BOHR AND NUCLEAR PHYSICS

I am grateful to H. H. Barschall, B. F. Bayman, A. P. French, Sir Rudolf Peierls, V. F. Weisskopf, and J. A. Wheeler for their comments on a draft of this article. I am also grateful to the American Council of Learned Societies, the National Science Foundation, and the Bush Foundation for support for this research.

1. Niels Bohr, "Reminiscences of the Founder of Nuclear Science and of Some Developments Based on His Work," *Proceedings of the Physical Society* (London) 78 (1961): 1083–1115; reprinted in Niels Bohr, *Essays, 1958–1962, on Atomic Physics and Human Knowledge* (New York: Wiley-Interscience, 1963).

2. Niels Bohr, "On the Theory of the Decrease of Velocity of Moving Electrified Particles on Passing through Matter," *Philosophical Magazine* 25 (1913): 10–31.

3. C. F. von Weizsäcker, "Zur Theorie der Kernmassen" [On the theory of nuclear masses], *Zeitschrift für* Physik 96 (1935): 431–458.

4. Niels Bohr, "Chemistry and the Quantum Theory of Atomic Constitution," *Journal of the Chemical Society* (London) (1932): 349–384.

5. Enrico Fermi, "Versuch einer Theorie der β-Strahlen" [A proposed theory of β-radiation], *Zeitschrift für* Physik 88 (1934): 161–177.

6. Niels Bohr, "Neutroneneinfang und Bau der Atomkerne," *Naturwissenschaften* 24 (1936): 241–245; idem, "Neutron Capture and Nuclear Constitution," *Nature* 137 (1936): 344–348, 351.

7. Gregory Breit and Eugene P. Wigner, "Capture of Slow Neutrons," *Physical Review* 49 (1936): 519–531.

8. Hans A. Bethe, "Nuclear Physics B: Nuclear Dynamics, Theoretical," *Reviews of Modern Physics* 9 (1937): 69–244; M. Stanley Livingston and Hans A. Bethe, "Nuclear Physics C: Nuclear Dynamics, Experimental," *Reviews of Modern Physics* 9 (1937): 245–390.

9. Niels Bohr and Fritz Kalckar, "On the Transmutation of Atomic Nuclei by Impact of Material Particles, I: General Theoretical Remarks," *Matematisk-Fysiske Meddelelser det Kongelige Danske Videnskabernes Selskab* [Mathematical-Physical Papers of the Royal Danish Academy of Sciences] 14, no. 10 (1937): 1–40.

10. Victor F. Weisskopf, "Statistics and Nuclear Reactions," *Physical Review* 52 (1937): 295–303.

11. Otto Hahn and Fritz Strassmann, "Über den Nachweis und das Verhalten der bei der Bestrahlung des Urans mittels Neutronen entstehenden Erdalkalimetalle" [On the detection and characteristics of alkaline-earth metals from the irradiation of uranium by neutrons], *Naturwissenschaften* 27 (1939): 11–15.

12. Lise Meitner and O. R. Frisch, "Disintegration of Uranium by Neutrons: A New Type of Nuclear Reaction," *Nature* 143 (1939): 239–240; O. R. Frisch, "Physical Evidence for the Division of Heavy Nuclei under Neutron Bombardment,"

Nature 143 (1939): 276; Niels Bohr, "Disintegration of Heavy Nuclei," *Nature* 143 (1939): 330.

13. Niels Bohr and John A. Wheeler, "The Mechanism of Nuclear Fission," *Physical Review* 56 (1939): 426–450.

14. For additional information, see: H. H. Barschall, "Three Decades of Fast-Neutron Experiments," *Physics Today* 22, no. 8 (1969): 54–59; Niels Bohr, *Essays, 1958–1962, on Atomic Physics and Human Knowledge* (New York: Wiley-Interscience, 1963); Joan Bromberg, "The Impact of the Neutron: Bohr and Heisenberg," *Historical Studies in the Physical Sciences* 3 (1971): 307–341; Laurie M. Brown, "The Idea of the Neutrino," *Physics Today* 31, no. 9 (1978): 23–28; Otto R. Frisch, *What Little I Remember* (Cambridge: Cambridge University Press, 1979); Otto R. Frisch and John A. Wheeler, "The Discovery of Fission," *Physics Today* 20, no. 11 (1967): 43–52; Fritz Krafft, *Im Schatten der Sensation: Leben und Wirken von Fritz Strassmann* [In the shadow of sensation: The life and work of Fritz Strassmann] (Weinheim: Verlag Chemie, 1981) [includes a documentary account of the discovery of fission]; Wolfgang Pauli, ed., *Niels Bohr and the Development of Physics: Essays Dedicated to Niels Bohr on the Occasion of His Seventieth Birthday* (London: Pergamon, 1955), esp. the essays by F. L. Friedman and V. F. Weisskopf, J. A. Wheeler, and J. Lindhard; Rudolf Peierls, "An Appreciation of Niels Bohr," *Proceedings of the Physical Society of London* 81 (1963): 793–799; idem, "Introduction," in Niels Bohr, *Collected Papers on Nuclear Physics* (forthcoming); Emilio Segrè, *Enrico Fermi, Physicist* (Chicago: University of Chicago Press, 1970) [treats the Rome experiments]; Roger H. Stuewer, ed., *Nuclear Physics in Retrospect: Proceedings of a Symposium on the 1930s* (Minneapolis: University of Minnesota Press, 1979); Roger H. Stuewer, "The Nuclear Electron Hypothesis," in William R. Shea, ed., *Otto Hahn and the Rise of Nuclear Physics* (Dordrecht, Netherlands: D. Reidel, 1983), pp. 19–68 [discusses Bohr's, Pauli's, others' contributions c. 1930]; idem, "Nuclear Physicists in a New World: The Émigrés of the 1930s in America," *Berichte zur Wissenschaftsgeschichte* 7 (1984): 23–40; idem, "Bringing the News of Fission to America" *Physics Today,* Oct. 1985; Spencer R. Weart, "Scientists with a Secret," *Physics Today* 29, no. 2 (1976): 23–30; idem, *Scientists in Power* (Cambridge, Mass.: Harvard University Press, 1979); John Archibald Wheeler, "Niels Bohr and Nuclear Physics," *Physics Today* 16, no. 10 (1963): 36–45.

PHYSICS IN COPENHAGEN IN 1934 AND 1935

1. This essay is based on a section of John A. Wheeler, "Some Men and Moments in the History of Nuclear Physics: The Interplay of Collegues and Motivations," in Roger H. Stuewer, ed., *Nuclear Physics in Retrospect* (Minneapolis: University of Minnesota Press, 1979).

SOME RECOLLECTIONS OF BOHR

1. Lev Landau and Rudolf E. Peierls, "Erweiterung des Unbestimmtheitsprinzip für die relativistische Quantentheorie" [Extension of the indeterminacy principle for relativistic quantum theory] *Zeitschrift für Physik* 69 (1931): 56–69.

2. See, however: Niels Bohr, Rudolf Peierls, and George Placzek, "Nuclear Reactions in the Continuous Region," *Nature* 144 (1939): 200–201.

REMINISCENCES FROM THE POSTWAR YEARS

1. Excerpted from Abraham Pais, "Reminiscences from the Postwar Years," in Stefan Rozental, ed., *Niels Bohr: His Life and Work as Seen by His Friends and Colleagues* (Amsterdam: North-Holland, 1967).

2. See "The Bohr-Einstein Dialogue" in Part III of this volume.

NIELS BOHR AS A POLITICAL FIGURE

1. See Peter Robertson, *The Early Years: The Niels Bohr Institute, 1921–1930* (Copenhagen: Akademisk Forlag, 1979), ch. 1.

2. This is how Frisch describes it in his article "The Interest Is Focussing on the Atomic Nucleus," in Stefan Rozental, ed., *Niels Bohr: His Life and Work as Seen by His Friends and Colleagues* (Amsterdam: North-Holland, 1967). But in his book *What Little I Remember* (Cambridge: Cambridge University Press, 1979), Frisch places this interview in London, where he spent the year 1933–34 before joining Bohr's institute.

3. The full text of the address, "Natural Philosophy and Human Cultures," can be found in *Nature* 143 (1939): 268–272. Reprinted in Niels Bohr, *Atomic Physics and Human Knowledge* (New York: Wiley, 1958).

4. *Danmarks Kultur ved Aar 1940* [Danish culture in the year 1940] (Copenhagen: Danske Forlag, 1941–1943).

5. The full text follows this article.

6. Vigo Kampmann, "Niels Bohr and the Danish Atomic Energy Establishment," in Rozental, *Niels Bohr*.

ENERGY FROM THE ATOM

1. Reprinted from *The Times* (London), 11 August 1945, p. 5.

NIELS BOHR AND NUCLEAR WEAPONS

1. Niels Bohr and John A. Wheeler, "The Mechanism of Nuclear Fission," *Physical Review* 56 (1939): 426–450.

2. Niels Bohr, "Resonance in Uranium and Thorium Disintegrations and the Phenomenon of Nuclear Fission," *Physical Review* 55 (1939): 418–419.

3. A Danish text of the lecture was subsequently published as "Nyere undersøgelser over atomkernernes omdannelser" [Recent investigations on the transmutation of atomic nuclei], *Fysisk Tidsskrift* 39 (1941): 3ff.

4. Unless otherwise noted, the quotations and statements of fact contained in this essay have been derived from Aage Bohr, "The War Years and the Prospects Raised by the Atomic Weapons," in Stefan Rozental, ed., *Niels Bohr: His Life and Work as Seen by His Friends and Colleagues* (Amsterdam: North-Holland, 1967); and Margaret Gowing, *Britain and Atomic Energy, 1939–1945* (London: Macmillan, 1964).

5. Richard G. Hewlett and Oscar E. Anderson, *The New World, 1939–1946* (University Park: Pennsylvania State University Press, 1962).

6. Donald C. Watt, "The Historiography of Nuclear Diplomacy," *Science* 194 (1976): 174–175.

7. Aage Bohr, private communication.

8. Niels Bohr, Memorandum dated 24 March 1945, as quoted in Bohr, *Open Letter to the United Nations* (Copenhagen: J. H. Schultz Forlag, 1950); reprinted in Rozental, ed., *Niels Bohr.*

9. The Russell-Einstein manifesto, issued in July 1955, was a peace appeal publicly addressed to the governments of the world. Its chief architects were Bertrand Russell and Albert Einstein.

MEETINGS IN WARTIME AND AFTER

1. S. Flügge, "Kann der Energieinhalt der Atomkerne technisch nutzbar gemacht werden?" [Can the energy content of the atomic nucleus be made technically useful?], *Naturwissenschaften* 23/24 (1939): 402–410.

OPEN LETTER TO THE UNITED NATIONS

1. The full text of the letter was originally published by J. H. Schultz Forlag, Copenhagen, and can also be found in Stefan Rozental, ed., *Niels Bohr: His Life and Work as Seen by His Friends and Colleagues* (Amsterdam: North-Holland, 1967).

THE PHILOSOPHY OF NIELS BOHR

1. This article is essentially the text of a talk prepared for Danish radio shortly after Bohr's death. The text was subsequently published in *Bulletin of the Atomic Scientists* 19 (September 1963): 8–14.

2. For a fuller discussion, see Aage Petersen, *Quantum Theory and the Philosophical Tradition* (Cambridge, Mass.: MIT Press, 1968). A more specific discussion of Bohr's ideas about complementarity can be found in Gerald Holton, "The Roots of Complementarity," *Daedalus* 99 (1970): 1015–1055; reprinted in Holton, *Thematic Origins of Scientific Thought* (Cambridge, Mass.: Harvard University Press, 1973), pp. 115–161. See also Léon Rosenfeld, "Niels Bohr's Contribution to Epistemology," *Physics Today* 16, no. 10 (1963): 47–54; and Abner Shimony, "Role of the Observer in Quantum Theory," *American Journal of Physics* 31 (1963): 755–773.

LIGHT AND LIFE

1. Reprinted in Niels Bohr, *Atomic Physics and Human Knowledge* (New York: Wiley, 1958), pp. 3–12.

2. Niels Bohr, "Light and Life Revisited," in Bohr, *Essays, 1958–1962, on Atomic Physics and Human Knowledge* (New York: Wiley-Interscience, 1963), pp. 23–29.

COMPLEMENTARITY AS A WAY OF LIFE

1. Niels Bohr, "The Quantum of Action and the Description of Nature," in *Atomic Theory and the Description of Nature* (Cambridge: Cambridge University Press, 1934), p. 96.

2. Max Jammer, *The Philosophy of Quantum Mechanics* (New York: McGraw-Hill, 1974), p. 102.

3. Bohr, "The Quantum of Action and the Description of Nature," pp. 99–100.

4. Max Born, "Physics and Metaphysics," *Memoirs of the Manchester Literary and Philosophical Society* 91 (1950): 35–53.

THE COMPLEMENTARITY PRINCIPLE AND EASTERN PHILOSOPHY

I would like to express my gratitude to Krishan Lal (National Physical Laboratory, New Delhi) and L. S. Kothari (University of Delhi) for their valuable suggestions during the preparation of this essay.

1. Niels Bohr, "Quantum Physics and Philosophy: Causality and Complementarity" (1958), in Bohr, *Essays, 1958–1962, on Atomic Physics and Human Knowledge* (New York: Wiley-Interscience, 1963).

2. Sri Aurobindo, *The Upanishads,* pt. 1 (Pondicherry, India: Sri Aurobindo Ashram, 1981), pp. 90–92.

3. Nathmal Tatia, *Studies in Jain Philosophy* (Banaras, India: Jain Cultural Research Society, 1951), p. 22.

4. P. C. Mahalanobis, "SANKHYA," *Indian Journal of Statistics* 18 (1957): 183ff.

5. J. B. S. Haldane, "SANKHYA," *Indian Journal of Statistics* 18 (1957): 195ff.

6. Bohr, *Essays, 1958–1962,* p. 7.

7. Léon Rosenfeld, *Physics Today* 16, no. 10, (1963): 47.

8. Werner Heisenberg, *Physics and Philosophy* (London: Allen & Unwin, 1959), p. 173.

9. Ibid., p. 182.

10. See "The Bohr-Einstein Dialogue" in Part III of this volume.

11. See "Delayed-Choice Experiments" in Part III of this volume.

12. John A. Wheeler, "Genesis and Observership," in Robert E. Butts and Jaakko Hintikka, eds., *Foundational Problems in the Special Sciences* (Dordrecht, Netherlands: D. Reidel, 1977), pp. 3–33.

13. D. S. Kothari, "Atom and Self," *Proceedings of the Indian National Science Academy* (Part A) 46 (1980): 1–28.

14. Idem, "Modern Physics and Syādvāda," in Krishan Lal, ed., *Synthesis, Crystal Growth and Characterization* (Amsterdam: North-Holland, 1984), pp. 555–568.

COMPLEMENTARITY AND MARXISM-LENINISM

1. J. Robert Oppenheimer, "Physics in the Contemporary World," Arthur D. Little Memorial Lecture, 1947 (Portland, Maine: Anthoensen Press, 1948). Reprinted in Oppenheimer, *The Open Mind,* (New York: Simon and Schuster, 1955), pp. 81–102.

2. Vladimir I. Lenin, *Materialism and Empirio-Criticism: Critical Comments on a Reactionary Philosophy* (Moscow, 1952).

3. Vladimir A. Fock, "Mozhno li schitat', chto kvantomekhanicheskoe opisanie fizicheskoi real'nosti iavliaetsia polnym?" [Can we consider the quantum-mechanical description of nature to be complete?], *Uspekhi fizicheskikh nauk* 16, no. 4 (1936): 436–457.

4. Mikhail E. Omel'ianovskii, *V. I. Lenin i fizika XX veka* [V. I. Lenin and Physics in the Twentieth Century] (Moscow, 1947).

5. Andrei A. Zhdanov, *Vystuplenie na diskussii po knige G. F. Aleksandrova "Istor-*

iia zapadnoevropeiskoi filosofii," 24 *iiunia 1947g* [Speech at the Discussion of G. F. Aleksandrov's Book "The History of West European Philosophy" of 24 June, 1947] (Moscow, 1951), p. 43.

6. Moisei A. Markov, "O prirode fizicheskogo znaniia" [On the Nature of Physical Knowledge], *Voprosy filosofii,* no. 2 (1947): 140–176.

7. Aleksandr A. Maksimov, "Ob odnoi filosofskom kentavre" [Concerning a Philosophical Centaur], *Literaturnaia gazeta,* April 10, 1948, p. 3.

8. Erwin Schrödinger, "Die gegenwärtige Situation in der Quantenmechanik" [The current situation in quantum mechanics], *Naturwissenschaften* 48 (1935): 807–812.

9. Aleksandr A. Maksimov, "Diskussiia o prirode fizicheskogo znaniia" [Discussion on the nature of physical knowledge], *Voprosy filosofii,* no. 3 (1948): 228.

10. Vladimir A. Fock, "Ob interpretatsii kvantovoi mekhaniki" [Concerning the interpretation of quantum mechanics], in Petr N. Fedoseev et al., eds., *Filosofskie problemy sovremennogo estestvoznaniia* [Philosophical Problems of Modern Science] (Moscow, 1959).

11. Ibid., p. 215.

12. For more details on Soviet interpretations of quantum mechanics, see Loren R. Graham, *Science and Philosophy in the Soviet Union* (New York: Knopf, 1972), pp. 69–110. An expanded second edition of this book, covering Soviet discussions until 1985, is being published by Columbia University Press.

A GLIMPSE OF THE OTHER SIDE

1. Wolfgang Pauli, "Die allgemeinen Prinzipien der Wellenmechanik" [The General Principles of Wave Mechanics] in *Handbuch der Physik,* 2nd. ser., vol. 24 (Berlin: Springer Verlag, 1933), pp. 83–272.

2. See, for example, Heinz Pagels, *The Cosmic Code* (New York: Simon and Schuster, 1982).

3. See, for example, Christian de Duve, *A Guided Tour of the Living Cell* (New York: Freeman, 1984).

SOME CLOSING REFLECTIONS

This essay owes a great deal to my many collaborative discussions with Samuel Y. Gibbon, Jr.

Glossary

Absorption edge A sharp transition between slight and strong absorption of X rays in a given substance as the X-ray wavelength is changed. Strong absorption sets in when the X-ray photon energy hv ($= hc/\lambda$) exceeds a certain threshold value, somewhat similar to the critical excitation of atoms by electrons. *See also* Franck-Hertz experiment.

Action A technical term, introduced into classical mechanics in 1747 by Pierre de Maupertuis, to denote a product of mass, velocity, and distance. Planck's constant, h, has these same dimensions and was identified as a "quantum of action" in the early development of atomic dynamics. *See also* Planck's constant.

Adiabatic principle This principle, formulated by Paul Ehrenfest in 1913, asserts in effect that one allowed motion of a system can be converted into another through a slow, smooth (adiabatic) change. In such a change, certain identifiable quantities (adiabatic invariants) may be conserved. For example, if the string of a pendulum is very slowly shortened, the ratio of energy to frequency remains constant. Such considerations were very important in the development of atomic mechanics.

Alpha particle (alpha rays) One of the three types of radiation (the others being beta rays and gamma rays) discovered in early studies of radioactivity around 1900. Rutherford showed that alpha particles were helium nuclei, which are made up of two protons and two neutrons tightly bound together.

Angstrom unit (Å) A unit of length equal to one ten-billionth (10^{-10}) of a meter. Radii of individual atoms, and interatomic distances in solids, are on the order of 1 Å. Anders J. Ångstrom (1814–1874) was a Swedish physicist whose spectroscopic data were a major source of information for Balmer, Rydberg, and others.

Angular momentum A fundamental dynamical quality in both classical and quantum mechanics. It is basically the measure of a linear momentum (mv) multiplied by its lever arm with respect to a certain center; thus, for a particle in a circular orbit we have $l = mvr$. In addition to such "orbital angular momentum," there may be intrinsic angular momentum, or "spin." In quantum mechanics the natural unit of angular momentum is $h/2\pi$, often denoted $\hbar$.

Antiparticle A particle of the same mass as a "normal" particle, but with the opposite sign of electric charge, magnetic moment, and so on. Every fundamental particle (such as an electron, proton, or neutron) has an antiparticle counterpart.

Artificial radioactivity The relatively short-lived radioactivity of unstable nuclei produced through nuclear reactions and bombardments. To be contrasted with the long-lived "natural radioactivity" of sequences of heavy elements descended from isotopes of uranium and thorium.

Atomic number The ordinal number of an atomic species in the periodic classification, starting with hydrogen as number 1. It is equal to the number (Z) of protons in the nucleus of the atom.

Atomic volume A relative measure of the size of an atom of a particular element, as given by the atomic weight divided by the density of the element in its fully condensed (solid or liquid) state. This ratio gives a measure of the total volume (in cubic centimeters or milliliters) occupied by 6×10^{23} atoms (Avogadro's number).

Atomic weight (A) This is the mass of an individual atom of a given chemical species, expressed as a multiple (not an integer) of a chosen "atomic mass unit" (a.m.u.). The current unit is one-twelfth of the mass of the carbon isotope C^{12} that contains 6 protons and 6 neutrons. (The *mass number* is the exact integer closest to A and equals the total number of protons and neutrons in the atom.)

Aufbauprinzip This German phrase, meaning "building-up principle" has become part of the universal language of physics. The *Aufbau* is the imagined process by which electron systems of massive atoms are constructed by successive addition of electrons to lighter atoms. At the same time, the central nuclear charge is increased to maintain the electrical neutrality of the atom as a whole.

Balmer series The series of lines in the optical spectrum of atomic hydrogen whose wavelengths were fitted to the formula $\lambda_n = \lambda_0[n^2/(n^2 - 4)]$ by Johann J. Balmer in 1885 and which gave Niels Bohr a vital clue in the development of his theory of atomic structure. Expressed in terms of the radiation frequencies $\nu \, (= c/\lambda)$, the Balmer formula becomes

$$\nu_n = R \left(\frac{1}{2^2} - \frac{1}{n^2} \right)$$

Bell's theorem This theorem, developed by John S. Bell in 1964, proves that observable discrepancies must exist between the predictions of the generally accepted form of quantum mechanics and the predictions from theories of local causality (hidden variables). The experimental evidence supports the standard quantum theory.

Beta decay The process by which an unstable nucleus is converted to a more stable one through the emission of a beta particle (negative electron) and a normally unobservable neutral particle (an antineutrino). The product nucleus is one unit higher in atomic number (nuclear charge) than the parent.

Beta particle (Beta rays) Early studies of natural radioactivity around 1900 revealed the emission of streams of negatively charged particles (as shown by their deflection by a magnet). They were subsequently shown to be identical with electrons. However, the term "beta particle" is kept to describe an electron emitted from a nucleus in the process of beta decay, in contrast to the extranuclear electrons that exist stably in all atoms.

Binding energy The energy that must be supplied to remove a particle completely from a system (for example, an electron from a neutral atom or a neutron from a nucleus).

Black-body radiation The continuous spectrum of radiation emitted by a body which, when cold, is a perfect absorber (perfectly black) at all wavelengths. The successful fitting of the observed spectrum with a theoretical formula by Max Planck in 1900 was the beginning of quantum theory.

Bohr atom The original form of atomic model devised by Niels Bohr, with electrons in circular orbits around a central nucleus.

Bohr-Sommerfeld theory The elaboration of the original Bohr model by Arnold Sommerfeld to include elliptic orbits and other characteristics.

Bose-Einstein statistics *See* Quantum statistics.

Boson A particle that obeys Bose-Einstein statistics. *See* Quantum statistics.

Breit-Wigner formula Probably the most famous formula in nuclear physics. Developed in 1936 by Gregory Breit and Eugene Wigner, it gives the theoretical variation of the cross-section (σ) for a nuclear process as a function of bombarding energy E when the compound nucleus has a well-defined resonance at a certain energy E_r:

$$\sigma \sim \frac{1}{v^2} \cdot \frac{1}{(E - E_r)^2 + \Gamma^2/4}$$

where v is the speed of the bombarding particle and Γ is the "width" of the resonant state. *See also* Cross-section; Resonance; Width.

Classical mechanics (Newtonian mechanics) The whole scheme of theoretical dynamics based on the concept of particles traveling in precisely definable trajectories according to Newton's laws of motion.

Cloud chamber The first device that made visible the tracks of fast-moving atomic or nuclear particles. Invented by C. T. R. Wilson in 1911, it exploited the fact that a supersaturated vapor condenses preferentially into droplets along the trail of ionization caused by a fast particle.

Combination principle A principle developed empirically around 1900 by J. R. Rydberg and Walter Ritz, stating that the wave-numbers (reciprocal wavelengths) of the lines in the spectrum of an element can all be expressed as differences between members of a limited number of "terms." With the development of Bohr's theory it became clear that these "terms" corresponded to the discrete energy levels of the atom.

Commute To commute is to exchange. In mathematics, certain operations commute — that is, the result is independent of the order in which they are taken — and others do not. For example, $(x^2)^3 = (x^3)^2$ so the operations of successively squaring and cubing commute. But $x(d/dx)\,[f(x)]$ is not equal to $(d/dx)\,[xf(x)]$, so the operations of multiplying by x and differentiating with respect to x do not commute. In quantum mechanics, observables that are represented by operators that commute can simultaneously have well-defined values; observables represented by noncommuting operators cannot. *See also* Observable.

Complementarity The existence of different aspects of the description of a physical system, seemingly incompatible but both needed for a complete description of the system. In particular, the wave-particle duality. *See also* Duality.

Compound nucleus A short-lived, excited state of the nucleus formed by the collision of two lighter particles (one of which may be a single neutron or proton). The compound nucleus may break up again into the particles that formed it, or disintegrate in some other way, or fall into its stable (ground) state through emission of gamma radiation.

Compton effect The collision of an X-ray photon with an electron as if both were simple particles. In the process, the electron picks up kinetic energy from the photon; the result is that the scattered X rays have lower quantum energy and hence longer wavelength than originally. The effect was discovered by Arthur H. Compton in 1922.

Compton wavelength A length that characterizes the change of X-ray wavelength in the Compton effect. It is given by h/mc, where h = Planck's constant,

m = electron mass, and c = speed of light. Numerically, it equals 2.4×10^{-10} cm.

Conductivity The ability of a material to transport heat or electricity (electric current). The rate of transport is characterized by a coefficient K (thermal conductivity) or σ (electrical conductivity). The larger the value of K or σ, the greater the rate of transport under a given temperature gradient or potential gradient (electric field).

Conservation In physics, the condition that certain properties of a system remain unaffected by internal changes in the system, provided that the system does not interact with its surroundings. Prime examples are conservation of momentum, total energy, and electric charge in a system of two colliding particles.

Correspondence principle The guiding principle, developed and extensively exploited by Bohr, according to which the predictions of quantum mechanics should not conflict with those of classical theory in the domain in which the latter are known to be valid. Typically this means situations involving high quantum numbers where the effects of quantization can become relatively unimportant.

Coulomb barrier *See* Potential barrier.

Coulomb force The basic force between electric charges, which decreases as the inverse square of the distance between them.

Cross-section The effective target area presented by a particle to an incoming beam of radiation or other particles. It is not a unique quantity; it depends on the details (for example, the energy and type of incident particles) of the process involved.

Curie's law This law, formulated by Pierre Curie in 1895, states that the magnetic susceptibility of a paramagnetic substance varies inversely as the absolute (Kelvin) temperature, T. *See also* Magnetism; Susceptibility.

De Broglie waves The waves, of wavelength h/p (where h = Planck's constant, and p = momentum), which Louis de Broglie, in 1924, hypothesized to be associated with any kind of particle.

Degree of freedom A dynamically independent coordinate in the description of motion. For example, a point particle has three degrees of freedom for translational motion; an extended rigid body has three additional degrees of freedom — three mutually orthogonal axes — for rotation.

Deuterium (deuteron) The isotope otherwise known as "heavy hydrogen." Its nucleus contains one proton and one neutron, which makes deuterium about twice as massive as the ordinary atomic hydrogen. A *deuteron* is the singly charged ion resulting from the removal of the electron from a neutral deuterium atom. It is a very important form of bombarding particle in nuclear physics.

Diamagnetism *See* Magnetism.

Direct reactions Nuclear reactions that take place without involving an intermediate compound-nucleus stage. *See also* Compound nucleus.

Dispersion The splitting of white light into different colors by a prism or other such device. This effect, due to a variation of refractive index with wavelength, can be used to make inferences about the number of electrons per atom; hence, it was important in the development of atomic theory.

Dispersion relations Mathematical relations connecting the scattering and absorptive components of the interaction of light (or other radiation) with matter. The relations involve considerations of causality in physical processes.

Duality The phenomenon by which, in the atomic domain, objects exhibit the properties of both particles and waves, which in classical, macroscopic physics are mutually exclusive categories.

Eigenvalue   A hybrid German-English word, meaning "proper value" or "characteristic value," that has become embedded in the language of mathematical physics. In quantum mechanics it denotes a specific allowed value of energy, angular momentum, and so on.

Eightfold way   A classification scheme for elementary particles, based on group theory and developed independently by Murray Gell-Mann and Yuval Ne'eman in 1961, that showed how selected groupings of eight or ten different particles could be understood as different combinations of three basic entities (quarks). *See also* Group theory; Quark.

Electric field   A measure of the condition, at any given point in space, that defines the magnitude of the force exerted on a stationary electrically charged particle at that point. The source of any such electric field is electrically charged particles elsewhere.

Electromagnetism   The study of the related phenomena of electricity and magnetism. The adjective "electromagnetic" refers to combined electric and magnetic effects, as in "electromagnetic wave." *See also* Maxwell's equations.

Electron   The fundamental negatively charged particle that is a constituent of all atoms. Its charge $(-e)$ is -1.6×10^{-19} coulomb and its mass is 9.1×10^{-31} kg. (1/1,840 of the mass of a hydrogen atom).

Electron radius   A length, often called the "classical electron radius," constructed from the values of electron charge (e), electron mass (m) and the speed of light (c). These quantities, in the combination e^2/mc^2, define a length $r_0 = 2.8 \times 10^{-13}$ cm, which features in the theoretical cross-section for scattering of electromagnetic waves by free electrons. (The electron is believed, however, to have negligible or zero true extension.) *See also* Cross-section.

Electron theory of metals   The theory that the characteristic properties of metals (especially their high electrical and thermal conductivities) can be interpreted in terms of vast numbers of freely moving electrons (roughly one per atom) inside the metal. Insulators do not have such free electrons.

Electron volt (eV)   An amount of energy equal to that acquired by an electron (or any other particle carrying one unit of the elementary charge e) in falling through an electric potential difference of one volt. Numerically, 1 eV = 1.6×10^{-19} Joule. The electron volt is a highly convenient energy unit in atomic physics. In nuclear physics the MeV (million electron volts) is more suitable as a unit for describing energy changes.

Energy density   The amount of energy per unit volume—in particular, the energy associated with electric or magnetic fields, or with their combination in electromagnetic radiation.

Energy level   A more or less sharply defined energy state of an atomic or nuclear system.

Equivalence principle   The famous principle, enunciated by Einstein in 1911, that over a limited region the effects of gravitation and of being in an accelerated frame are indistinguishable. This was the basis of his general theory of relativity.

Even-even nucleus   A nucleus containing even numbers of both protons and neutrons.

Even-odd nucleus   A nucleus containing an even number of protons and an odd number of neutrons. (In an "odd-even" nucleus, there are an odd number of protons and an even number of neutrons.)

Exchange force A type of force describable in terms of the continual exchange of some entity between two interacting particles. (A crude macroscopic example would be a situation in which a ball is thrown back and forth between two ice-skaters, who would be pushed apart as a result.)

Excitation The process of raising an atomic or nuclear system above its state of lowest energy—the "ground state." *Excitation energy* is the energy needed to raise a system (atom, nucleus, and so on) from its ground state to a given excited state. An *excited state* is any state of an atom or nucleus having a higher energy than the ground state.

Exclusion principle The principle (first arrived at empirically by Wolfgang Pauli in 1925 from studies of spectral data) that no two electrons can occupy the same quantum state and hence have all their quantum numbers in common.

Faraday rotation The rotation of the plane of polarization of light that passes through a transparent medium placed in a strong magnetic field along the direction of the beam. Discovered by Michael Faraday in 1845.

Fermi-Dirac statistics *See* Quantum statistics.

Fermion A particle that obeys Fermi-Dirac statistics. *See also* Quantum statistics.

Field theory A generic name for any theory that describes interactions between particles in terms of fields of force that exist throughout space (as contrasted with action-at-a-distance theories).

Fine structure What may appear, in a spectroscope with low resolution, to be light of a single wavelength will often (usually) be seen under higher resolution to consist of several closely spaced components, with wavelengths differing by a few parts in a thousand. A prime cause of this "fine structure" is a separation of the energy levels having the same "principal quantum number" n but different subsidiary quantum numbers. *See also* Hyperfine structure; Quantum number.

Fine-structure constant The fine-structure constant, α, is a dimensionless quantity (a pure number) given by the combination $2\pi e^2/hc\ (= e^2/\hbar c)$ of fundamental constants. Its numerical value is close to $1/137$.

Fission The process in which the nucleus of a heavy atom (near the top end of the periodic table) splits, either spontaneously or under external stimulus, into two (usually unequal) fragments plus other debris (notably one or two surplus neutrons). *See also* Width.

Fourier coefficient A number (fraction) giving the relative strength of a particular harmonic when a complicated periodic phenomenon (for example, a given note as played on a piano) is analyzed into a fundamental and a set of harmonics. Jean Baptiste Fourier (1772–1837) showed that any periodic phenomenon can be analyzed in this way.

Franck-Hertz experiment The classic experiment by James Franck and Gustav Hertz (1914) that revealed the existence of discrete energy levels in atoms. Electrons with less than a certain threshold energy were shown not to lose any energy when they collided with mercury atoms.

Free energy The maximum amount of energy that is available in the form of work from a given thermodynamic system.

Frequency The number of cycles per unit time of a periodic phenomenon. Usually measured in reciprocal seconds (the unit now known as the Hertz).

Fundamental frequency The lowest characteristic frequency of a physical system.

Gamma rays The name originally given to the most penetrating of the three types of radiation found to be emitted from radioactive substances. Gamma rays

Glossary

are electromagnetic radiation (like light) but are shorter in wavelength by factors ranging from thousands to millions. The term is still kept to describe electromagnetic radiation emitted from an atomic nucleus, in contrast to the light or X rays emitted in quantum jumps of extranuclear electrons. *See also* Alpha particle; Beta particle.

Geiger-Müller counter An early form of particle detector, developed by Hans Geiger and Walther Müller in 1928, that undergoes a measurable electric discharge when a charged particle passes through it.

Grand unified theories (GUTs) Theories aiming to establish that the strong interactions, the electromagnetic interactions, and the weak interactions are different aspects of one fundamental force. Their ultimate goal would be to incorporate the gravitational interaction in this same scheme.

Group theory A generic name for theories based upon self-contained groups of mathematical operations—basically symmetry operations such as reflection or rotation—which correspond to the symmetry properties of physical systems. A group may have a very direct geometrical interpretation (for example, in crystallography) or be quite abstract (as in the theoretical classification of elementary particles).

Hamilton-Jacobi theory A particular formulation of the general equations of particle dynamics, developed by W. R. Hamilton (1805–1865) and C. G. J. Jacobi (1804–1851). The quantity "action" figures prominently in this approach, which helps make the connection between classical and quantum mechanics. *See also* Action.

Harmonic oscillator A basic physical system, typified classically by a mass attached to a spring, that oscillates harmonically (sinusoidally) with time.

Heavy hydrogen The hydrogen isotope of mass 2, also called deuterium (D).

Heavy water Water in which the hydrogen atoms are deuterium (heavy hydrogen) instead of ordinary hydrogen. It is given the chemical formula D_2O (as contrasted with H_2O).

Hidden variables Possible physical variables that may help define the state of a system but whose existence has not been demonstrated. Postulated by those, in particular David Bohm, who have shared Einstein's dislike of a seemingly inescapable random element in atomic processes.

Hole theory One aspect of P. A. M. Dirac's relativistic theory of electrons, according to which a hole in the otherwise completely filled "sea" of negative energy states behaves as a positive particle. *See also* Positron.

Homologue In chemistry, one of a number of substances having similar properties but a regular difference in mass or structure (for example, the hydrocarbon series methane, ethane, propane, and so on). Also used to describe elements belonging to the same group of the periodic system. *See also* Periodic system.

Hyperfine structure A subdivision of a spectral line into several components on a still smaller scale than that of the fine structure. A prime cause of hyperfine structure is the magnetism of nuclei. *See also* Fine structure.

Ideal gas A gas for which the pressure multiplied by volume divided by absolute temperature (pV/T) is a constant for a given mass of the gas. This relation is to be expected theoretically for a system of identical, perfectly elastic, pointlike particles that have no long-range interaction with one another.

Induced emission (stimulated emission) The emission of radiation by an excited atom under the influence of other radiation falling upon it. This process is the basis of the action of a *laser* (*l*ight *a*mplification by *s*timulated *e*mission of radiation).

Interference   The process by which waves of the same wavelength and frequency, coming from two or more sources that have a defined phase relationship, will reinforce one another at some places and will tend to cancel one another at other places. *Interference fringes* are the pattern of regions of enhanced and diminished intensity resulting from interference.

Ionization   The process of removing one or more electrons from a neutral atom, converting the latter into a positively charged "ion."

Isomer   In chemistry, one of two or more compounds having identical chemical compositions but different properties (for example, the sugars fructose and glucose). *Nuclear isomers* are nuclei having the same numbers of neutrons and protons but different radioactive decay modes. Essentially they are just different energy states of the same nucleus, but happen both to be fairly long-lived.

Isotope   From the Greek *iso* (same) and *topos* (place). The name given to different atomic forms that have the same atomic number (Z) and are essentially identical in chemical behavior. They are atomic species of the same element, differing only in the number of neutrons in the nucleus.

Kinetic energy   The energy that an object has by virtue of its motion, over and above any energy that it has at rest.

Klein's paradox   A paradox proposed by Oskar Klein in 1929 after P. A. M. Dirac had published his relativistic quantum theory of electrons, which implied the existence of seemingly unreal negative energy states. Klein argued that, if such states existed, electrons in positive energy states could fall into them. It was later concluded that such states did exist but were normally all filled and therefore unavailable. *See also* Exclusion principle.

Lamb shift   A small correction required by quantum electrodynamics (QED) to the energies of certain states in atomic hydrogen as calculated from ordinary (nonrelativistic) quantum mechanics. The verification of this shift by W. E. Lamb and R. C. Retherford in 1946 was a clear demonstration of the correctness of QED. *See also* Quantum electrodynamics.

Laser   *See* Induced emission.

Lifetime   The average time for which an unstable atomic or nuclear state survives before it decays by disintegration or emission of radiation.

Liquid-drop model   A theoretical model of the nucleus that likens it to a droplet of liquid. The forces between neutrons and protons have many remarkable resemblances to the forces between the molecules in a liquid, even though the scales of distance and force strength are very different.

Lorentz force   The force exerted on a charged particle when it moves in a magnetic field. The force acts perpendicularly to the particle's velocity and is proportional to the magnitude of the velocity.

Magnetic field   A measure of the condition, at any given point in space, that defines the magnitude of the force exerted on an element of electric current at that point (and which vanishes if the current stops). The source of any such magnetic field is electric charges in motion elsewhere.

Magnetic moment   A measure of the strength of the equivalent bar magnet represented by the circulating and/or spinning electric charges in an atomic or nuclear system.

Magnetism   A property of matter that describes its ability to generate a magnetic field (as in a bar magnet) or its response to a magnetic field externally applied. *Diamagnetism* is a fundamental tendency of all atoms to be repelled by a magnet. *Paramagnetism* is a tendency of *some* kinds of atoms to be attracted by a magnet. Where paramagnetism exists, it completely swamps the diamagnetic tendency.

Magneton (Bohr magneton) A natural atomic unit of magnetism equal to $eh/4\pi mc$ (where e = elementary charge, h = Planck's constant, m = electron mass, c = speed of light).

Malus's law An originally empirical result in optics: the intensity of light emerging from two polarizing devices in sequence is proportional to the square of the cosine of the angle between their polarization axes. Discovered by E. L. Malus in 1808.

Mass defect Not an imperfection, but simply a technical term for the arithmetical difference between the mass of an isotope (measured in atomic mass units) and its *mass number, A*, an integer equal to the total number of neutrons and protons in the atom.

Matrix mechanics The particular formulation of quantum mechanics, based on the properties of mathematical matrices, that was developed by Werner Heisenberg beginning in 1925.

Maxwell-Boltzmann law A theoretical law, developed by James Clerk Maxwell (1831–1879) and Ludwig Boltzmann (1844–1906), governing the distribution of speeds of gas molecules at different temperatures. The law states that the distribution of energies for a collection of particles in thermal equilibrium is governed by the factor exp $(-E/kT)$, where E is the energy, T the absolute temperature, and k the Boltzmann constant. In the original application of the law to gases, E was simply the kinetic energy $\frac{1}{2}\,mv^2$ of molecular motion, but the Boltzmann factor applies more generally. For example, in the analysis of the distribution of orientations of magnetic atoms in a magnetic field, one has $E = -\mu B \cos\theta$, where μ is the magnetic moment, B the magnetic field strength, and θ the angle of the magnetic moment to the field direction.

Maxwell's demon A hypothetical hobgoblin, invented by James Clerk Maxwell, which by operating a frictionless shutter separating two halves of a container of gas could cause faster molecules to accumulate on one side and slower molecules on the other, thus creating a temperature difference without doing work, in violation of the second law of thermodynamics (the law immortalized by C. P. Snow in his *Two Cultures*).

Maxwell's equations A set of four equations, embodying all the fundamentals of classical electromagnetism, presented in organized form by James Clerk Maxwell in 1864. They describe (1) Coulomb's law of electrostatics; (2) the absence of magnetic monopoles; (3) Faraday's law of electromagnetic induction (the association of an electric field with a changing magnetic field); and (4) "Maxwell's induction law" (the association of a magnetic field with a changing electric field). Using these, Maxwell proved that light is an electromagnetic wave.

Mean free path The average distance between successive collisions for particles traveling through a region occupied by other particles (such as molecules in a gas, or free electrons in a metal).

Measurement theory The detailed theory of actual measurements in quantum mechanics, as distinct from the probabilistic statements that apply to physical systems in the absence of an act of observation.

Microwaves Electromagnetic radiation in the wavelength range from about 1 mm to 10 cm. The cosmic background radiation, believed to be the remnant of the "big bang" in which our universe originated, has a mean wavelength of about 1 mm, corresponding to a current temperature of about 3° Kelvin.

Model In physics, the word "model" normally means a particular mathematical or theoretical representation of a physical system, since in atomic and nuclear physics our macroscopic pictures of objects have little or no application.

Moderator A body of liquid or solid material in which neutrons are slowed down by multiple collisions. *See also* Neutron.

Momentum (linear momentum) The product of mass and velocity for a moving object.

Monochromatic Being all of the same wavelength or frequency (as applied particularly to light or other electromagnetic radiation).

Monomer A chemical unit of relatively low molecular weight from which polymers may be formed. *See also* Polymer.

Muon A charged particle about 200 times as massive as the electron, but otherwise closely resembling it as a carrier of the weak nuclear interaction. *See also* Weak interaction.

Neutrino A neutral particle of negligible (or perhaps zero) mass, postulated by Wolfgang Pauli in 1931 as an unobservable companion of the electron in beta decay, and whose existence would allow energy and angular momentum to be conserved in this process. It was finally detected unambiguously in 1956 by Clyde Cowan and Frederick Reines.

Neutron An electrically neutral particle of almost the same mass as the proton (but slightly heavier). If one chooses to ignore their substructure, neutrons and protons can be regarded as the fundamental constituents of all atomic nuclei. *Fast neutrons,* in the accepted idiom of nuclear physics, are neutrons with kinetic energies on the order of MeV (millions of electron volts). *Slow neutrons* are neutrons with energies on the order of eV or less. *Thermal neutrons* are neutrons reduced to thermal energies (on the order of 1/40 eV) through multiple collisions with atoms in a body of material (a moderator) at room temperature.

Newton's bucket The subject of a famous experiment performed by Isaac Newton to demonstrate (as he believed) the absolute character of rotation, from the fact that the surface of the water in the bucket, initially flat, became concave when it took up the rotation.

Nicol prism A device, constructed from a crystal of calcite, that acts like a piece of Polaroid sheet in producing linearly polarized light from an unpolarized beam. Invented by William Nicol (1768–1851).

Noble gas A member of the series helium, neon, argon, krypton, xenon, radon (also known as the "rare" or "inert" gases). Called "noble" because, possessing fully completed electron shells, these elements form almost no associations through chemical combination with other atoms.

Nucleosynthesis The process of building up heavier nuclei from lighter ones as a result of thermonuclear reactions in the interiors of stars.

Nucleus The central core of an atom. It is composed of neutrons and protons and carries more than 99.95 percent of the mass of any type of atom, but is confined to a region representing less than about 1/10,000 of the diameter of the atom.

Observable A physically observable (measurable) attribute of a system, associated in quantum mechanics with a specific operator. *See also* Operator.

Operator Fundamentally, any mathematical entity that specifies some manipulative procedure. The signs $+$, $-$, and $\times$ are operators, whose use we understand in the expressions $a + b$, $a - b$, $a \times b$. In quantum mechanics the most common operators are (a) in wave mechanics, differential operators such as d/dx or d/dt; (b) in matrix mechanics, matrix operators of the kind used in coordinate transformations in analytic geometry.

Packing fraction A quantity equal to 10,000 times an isotopic mass defect, divided by the isotopic mass number A. It is a relative measure of the average binding energy per neutron or proton in a nucleus. *See also* Binding energy; Mass defect.

Pair production The creation of an electron-positron pair (or some other combination of particle and antiparticle) from the energy of a photon — possible if the photon has energy greater than $2mc^2$.

Paramagnetism *See* Magnetism.

Periodic system (periodic table) A classification of the chemical elements in order of ascending atomic weight (with one or two exceptions), arranged so as to exhibit a periodic recurrence of similar chemical properties (for example, the sequence fluorine, chlorine, bromine, iodine). The first reasonably complete version of the periodic classification was devised by Dmitri Mendeleev in 1869.

Phase A specific stage in a periodic phenomenon (as in the phases of the moon). In physics, the phase is measured by an angle between zero and 2π (plus perhaps an integral multiple of 2π), since 2π radians corresponds to one complete cycle.

Phase space A mathematical space whose coordinates are the positions (x) and momenta (p) of particles. For a single particle moving in one dimension, the particle can be considered as located within a "cell" of phase space of area $\Delta p \cdot \Delta x = h$, according to the uncertainty principle. *See also* Uncertainty principle.

Photodisintegration The disintegration of a nucleus by energetic photons (gamma rays or X rays of very short wavelength and energies on the order of several MeV).

Photoelectric effect The ejection of electrons from individual atoms or from matter in bulk by incident light or other electromagnetic radiation.

Photon A quantum of electromagnetic radiation, with energy equal in magnitude to $h\nu$, where ν is the frequency. This is the smallest possible amount of radiant energy at the given frequency.

Pickering series A series spectrum of singly ionized helium (He^+) made up of transitions ending on the $n = 4$ orbit. The spectrum was discovered by E. C. Pickering in 1896 in the light from a star, and was ascribed by him to hydrogen. Bohr identified its true origin in the second paper of his Trilogy, in 1913.

Planck radiation formula The theoretical formula, involving Planck's constant h in an essential way, that Max Planck devised in 1900 to fit the spectrum of black-body radiation. The formula states that the energy density of radiation inside an enclosure at absolute temperature T varies with frequency ν according to the equation

$$\rho(\nu, T) = \frac{8\pi\nu^2}{c^3} \cdot \frac{h\nu}{\exp(h\nu/kT) - 1}.$$

Planck's constant (h) This constant, the "quantum of action," is the quantity that underlies all of quantum physics. It has the dimensions of momentum $\times$ distance or energy $\times$ time. It was discovered by Max Planck (1900) in his attempt to explain the spectrum of the radiation from a hot body. Its magnitude is 6.625×10^{-34} joule-sec. *See also* Action.

Plasma An almost completely ionized state of matter, developed in electrical discharges in gases at low pressure and in matter under similar conditions in interstellar space.

Pockels cell An optical shutter for polarized light, whose action depends on the birefringence (double refraction) that can be induced in certain crystals by an applied electric field.

Polarization A state involving some kind of specific orientation in a physical system. A substance is said to be electrically polarized if it has a separation of positive and negative charge along a particular direction. As applied to light or other electromagnetic waves, it refers to the direction of transverse oscillation of the electric field of the wave.

Polymer A substance made up of very large molecules which are, at least approximately, multiples of units of lower molecular weight (monomers). A polymer molecule may be built of hundreds or thousands of monomers. Examples are rubber, plastics, and many animal and plant tissues.

Positive rays A name for the positively charged particles that were observed to emerge through a hole in the cathode in a gas discharge tube at low pressure. They were identified as atoms stripped of one or more electrons, and "positive ray analysis" provided the basis of mass spectroscopy.

Positron The antiparticle of the electron. It has the same mass but carries a positive charge $(+e)$. *See also* Antiparticle.

Potential barrier A region of high potential energy that inhibits or prevents particles from passing from one side to the other of that region. A particularly important case is the *Coulomb barrier* around a nucleus, due to the repulsion between the charge on the nucleus and the positive charge in an incident particle (for example, a proton or an alpha particle).

Proton The fundamental particle that is the nucleus of the hydrogen atom. It has charge $+e$ $(= 1.6 \times 10^{-19}$ coulomb), diameter about 10^{-15} m, and mass 1.66×10^{-27} kg (about 99.95 percent of the total mass of a hydrogen atom). Protons, with neutrons, are the basic constituents of all atomic nuclei.

Quantization The restriction of the magnitudes of certain quantities (especially energy and angular momentum) to values defined in terms of Planck's constant.

Quantum A characteristic amount of energy, momentum, angular momentum, and so on defined naturally by the finite (nonzero) magnitude of Planck's constant, h. The discreteness represented by the quantum is quite negligible in most macroscopic systems but is crucial in atomic dynamics.

Quantum electrodynamics (QED) As the name implies, this is a theory that combines quantum theory with electromagnetism. It provides a rigorously correct description of all extranuclear atomic phenomena, which depend only on electric and magnetic interactions.

Quantum field theory The theory that deals with the fact that electric, magnetic, and other fields, which classically are continuous, must be treated as quantized systems with definite energy states, analogously to atoms.

Quantum number An integer (or sometimes a half-integer) that characterizes some feature of a quantized atomic state. The first example was the quantum number n defining the allowed values $(nh/2\pi)$ of the orbital angular momentum of the electron in Bohr's original theory of the hydrogen atom. In atomic physics the *principal quantum number* (*n*) is the chief determinant of the energy of a state. The *azimuthal quantum number* (*k*) labels the allowed orbits of different ellipticity for the same *n* in the semiclassical Bohr-Sommerfeld model. The *inner quantum number* (*j*) is related to the total angular momentum of an orbiting electron.

Quantum state   A specific state of an atomic or nuclear system, characterized by specific values of energy, momentum, quantum numbers, and so on.

Quantum statistics   The rules applying to a large collection of identical atomic or nuclear particles. There are two distinct categories: (1) *Bose-Einstein statistics,* in which there is no restriction on the number of particles that can occupy the same quantum state (that is, have the same energy and the same set of quantum numbers). Photons and neutral helium-4 atoms are in this class. (2) *Fermi-Dirac statistics,* in which no two particles can occupy the same quantum state. Such particles obey the Pauli exclusion principle. Electrons, protons, and neutrons are in this category. *See also* Exclusion principle.

Quantum theory   Originally restricted to Max Planck's hypothesis (1900) that radiant energy can be transferred only in quanta of magnitude $h\nu$ (ν = frequency); now applied to the whole of atomic mechanics.

Quark   A family name for the various types of particles that are believed to be the ultimate building blocks for all other particles except electrons, muons and neutrinos. Quarks are believed to be pointlike particles with electric charges of $+e/3$ or $-2e/3$.

Radiation   In its broadest sense, anything that is emitted from a source. In atomic physics it means electromagnetic radiation (X rays, ultraviolet radiation, visible light, infrared radiation, microwaves, radio waves). In nuclear physics, besides electromagnetic waves (gamma rays) it includes alpha rays (helium nuclei) and beta rays (electrons).

Radiation field   The electromagnetic field produced by the oscillatory motion of electric charges in a source.

Radioactivity   The phenomenon of spontaneous emission of various radiations (alpha, beta, or gamma rays) from unstable nuclei.

Rare earths   *See* Transition group.

Red shift   The shift toward longer wavelengths exhibited by light or other electromagnetic waves emitted from a source that is very massive (gravitational red shift) or is receding from the observer (Doppler red shift).

Relativistic quantum mechanics   The form of quantum mechanics needed when the modifications of mechanics due to special relativity must be taken into account. The original quantum mechanics of Heisenberg and Schrödinger was "nonrelativistic"; the extension to the relativistic domain was pioneered by Dirac in 1928.

Renormalization   A systematic procedure for subtracting away mathematically infinite self-energies in atomic and nuclear theory so as to obtain finite and physically significant values of various quantities. *See also* Self-energy.

Resonance   In general, a phenomenon by which an external influence is exactly matched to the natural motion of a system; the result is a particularly large response (for example, "pumping" a swing). In atomic and nuclear physics, a resonance occurs when the total energy of two colliding particles corresponds to the energy of a quantized state of the combined system. The resonance is characterized by a high probability of interaction at or near that energy. *See also* Breit-Wigner formula.

Rest mass The intrinsic mass of an object as measured in its own frame of reference. Because mass and energy are equivalent (through Einstein's equation $E = mc^2$), a moving object is more massive because it possesses kinetic energy. (In theoretical physics, however, the word "mass" is usually reserved for rest mass.)

Riemann surface An abstract type of surface devised by Bernhard Riemann (1826–1866) for the representation of many-valued functions $f(z)$ of a complex variable. For example, if $f(z) = r\,e^{i\theta}$ ($i = \sqrt{-1}$, θ = angle in radians), every change of θ by 2π brings one back to the same point in the complex plane. But by making a stack of planes (sheets) for successive intervals of θ ($0 - 2\pi$, $2\pi - 4\pi$, and so on) and joining them along suitably chosen cuts, one obtains a continuous surface (like a spiral ramp) on which each of the identical values of $f(z)$ is associated with a different point on the surface, thus dissolving the many-valuedness.

Rigid rotator An idealized prototype physical system in both classical and quantum mechanics. A classical example would be a dumbbell rotating about its center; an atomic example would be a diatomic molecule doing likewise.

Rydberg constant (R) The constant, dimensionally a reciprocal length, that was introduced in 1890 by J. R. Rydberg in the numerical representation of the wavelengths of series spectra. Its value is about 109,700 cm^{-1}.

Scattering A general expression for the process of collision of one atomic or nuclear particle with another. *Rutherford scattering* is the deflection of an incident charged particle by the electric field of an atomic nucleus. *Double scattering* is the scattering of a beam of particles at two separate "targets" in succession.

Selection rules Specific rules, derivable from quantum-mechanical analysis, that place definite limitations on the possible changes of quantum numbers for a system undergoing a change from one state to another—for example, the transition of an atom from a higher state to a lower state through the emission of a photon.

Self-energy The energy possessed by a particle in isolation from all others.

Shell structure The tendency of electrons outside the nucleus, or of neutrons and protons inside the nucleus, to arrange themselves in "shells" corresponding classically to the same orbit radius and the same energy for each particle in the shell.

Spectrum The distribution in wavelength of the radiation of different wavelengths emitted by a given type of excited atom or nucleus (or by a collection of atoms, as in an incandescent solid). A *continuous spectrum* is a smooth distribution of wavelengths such as is represented by white light or by the thermal radiation from a hot body. A *line spectrum* is a characteristic set of sharply defined wavelengths emitted by individual atoms of a particular element. A *series spectrum* is a set of wavelengths that are described by a single formula with one variable parameter (a quantum number). The most famous example is the Balmer series. *See also* Balmer series.

Spin The intrinsic angular momentum of a particle such as an electron or a proton. *See also* Angular momentum.

Spin flip The process of inverting the spin angular momentum of particles such as electrons, protons, and neutrons, which have only two quantized orientations in a magnetic field.

Spontaneous emission The emission of radiation from isolated atoms in an excited state (that is, a state of higher energy than the ground state). *See also* Induced emission.

Standard deviation A term from mathematical statistics that describes the spread of a set of independent measurements of a quantity about its average value.

Stark effect The splitting of a single spectral line into components of slightly different wavelengths as a consequence of placing the source of the light in an intense electric field. The effect was first demonstrated in 1913 by Johannes Stark.

State vector A vector in an abstract space (hyperspace) of arbitrarily many dimensions that completely characterizes the state of a system in quantum mechanics. Basically it plays the same role as a wave function but is expressed in mathematically very different terms. *See also* Wave function.

Stationary state A state of an atomic or nuclear system that exists for a long time (as measured on the natural time scale for atomic or nuclear processes) and hence has a fairly sharply defined energy in accordance with the uncertainty principle: $\Delta E \approx h/$Lifetime. *See also* Uncertainty principle.

Statistical mechanics The field of theoretical physics that treats of the behavior and properties of very large numbers of identical particles.

Stern-Gerlach experiment A crucial experiment, performed in 1921 by Otto Stern and Walther Gerlach, showing that neutral atoms are limited to discrete (quantized) orientations with respect to a magnetic field.

Stopping power A quantitative measure of the effectiveness of a material in removing energy (by collisions) from a fast nuclear particle traveling through it.

Strong interaction The short-range interaction that represents the main force that holds nuclei together.

Superconductivity The property possessed by a number of materials (especially metals) of losing their electrical resistance completely below a certain temperature, so that, for example, a current once started will continue to flow indefinitely without need for a battery.

Superluminal Faster than the speed of light in a vacuum (c). Physicists generally believe that (in accordance with Einstein's relativity theory) no real influence can travel at a speed greater than c.

Superposition A basic property of waves, such that the net effect of two or more waves is obtained by simply adding (superposing) their separate effects at a given point. *See also* Interference.

Surface tension The tendency of a liquid surface to contract to its minimum possible area. It arises from the force of attraction between molecules.

Susceptibility A measure of the degree to which a substance can be electrically or magnetically polarized by external electric or magnetic fields.

Symmetry In quantum mechanics, a mathematical property of wave functions, which may remain unchanged (symmetric) or reverse sign (antisymmetric) under some imagined symmetry operation (for example, replacing x by $-x$). The wave function for a pair of identical particles must be either symmetric or antisymmetric with respect to an imagined interchange of the particles — symmetric for bosons and antisymmetric for fermions; the latter case embodies the exclusion principle. *See also* Boson; Fermion.

Thermodynamic equilibrium The dynamic equilibrium of a physical system at a given temperature.

Thermoelectricity   The generation of electric voltages and currents through temperature differences in a physical system.

Transition   A general term for the process by which a physical system changes from one state to another.

Transition group   A sequence of consecutive elements in the periodic system having similar chemical and other properties. Such sequences occur in the *Aufbau* (building up of atoms), when inner electron shells are being filled while the outer (valence) electron structure remains constant. Two notable transition groups are the transition metals, between scandium ($Z = 21$) and nickel ($Z = 28$), and the rare earths, between lanthanum ($Z = 57$) and lutecium ($Z = 71$). *See also* Periodic system.

Transition probability   The probability or relative probability that a transition between one quantum state and another will take place. Often measured or calculated as a probability per unit time.

Transuranic element   An element of larger mass and atomic number than the heaviest of the naturally occurring elements (uranium). All such elements are relatively short-lived; they are produced in nuclear reaction processes.

Tunneling   The ability of particles to traverse regions in which their kinetic energy would have a negative value — impossible in classical mechanics. The effect depends on the wave property of particles.

Uncertainty principle   This principle, enunciated in 1927 by Werner Heisenberg, asserts the fundamental impossibility of simultaneously obtaining arbitrarily precise measures of particular pairs of quantities, most notably momentum and position or energy and time, in the description of atomic systems. Also known as the *indeterminacy principle*.

Uncertainty relations   The quantitative expression of the uncertainty principle in such inequalities as $\Delta p \cdot \Delta x \gtrsim h/2\pi$, or $\Delta E \cdot \Delta t \gtrsim h/2\pi$ (where p = momentum, x = position, E = energy, t = time, h = Planck's constant).

Vacuum fluctuation   The existence of a fluctuating electromagnetic field in empty space, even in the absence of an applied field, as a result of the quantization of electromagnetic field energy.

Vacuum polarization   A key feature of quantum electrodynamics according to which the vacuum is not empty but contains "virtual" electron-positron pairs that can be separated (polarized) in the vicinity of a single electron of positive energy. *See also* Quantum electrodynamics.

Valence   The effective number of units of elementary charge ($\pm e$) that an atom has available in processes of chemical combination or bonding.

Van der Waals force   The basic force between neutral atoms and molecules, repulsive at very short distances but attractive at somewhat longer range. First explored by J. D. Van der Waals (1837 – 1923).

Wave equation   A mathematical equation, involving the space and time derivatives of some variable, whose solutions include the possibility of progressive waves.

Wave function   A mathematical function of space and time, usually denoted ψ (psi), that is the solution to Schrödinger's wave equation in quantum mechanics. The square of the magnitude of ψ was recognized by Max Born in 1926 as a measure of the probability of finding a particle at a given point.

Wave mechanics   The form of nonrelativistic quantum theory developed by Erwin Schrödinger in 1925, based upon the fact that particles of a given momentum p have an equivalent wavelength (the de Broglie wavelength) given by $\lambda = h/p$, where h is Planck's constant.

Wave packet   A wave disturbance of limited extent in space. It spreads as it travels.

Weak interaction   The nuclear interaction that is responsible for the process of beta decay. Named in contrast to the "strong interaction," which is billions of times stronger. *See also* Beta decay; Strong interaction.

Width   Basically, the finite (nonzero) spread in energy of any atomic or nuclear state that is not completely stable. The width, usually denoted Γ, is related to the mean lifetime τ for decay of a state through the uncertainty relation $\Gamma = h/\tau$. *See also* Breit-Wigner formula.

X rays   A name given to the form of penetrating radiation (discovered by Wilhelm Roentgen in 1896) produced when energetic electrons are suddenly brought to rest by collision with a "target." In 1912, X rays were identified as electromagnetic waves, similar to light but thousands of times shorter in wavelength (on the order of one angstrom). *Characteristic X rays* are line spectra arising from quantum jumps of electrons in the inner shells of atoms.

Zeeman effect   The broadening (or splitting into several components) of a single spectral line when the source of the light is placed between the poles of a strong magnet. The effect was first observed in 1896 by Pieter Zeeman.

Zero-point energy   An amount of energy that is retained by an atomic oscillator (say, a diatomic molecule) even at the absolute zero of temperature, where classically all motion would cease.

Works by Niels Bohr

1. "Determination of the Surface-Tension of Water by the Method of Jet Vibration," *Philosophical Transactions of the Royal Society*, ser. A, 209 (1909): 281–317.

2. "On the Determination of the Tension of a Recently Formed Water-Surface," *Proceedings of the Royal Society*, ser. A, 84 (1910): 395–403.

3. "Studier over metallernes elektrontheori" [Studies in the electron theory of metals], dissertation, Copenhagen, 1911.

4. "Note on the Electron Theory of Thermo-electric Phenomena," *Philosophical Magazine* 23 (1912): 984–986.

5. "On the Theory of the Decrease of Velocity of Moving Electrified Particles on Passing through Matter," *Philosophical Magazine* 25 (1913): 10–31.

6. "On the Constitution of Atoms and Molecules": Part 1—"On the Constitution of Atoms and Molecules," Part 2—"Systems Containing Only a Single Nucleus," Part 3—"Systems Containing Several Nuclei," *Philosophical Magazine* 26 (1913): 1–25, 476–502, 857–875.

7. "The Spectra of Helium and Hydrogen," *Nature* 92 (1913): 231–232.

8. "Om brintspektret" [On the hydrogen spectrum], *Fysisk Tidsskrift* 12 (1914): 97–114.

9. "Atomic Models and X-Ray Spectra," *Nature* 92 (1914): 553–554.

10. "On the Effect of Electric and Magnetic Fields on Spectral Lines," *Philosophical Magazine* 27 (1914): 506–524.

11. "On the Series Spectrum of Hydrogen and the Structure of the Atom," *Philosophical Magazine* 29 (1915): 332–335.

12. "The Spectra of Hydrogen and Helium," *Nature* 95 (1915): 6–7.

13. "On the Quantum Theory of Radiation and the Structure of the Atom," *Philosophical Magazine* 30 (1915): 394–415.

14. "On the Decrease of Velocity of Swiftly Moving Electrified Particles in Passing through Matter," *Philosophical Magazine* 30 (1915): 581–612.

15. "On the Quantum Theory of Line Spectra": Part 1—"On the General Theory," Part 2—"On the Hydrogen Spectrum," Part 3—"On the Spectra of Elements of Higher Atomic Structure," *Det Kongelige Danske Videnskabernes Selskab. Skrifter. Naturvidenskabelig og Matematisk Afdeling* 8, no. 4 (1918–1922). Published in German as *Über die Quantentheorie der Linienspektren* (Braunschweig: Vieweg, 1923).

16. *On the Model of a Triatomic Hydrogen Molecule* (Uppsala: Almqvist and Wiksells, 1919).

17. "Über die Serienspektra der Elemente" [On the series spectra of the elements], *Zeitschrift für Physik* 2 (1920): 423–469.

18. "Atomic Structure," *Nature* 107 (1921): 104–107.

19. "Zur Frage der Polarisation der Strahlung in der Quantentheorie" [On the question of the polarization of radiation in the quantum theory], *Zeitschrift für Physik* 6 (1921): 1–9.

20. "Atomic Structure," *Nature* 108 (1921): 208–209.

21. *Abhandlungen über Atombau aus den Jahren 1913–1916* [Papers on atomic structure from the years 1913–1916] (Braunschweig: Vieweg, 1921). Contains works listed here as nos. 6, 7, 9, 10, 11, 12, and 13. Also contains "Die Anwendung der Quantentheorie auf periodische Systeme" [Application of quantum theory to the periodic system].

22. "Atomernes bygning og stoffernes fysiske og kemiske egenskaber" [Atomic structure and the physical and chemical properties of matter], *Fysisk Tidsskrift* 19 (1921): 153–220. Published in book form under the same title by Gjellerups Forlag, Copenhagen, 1922. Published in German as "Der Bau der Atome und die physikalischen und chemischen Eigenschaften der Elemente," *Zeitschrift für Physik* 9 (1922): 1–67.

23. "The Effect of Electric and Magnetic Fields on Spectral Lines," Seventh Guthrie Lecture, *Proceedings of the Physical Society* 35 (1922): 275–302.

24. *The Theory of Spectra and Atomic Constitution* (Cambridge: Cambridge University Press, 1922; 2nd ed., 1924). Contains works listed here as nos. 8, 17, and 22. Published in German as *Drei Aufsätze über Spektren und Atombau* [Three articles on spectra and atomic structure] (Braunschweig: Vieweg, 1922). Published in French as *Les Spectres et la Structure de l'Atome* [Spectra and atomic structure] (Paris: J. Hermann, 1923).

25. "On the Selection Principle of the Quantum Theory," *Philosophical Magazine* 43 (1922): 1112–16.

26. "Om atomernes bygning" [On atomic structure], Nobel Lecture, delivered December 1922, in *Les Prix Nobel* (Stockholm 1921–22). Published in Danish by Gjellerups Forlag, Copenhagen, 1923); and also in *Fysisk Tidsskrift* 21 (1923): 6–44. Published in German as "Über den Bau der Atome," *Naturwissenschaften* 11 (1923): 606–624; and in book form under the same title by Springer Verlag, Berlin, 1924. Published in English as "The Structure of the Atom," *Nature* 112 (1923): 29–44.

27. "Röntgenspektren und periodisches System der Elemente" [X-ray spectra and the periodic system of the elements] (with Dirk Coster), *Zeitschrift für Physik* 12 (1923): 342–374.

28. "Linienspektren und Atombau" [Line spectra and atomic structure] *Annalen der Physik* 71 (1923); 228–288.

29. "Über der Anwendung der Quantentheorie auf den Atombau" [On the application of quantum theory to atomic structure], Part 1—"Die Grundpostulate der Quantentheorie" [The basic postulates of quantum theory], *Proceedings of the Cambridge Philosophical Society* 22, supplement (1924): 1–42.

30. "Über die Quantentheorie der Strahlung" [The quantum theory of radiation] (with H. A. Kramers and J. C. Slater), *Zeitschrift für Physik* 24 (1924): 69–87. Published in English as "The Quantum Theory of Radiation," *Philosophical Magazine* 47 (1924): 785–802.

31. "Zur Polarisation des Fluorescenzlichtes" [The polarization of fluorescence radiation], *Naturwissenschaften* 12 (1924): 1115–17.

32. "Über die Wirkung von Atomen bei Stossen" [The excitation of atoms by collisions], *Zeitschrift für Physik* 34 (1925): 142–147.

33. "Atomic Theory and Mechanics," *Nature* 116 (1925): 845–852. Published in German as "Atomtheorie und Mechanik," *Naturwissenschaften* 14 (1926): 1–10.

34. "Spinning Electrons and the Structure of Spectra," *Nature* 117 (1926): 265.

35. "Sir Ernest Rutherford," *Nature* 118, supplement (1926): 51–52.

36. "Sir J. J. Thomson's Seventieth Birthday," *Nature* 118 (1926): 879.

37. "Atom," *Encyclopaedia Britannica,* 13th ed. (1926).

38. "The Quantum Postulate and the Recent Development of Atomic Theory," in *Atti del Congresso Internazionale dei Fisici* (Como, 1927). Also published in *Nature* 121, supplement (1928): 580–590. Published in German as "Das Quantenpostulat und die neuere Entwicklung der Atomistik," *Naturwissenschaften* 16 (1928): 245–257.

39. "Sommerfeld und die Atomtheorie" [Sommerfeld and atomic theory] *Naturwissenschaften* 16 (1928): 1036.

40. "Wirkungsquantum und Naturbeschreibung" [The quantum of action and the description of nature], *Naturwissenschaften* 17 (1929): 483–486.

41. *Atomteori og Naturbeskrivelse* [Atomic theory and the description of nature], Festschrift (University of Copenhagen, 1929). Contains works listed here as nos. 33, 38, and 40.

42. "Atomteorien og grundprincipperne for naturbeskrivelsen" [Atomic theories and the principles for the description of nature], *Fysisk Tidsskrift* 27 (1929): 103–114. Published in German as "Die Atomtheorie und die Prinzipien der Naturbeschreibung," *Naturwissenschaften* 18 (1930): 73–78.

43. *Atomic Theory and the Description of Nature* (Cambridge: Cambridge University Press, 1934; rpt. 1961). Contains works listed here as nos. 33 and 38. Also contains item no. 40 in English translation: "The Atomic Theory and the Fundamental Principles Underlying the Description of Nature." Published in German as *Atomtheorie und Naturbeschreibung* (Berlin: Springer, 1931); this volume contains items 33, 38, 40, and 42. Published in French as *La Théorie Atomique et la Description des Phénomènes* (Paris: Gauthier-Villars, 1932). Published in Danish as *Atomteori og Naturbeskrivelse* (Copenhagen: Schultz Forlag, 1958).

44. "Maxwell and Modern Theoretical Physics," *Nature* 128 (1931): 691–692.

45. "Chemistry and the Quantum Theory of Atomic Constitution," Faraday Lecture, *Journal of the Chemical Society* (1932): 349–384.

46. "Atomic Stability and the Conservation Laws," in *Atti del Convegno di Fisica Nucleare della "Fondazione Alessandro Volta,"* Como (1932).

47. "Light and Life," Address delivered in 1932 to the 11ème Congrès International de la Lumière, Copenhagen, *Nature* 131 (1933): 421–423, 457–459. Published in Danish as "Lys og liv," *Naturens Verden* 17 (1933): 49–54. Published in German as "Licht und Leben," *Naturwissenschaften* 21 (1933): 245–250.

48. "Zur Frage der Messbarkeit der elektromagnetischen Feldgrössen," [The measurability of electromagnetic field magnitudes] (with L. Rosenfeld), *Matematisk-Fysiske Meddelelser det Kongelige Danske Videnskabernes Selskab* 12, no. 8 (1933): 1–65.

49. "Sur la méthode de correspondance dans la théorie de l'électron" [The correspondence method in the theory of the electron], in *Structure et Propriétés des Noyaux Atomiques: Rapports et Discussions du Septième Conseil de Physique* (Paris: Gauthiers-Villars, 1934), pp. 216–228.

50. "Friedrich Paschen zum siebzigsten Geburtstag" [For the seventieth birthday of Friedrich Paschen], *Naturwissenschaften* 23 (1935): 73.

51. "Zeeman Effect and Theory of Atomic Constitution," *Zeeman Verhandelingen* (1935): 131ff.

52. "Quantum Mechanics and Physical Reality," *Nature* 136 (1935): 65. Published in expanded form as "Can Quantum-Mechanical Description of Reality Be Considered Complete?" *Physical Review* 48 (1935): 696–702.

53. "Neutron Capture and Nuclear Constitution," *Nature* 137 (1936): 344–348, 351. Published in German as "Neutroneneinfang und Bau der Atomkerne," *Naturwissenschaften* 24 (1936): 241–245.

54. "Conservation Laws in Quantum Theory," *Nature* 138 (1936): 25–26.

55. "Atomkernenes egenskaber" [The properties of the atomic nucleus], *Fysisk Tidsskrift* 34 (1936): 186ff.

56. "Kausalität und Komplementarität" [Causality and complementarity] *Erkenntnis* 6 (1937): 293ff. Published in Danish as "Kausalitet og komplementaritet," *Naturens Verden* 21 (1937): 113ff.

57. "On the Transmutation of Atomic Nuclei by Impact of Material Particles, I: General Theoretical Remarks" (with F. Kalckar), *Matematisk-Fysiske Meddelelser det Kongelige Danske Videnskabernes Selskab* 14, no. 10 (1937): 1–40.

58. "Transmutations of Atomic Nuclei," *Science* 86 (1937): 161–165.

59. "Tribute to the late Lord Rutherford," *Nature* 140, supplement (1937): 1048–49.

60. "Biology and Atomic Physics," *Proceedings of the Galvani Congress* (Bologna, 1937). Published in Danish as "Biologi og atomfysik," *Naturens Verden* 22 (1938): 433ff.

61. "Wirkungsquantum und Atomkern" [The quantum of action and the atomic nucleus], *Annalen der Physik* 32 (1938): 5–19. Published in Danish as "Virkningskvantum og atomkerne," *Fysisk Tidsskrift* 36 (1938): 69ff.

62. "Analysis and Synthesis in Science," *International Encyclopedia of Unified Science* 1, no. 1 (1938).

63. "Nuclear Photo-Effect," *Nature* 141 (1938): 326–327.

64. "Resonance in Nuclear Photo-Effects," *Nature* 141 (1938): 1096–97.

65. "Natural Philosophy and Human Cultures," *Comptes Rendus du Congrès International de Science, Anthropologie et Ethnologie* (Copenhagen, 1938). Also published in *Nature* (London) 143 (1939); 268–272. Published in Danish as "Fysikkens erkendelseslaere og mennenskulturerne" [Physical epistemology and human cultures], *Tilskueren* (Jan, 1939).

66. "Disintegration of Heavy Nuclei," *Nature* 143 (1939): 330.

67. "Resonance in Uranium and Thorium Disintegrations and the Phenomenon of Nuclear Fission," *Physical Review* 55 (1939): 418–419.

68. "Nuclear Reactions in the Continuous Energy Region" (with R. E. Peierls and G. Placzek), *Nature* 144 (1939): 200–201.

69. "The Mechanism of Nuclear Fission" (with John A. Wheeler), *Physical Review* 56 (1939): 426–450.

70. "The Fission of Protoactinium" (with John A. Wheeler), *Physical Review* 56 (1939): 1065–66.

71. "The Causality Problem in Atomic Physics," *Conference on New Theories in Physics* (Warsaw, 1938). Published in French as "Le Problème Causal en Physique Atomique," *Réunion sur les Nouvelles Théories de la Physique, Varsovie, 1938* (Paris, 1939).

72. "Scattering and Stopping of Fission Fragments," *Physical Review* 58 (1940): 654–655.

73. "Velocity-Range Relation for Fission Fragments" (with J. K. Bøggild, K. J. Brostrøm, and T. Lauritsen), *Physical Review* 58 (1940): 839–840.

74. Successive Transformations in Nuclear Fission," *Physical Review* 58 (1940): 864–866.

75. "Velocity-Range Relation for Fission Fragments," *Physical Review* 59 (1941): 270–275.

76. "Nyere undersøgelser over atomkernernes omdannelser" [Recent Investigations on the Transmutation of Atomic Nuclei], *Fysisk Tidsskrift* 39 (1941): 3ff.

77. "Mechanism of Deuteron-Induced Fission," *Physical Review* 59 (1941): 1042. Also published in *Nature* 148 (1941): 229.

78. "Universitetet og forskningen" [Universities and research], *Politiken*, 3 June 1941.

79. "Dansk kultur" [Danish culture], in *Danmarks Kultur ved Aar 1940* [Danish culture in the year 1940] (Copenhagen: Danske Forlag, 1941–43).

80. "Ole Chievitz," *Ord och Bild* [Words and pictures] (Stockholm, 1944).

81. "Science and Civilization," *The Times* (London), 11 August 1945.

82. "A Challenge to Civilization," *Science* 102 (1945): 363–364.

83. "Newton's Principles and Modern Atomic Mechanics," *Royal Society, Newton Tercentenary Celebrations* (Cambridge: Cambridge University Press, 1946), pp. 56–61.

84. "Om maalingsproblemet i atomfysikken" [Measurement problems in atomic physics], *Matematisk Tidsskrift*, ser. B (1946): 163ff.

85. "Atomic Physics and International Cooperation," *Proceedings of the American Philosophical Society* 91 (1947): 137ff. Published in German as "Atomphysik und Internationale Zusammenarbeit," *Universitas* 6 (1951): 547ff.

86. "Problems of Elementary-Particle Physics," in *Report of an International Conference on Fundamental Particles and Low Temperatures,* vol. 1: *Fundamental Particles* (London: Physical Society, 1947), pp. 1–4.

87. "The Penetration of Atomic Particles through Matter," *Matematisk-Fysiske Meddelelser det Kongelige Danske Videnskabernes Selskab* 18, no. 8 (1948): 1–144.

88. "On the Notions of Causality and Complementarity," *Dialectica* 2 (1948): 312–318.

89. "Discussion with Einstein on Epistemological Problems in Atomic Physics," in *Albert Einstein: Philosopher-Scientist,* ed. P. A. Schilpp, (Evanston, Ill.: Library of Living Philosophers, 1949), pp. 201–241. Published in German as "Diskussionen mit Einstein über erkenntnistheoretische Probleme in der Atomphysik," in *Albert Einstein als Philosoph und Naturforscher* (Stuttgart: W. Kohlhammer, 1955), pp. 115ff. Published in Russian as "Diskussii s Einsteinom o problemakh teorii poznaniye v atomnoi fizike," *Uspekhi Fizicheskikh Nauk* 66 (1950): 571–598.

90. *Open Letter to the United Nations, June 9, 1950* (Copenhagen: J. H. Schultz Forlag, 1950). Published in Danish as *Åbent Brev til de Forenede Nationer, 9 Juni 1950* (Copenhagen: J. H. Schultz Forlag, 1950).

91. "Field and Charge Measurements in Quantum Electrodynamics" (with L. Rosenfeld), *Physical Review* 78 (1950): 794–798.

92. "On the Notions of Causality and Complementarity," *Science* 111 (1950): 51–54.

93. "Medical Research and Natural Philosophy," *Acta Medica Scandinavica* 142, supplement 266 (1952): 967ff.

94. "Hendrik Anthony Kramers," *Nederlands Tijdsschrift voor Natuurkunde* 18 (1952): 161ff.

95. "Ved Hendrik Anton Kramers' død" [On the death of Hendrik Anton Kramers], *Politikens kronik,* 27 April 1952).

96. "Physical Science and the Study of Religions," *Studia Orientalia Ioanni Pedersen septuagenario A.D. VII id. Nov. Anno MCMLIII* (Copenhagen: Ginar Mimles-Gaard, 1953), p. 385ff.

97. "Electron Capture and Loss by Heavy Ions Penetrating through Matter" (with J. Lindhard), *Matematisk- Fysiske Meddelelser det Kongelige Danske Videnskabernes Selskab* 28, no. 7 (1954): 1–31.

98. "Det fysiske grundlag for industriel udnyttelse af atomkerneenergien," [The physical basis for industrial use of nuclear energy], *Tidsskrift for Industri,* nos. 7–8 (1955): 168ff.

99. "Physical Science and Man's Position," *Proceedings of the International Conference on Peaceful Uses of Atomic Energy,* vol. 16 (Geneva: United Nations, 1956), pp. 57ff. Also published in *Philosophy Today* (1957): 65ff.

100. "Rydberg's Discovery of the Spectral Law," *Lunds Universitets Aarsskrift, N.F. (Part 2)* 50 (1955): 15ff.

101. "Albert Einstein: 1879–1955," *Scientific American* 192, no. 6 (1955): 31–32.

102. *The Unity of Knowledge* (New York: Doubleday, 1955).

103. "Mathematics and Natural Philosophy," *Scientific Monthly* 82 (1956): 85–88.

104. "On Atoms and Human Knowledge," *Daedalus* 87, no. 2 (1958): 164–175. Published in Danish as "Atomerne og den menneskelige erkendelse," *Oversigt over Det Kongelige Danske Videnskabernes Selskabs Virksomhed, 1955–1956* (Copenhagen, 1956).

105. *Atomic Physics and Human Knowledge* (New York: Wiley, 1958). Contains works listed here as nos. 47, 60, 65, 89, 102, and 104, as well as the essay "Physical Science and the Problem of Life." Published in German as *Atomphysik und menschliche Erkenntnis* (Braunschweig: Vieweg, 1958). Published in Swedish as *Atomfysik och mänskligt vetande* (Stockholm: Bonniers, 1959). Published in French as *Physique atomique et connaissance humaine* (Paris: Gauthier-Villars, 1961). Published in Russian as *Atomnaya Fizika i Chelovecheskoye Poznaniye* (Moscow: IL, 1961); this volume also contains items 106 and 108. Published in Italian as *Teoria dell'atomo e conoscenza umana* (Turin: Boringhieri, 1961); this volume also contains items 21, 24, and 43, but lacks 89.

106. "Quantum Physics and Philosophy: Causality and Complementarity," in *Philosophy in Mid-Century: A Survey,* ed. R. Klibansky (Florence: La Nuova Italia Editrice, 1958), pp. 308ff. Published in Russian as "Kvantovaya Fizika i Filosofiya," *Uspekhi Fizicheskikh Nauk* 67 (1959): 37ff.

107. "Über Erkenntnisfragen der Quantenphysik" [On questions of perception in quantum physics], *Max Planck Festschrift* (Berlin, 1958), pp. 169ff.

108. "Quantum Physics and Biology," in *Models and Analogues in Biology* (Bristol: Society for Experimental Biology, 1960).

109. "Physical Models in Living Organisms," in *Light and Life,* ed. W. D. McElroy and Bentley Glass (Baltimore: Johns Hopkins University Press, 1961).

110. "The Unity of Human Knowledge," Address given in 1960 at the congress in Copenhagen arranged by La Fondation Européenne de la Culture. Published in German as "Über die Einheit unseres Wissens," *Universitas* 16 (1961): 835ff; and as "Die Einheit menschlicher Erkenntnis," *"Europa," Monatzeitschrift für Politik, Wirtschaft und Kultur* (August 1961). Published in Danish as "Den menneskelige erkendelses enhed," *Berlingske Tidende,* 22 October 1960. Published in Italian as "Unità della conoscenza umana," *Responsabilità del Sapere* 57 (1961): 7ff.

111. "Atomvidenskaben og menneskehedens krise" [Atomic science and mankind's crisis], *Politiken,* 20 April 1961.

112. "Die Entstehung der Quantenmechanik" [The genesis of quantum mechanics], in *Werner Heisenberg und die Physik unserer Zeit* [Werner Heisenberg and physics in our time] (Braunschweig: Vieweg, 1961).

113. "Reminiscences of the Founder of Nuclear Science and of Some Developments Based on His Work," Rutherford Memorial Lecture, delivered in 1958, *Proceedings of the Physical Society* 78 (1961): 1083–1115.

114. "The Solvay Meetings and the Development of Quantum Physics," in *La Théorie quantique des champs* (New York: Interscience, 1962).

115. *Essays, 1958–1962, on Atomic Physics and Human Knowledge* (New York: Wiley, 1963). Contains works listed here as nos. 106, 110, 113, and 114. Also contains "The Connection between the Sciences," delivered as an address in 1960, and "Light and Life Revisited," delivered as an address in 1962, as well as item 112 in English translation: "The Genesis of Quantum Mechanics."

See also Niels Bohr, *Collected Works* (Amsterdam: North-Holland; New York: American Elsevier): Vol. 1 (1972) — *Early Work (1905–1911)*; Vol. 2 (1976) — *Work on Atomic Physics (1912–1917)*; Vol. 3 (1977) — *The Correspondence Principle (1918–1923)*; Vol. 4 (1977) — *The Periodic System (1918–1923)*. Other volumes in preparation.

Credits

TEXT

"A Personal Memoir," by James Franck, is a translated excerpt from Franck's obituary article "Niels Bohrs Persönlichkeit," *Naturwissenschaften* 9 (1963): 341–343, and is printed by permission of the publishers, Springer-Verlag, Heidelberg.

"Niels Bohr, the Quantum and the World" is similar to an essay of the same title in V. F. Weisskopf, *Physics in the Twentieth Century*, published by MIT Press, which has given permission to print this adaptation.

In the introduction to "The Trilogy" the translated letter from Bohr to his brother is reprinted by permission of the Niels Bohr Institute. The excerpts from Bohr's Rutherford Memorial Lecture are reprinted by permission of the Niels Bohr Institute and the Institute of Physics, publishers of *Proceedings of the Physical Society*. The extended excerpts from Bohr's three papers "On the Constitution of Atoms and Molecules" are reprinted by permission of the Niels Bohr Institute and of Taylor & Francis, Ltd., publishers of *Philosophical Magazine*.

The excerpts from Bohr's Nobel Lecture are printed by permission of the Nobel Foundation. The text of the lecture is taken from *Les Prix Nobel en 1921–1922*, published by the Foundation.

The excerpts from Bohr's "Discussion with Einstein" are reprinted, by permission of the Niels Bohr Institute and Open Court Publishing Co., from Paul A. Schilpp, ed., *Albert Einstein: Philosopher-Scientist* (copyright © 1969 by Library of Living Philosophers, Inc.).

"On Bohr's Views Concerning the Quantum Theory" is reprinted by permission of Professor David Bohm and Cambridge University Press, publishers of *Quantum Theory and Beyond*, copyright © Cambridge University Press, 1971.

"Waves and Particles: 1923–1924" is excerpted from Ch. 2 of John C. Slater, *Solid-State and Molecular Theory: A Scientific Biography*, © John Wiley & Sons, Inc., 1975, and is reprinted by permission of the publishers.

"Reminiscences from 1926 and 1927" by Werner Heisenberg is taken, with deletions indicated by ellipses, from *Physics and Beyond: Encounters and Conversations*, by Werner Heisenberg (translated by Arnold J. Pomerans), Vol. 42, World Perspective Series planned and edited by Ruth Nanda Anshen, copyright © 1971 by Harper & Row, Publishers, Inc.

"Physics in Copenhagen in 1934 and 1935," by John A. Wheeler, is based mainly on an excerpt from Professor Wheeler's article, "Some Men and Moments in Nuclear Physics," in *Nuclear Physics in Retrospect* (ed. Roger H. Stuewer), © University of Minnesota Press, 1979. It is reprinted here by permission of the publishers.

Bohr's article, "Transmutations of Atomic Nuclei," from *Science*, 86 (1937): 161–165, is reproduced by permission of the publishers.

The excerpt from "The Mechanism of Nuclear Fission," *Physical Review* 56 (1939): 426–430, is reproduced by permission of Professor John A. Wheeler and the American Physical Society.

"Reminiscences from the Postwar Years," by Abraham Pais, is the text of the major part of an article of the same title in *Niels Bohr: His Life and Work,* © North-Holland Publishing Co., 1967. It is reprinted by permission of the author and publishers.

Bohr's article, "Energy from the Atom," originally published in *The Times* (London), 11 August, 1945, is reprinted by permission of Times Newspapers, Ltd.

The abridged version of Bohr's "Open Letter to the United Nations," with deletions indicated by ellipses, was prepared for this book by the Niels Bohr Institute and is printed with their permission.

"The Philosophy of Niels Bohr," by Aage Petersen (*Bulletin of the Atomic Scientists,* September, 1963), copyright © 1963 by The Educational Foundation for Nuclear Science, Chicago, is reprinted by permission of the author and publishers.

Bohr's essay, "Light and Life," is reprinted by permission of the Niels Bohr Institute.

The marginal quotations contain a number of short excerpts from Otto Frisch, *What Little I Remember,* © Cambridge University Press, 1979. We thank the publishers for permission to reprint them in this book.

ILLUSTRATIONS (PAGE NUMBERS AT LEFT)

Century Speculations on the Complexity of Atoms," *Annals of Science* 39 (1982): 54.

57 Right: Reproduced from *Nobel Lectures: Physics, 1922–1941*, by permission of the Nobel Foundation.

58 Reproduced from E. N. da C. Andrade, *The Structure of the Atom* (London: G. Bell and Sons, 1934).

59 Reproduced from Helge Kragh, "Niels Bohr's Second Atomic Theory," *Historical Studies in the Physical Sciences* 10 (1979): 145.

60 Reproduced from *Nobel Lectures: Physics, 1922–1941*, by permission of the Nobel Foundation.

62 *Fysisk Tydsskrift* 22 (1924): 18. Photo by Dirk Coster.

63 Niels Bohr Institute.

69 AIP Niels Bohr Library, Margrethe Bohr Collection.

72 Photograph kindly supplied by Professor Gernot Born. Photographer unknown.

73 AIP Niels Bohr Library, Landé Collection.

78 Niels Bohr Institute.

95 Reproduced from *Nobel Lectures: Physics, 1922–1941*, by permission of the Nobel Foundation.

96 Bohr Scientific Correspondence, courtesy AIP Niels Bohr Library.

109 Niels Bohr Institute.

111 Niels Bohr Institute. Photo by Paul Ehrenfest.

124, 126, 127, 129, 130, 132, 133 Drawings from Paul A. Schilpp, ed., *Albert Einstein: Philosopher-Scientist*. Reproduced by permission of the Open Court Publishing Co., La Salle, Illinois, copyright © 1949, 1951 and copyright © 1969 by the Library of Living Philosophers, Inc.

125 AIP Niels Bohr Library. From negatives provided by Martin J. Klein, restored by William R. Whipple.

134 From *My World Line*, by George Gamow. Copyright © 1970 by the Estate of George Gamow. Reprinted by permission of Viking Penguin, Inc.

147 Montage by P. J. Kennedy.

150 Drawings by William Minty.

151 Prepared by Field Gilbert of Austin, Texas, for John Wheeler.

154, 155, 156 Drawings from David Bohm, "Bohr's Views Concerning the Quantum Theory," in *Quantum Theory and Beyond*, ed. Ted Bastin (Cambridge: Cambridge University Press, 1971).

167 Photograph kindly supplied by Det Kongelige Bibliotek, Copenhagen. Photographer unknown.

168 Photographs by P. Ehrenfest, Jr., AIP Niels Bohr Library, Weisskopf Collection.

173 Margrethe Bohr Collection (Copenhagen).

177 AIP Niels Bohr Library. Photo by P. Ehrenfest, Jr.

179 Drawings by William Minty.

188 Niels Bohr Institute.

189 Drawings by George Gamow. Reproduced from George Gamow, *Thirty Years That Shook Physics* (New York: Doubleday Anchor, 1966), by kind permission of the Estate of George Gamow.

192 Photo courtesy of Dr. Fabio Bevilacqua.

194 Reproduced by courtesy of the publishers of *Nature* (Macmillan Journals Ltd.).

202 Photograph kindly supplied by the Institut Solvay de Physique. Photographer unknown.

203 Photograph by S. A. Goudsmit. Courtesy of the AIP Niels Bohr Library, Goudsmit Collection.

206 Upper: From *Science* 86 (1937): 162. Reproduced by permission of the publishers.

206 Lower: Photograph by P. Ehrenfest, Jr., AIP Niels Bohr Library, Weisskopf Collection.

210 AIP Niels Bohr Library, Landé Collection.

213 Copyright © 1985 by The New York Times Company. Reprinted by permission of the publishers.

215 Reproduced by kind permission of Mrs. Ulla Frisch and the Niels Bohr Institute.

218 AIP Niels Bohr Library, Margrethe Bohr Collection.

224 AIP Niels Bohr Library, Margrethe Bohr Collection.

228 From George Gamow, *Thirty Years That Shook Physics* (New York: Doubleday Anchor, 1966). Reproduced by kind permission of the Estate of George Gamow.

230 Left: Niels Bohr Institute.

230 Right: Niels Bohr Institute.

248 Niels Bohr Institute, courtesy AIP Niels Bohr Library.

262 Reproduced by permission of Times Newspapers, Ltd.

269 Niels Bohr Institute, courtesy AIP Niels Bohr Library.

279 Niels Bohr Institute.

281 From *Merchant Airmen* (London: Her Majesty's Stationery Office, 1964). Reproduced with the permission of the Controller of Her Majesty's Stationery Office.

283 Kindly supplied by Professor R. V. Jones.

286 Niels Bohr Institute.

289 Niels Bohr Institute, courtesy AIP Niels Bohr Library.

304 Clockwise from upper left: AIP Niels Bohr Library; Princeton University, courtesy AIP Niels Bohr Library; AIP Niels Bohr Library, Fermi Film Collection; AIP Niels Bohr Library, Margrethe Bohr Collection; AIP Niels Bohr Library, Margrethe Bohr Collection.

327, 328, 336 Drawings by William Minty.

347 Photo by Lotte Meitner-Graf, London. Courtesy of the AIP Niels Bohr Library.

Index

Acheson, Dean, 233

Albert Einstein: Philosopher-Scientist, 121

Alpha decay, 93, 197; theory of, 173, 198, 209

Alpha particles: penetration through matter, 7, 39, 198; causing nuclear transmutations, 204

American Physical Society, 208, 216, 225

Anderson, Sir John (Lord Waverley), 268, 270, 271, 273, 274

Angular momentum quantization, 39, 46, 64, 85, 90. *See also* Spin

Aspect, Alain, 142, 145–146, 149

Atom, 22, 23; Rutherford model of, 6, 37, 40, 41, 42, 77, 80, 92, 197; nucleus of, 6, 38, 40, 77, 92–93, 113, 115, 186, 194, 197–210, 235–239, 263; Bohr model of, 7–8, 16, 20–21, 39–47, 80–86, 92–93; postage stamps on theme of, 10–11; hydrogen, 20, 38, 50, 83–88, 161, 178, 194; Thomson model of, 37–39, 40; Saturnian, 37, 40; in Greek thought, 303; early view of, 351

Atomic bomb: Bohr's work on, 13, 18, 27, 233, 269–270; in Germany, 257–258, 268, 278; Bohr's warnings about, 259; and Russia, 270–276, 284; and Churchill, 271–273, 282, 284–285; and Roosevelt, 272

Atomic structure: and quantum theory, 19–21, 24; theories of (ca. 1910), 37–39; Bohr's theories of (1912–1914), 39–49, 50, 66–67; Bohr's theories of (1921–1923), 51–60. *See also* Nobel Prize lecture; Particles; Trilogy

Atomkerne, Die (von Weizsäcker), 209

Atommechanik I (Born), 74

Atoms for Peace award (1957), 260

Atoms for Peace Conference (Geneva), 260

Auger, Pierre, 28

Balmer, J. J., 44

Balmer series, 17, 43, 44, 45, 46, 47, 84, 94, 95

Bell, J. S., 143, 147, 152

Bell's theorem, 144–145, 149

Beta decay, 93, 114, 197, 200; theory of, 115, 201, 203–204

Bethe, Hans A.: and quantum theory, 174, 225; single-particle reaction theory of, 204, 207; on nuclear dynamics, 208; and liquid-drop model, 209–210; and Bohr Institute, 221, 223, 232–234, 345

Biology, Bohr's views on, 26, 307–308, 315–319, 348–349

B-K-S paper. *See* Bohr-Kramers-Slater paper

Blackett, P. M. S., 14, 202

Bloch, Felix, 14, 22, 174, 190, 201

Bohm, David, 142, 145, 339

Bohr, Aage (son), 14, 220, 247, 257, 267–268, 269, 282; at Los Alamos, 233, 346; as Bohr's assistant, 259, 304

Bohr, Christian (father), 3, 253

Bohr, Christian (son), 204, 222

Bohr, Ellen Adler (mother), 3

Bohr, Hans (son), 207

Bohr, Harald (brother), 3–4, 6 253, 257, 280; letters from Niels to, 76, 102

Bohr, Jenny (sister), 3

Bohr, Margrethe Nørlund (wife), 6, 7, 9, 15, 197, 245, 257, 258, 266, 280

Bohr-Festspiele (Göttingen, 1922), 71–74

Bohr-Kramers-Slater paper, 9, 66, 105, 106–107, 115, 200

Bohr-Rosenfeld theory, 115

Bohr-Sommerfeld theory, 48, 103–104, 161

"Bonzenfreie Kolloquium" (Berlin, 1920), participants in, 72

Born, Max, 59, 67, 110, 144, 172, 198, 333, 339; at Göttingen, 71, 73, 75, 167, 174; and quantum mechanics, 75, 104, 106; and matrix mechanics, 194, 283; letter from Rutherford, 207; on complementarity, 324

Index

Index